W9-AQW-267

The Biochemical Basis of Neuropharmacology

The Biochemical Basis of Neuropharmacology

FIFTH EDITION

JACK R. COOPER, Ph.D.
Professor of Pharmacology
Yale University School of Medicine

FLOYD E. BLOOM, M.D.
Director, Division of Preclinical Neuroscience and Endocrinology
Research Institute of Scripps Clinic
La Jolla, California

ROBERT H. ROTH, Ph.D.
Professor of Pharmacology and Psychiatry
Yale University School of Medicine

New York Oxford
OXFORD UNIVERSITY PRESS
1986

Oxford University Press
Oxford New York Toronto
Delhi Bombay Calcutta Madras Karachi
Petaling Jaya Singapore Hong Kong Tokyo
Nairobi Dar es Salaam Cape Town
Melbourne Auckland

and associated companies in
Beirut Berlin Ibadan Nicosia

Published by Oxford University Press, Inc.,
200 Madison Avenue, New York, New York 10016

Library of Congress Cataloging-in-Publication Data

Cooper, Jack R., 1924–
The biochemical basis of neuropharmacology.

Bibliography: p.
Includes index.
1. Neurochemistry. 2. Neuropharmacology.
I. Bloom, Floyd E. II. Roth, Robert H., 1939–
III. Title. [DNLM: 1. Autonomic Agents. 2. Central
Nervous System—drug effects. 3. Nerve Tissue—
analysis. 4. Nerve Tissue—physiology. 5. Psychopharmacology. QV 77 C777b]
QP356.C66 1986 615'.78 85-29803
ISBN 0-19-504035-X
ISBN 0-19-504036-9 (pbk.)

Cover: Darkfield photomicrograph of a cluster of dopamine neurons in the A8 dopamine cell group of the rat. The dopaminergic neurons and their processes are visualized by immunohistochemically staining for tyrosine hydroxylase (TH), the catecholamine synthetic enzyme. TH is localized to noradrenergic and adrenergic neurons as well as dopamine-containing cells. Identification of these neurons as dopaminergic was made on the basis of lack of staining of A8 neurons for dopamine-beta-hydroxylase, the enzyme which catalyzes the oxidation of dopamine to norepinephrine. Both perikarya and processes of the dopamine neurons can be clearly visualized, and varicosities present in the axons of TH-positive axons can be observed. (Unpublished photomicrograph courtesy of Dr. Ariel Y. Deutch, Yale University School of Medicine) × 400

Printing (last digit): 9 8 7 6 5 4 3 2 1

Printed in the United States of America
on acid-free paper

This book is dedicated to the memory
of Nicholas J. Giarman, colleague and dear friend

Preface to the Fifth Edition

Considerable revision has occurred with this edition. Two new chapters have been added, one on molecular neurobiology, the other on modulation of synaptic transmission. We have also eliminated the separate chapter on cyclic nucleotides, incorporating these agents with other second messenger systems in the modulation chapter. The neuroactive peptides and the endorphins have been combined into one chapter. Other chapters have been updated with new information. Despite some complaints about the paucity of references, we still hold to our original aim of presenting overviews rather than specific articles so that only selected papers, which would contain useful references, are cited along with recent reviews.

We are grateful to those readers who took the time to offer useful suggestions. Finally, we once again with pleasure acknowledge the efficient and tireless (but never tiresome) help of our Oxford University Press editor, Jeffrey House.

July 1985

J.R.C.
F.E.B.
R.H.R.

Contents

1 | Introduction, 3

2 | Cellular Foundations of Neuropharmacology, 9
CYTOLOGY OF THE NERVE CELL, 9
BIOELECTRIC PROPERTIES OF THE NERVE CELL, 15
APPROACHES, 34
IDENTIFICATION OF SYNAPTIC TRANSMITTERS, 35
ANALYSIS OF MEMBRANE ACTIONS OF DRUGS AND TRANSMITTERS IN VITRO, 40
THE STEPS OF SYNAPTIC TRANSMISSION, 43

3 | Molecular Foundations of Neuropharmacology, 48
FUNDAMENTAL MOLECULAR INTERACTIONS, 53
MOLECULAR STRATEGIES IN NEUROPHARMACOLOGY, 59
GENERAL STRATEGIES FOR CLONE SCREENING AND SELECTION, 61
BEYOND THE CLONES, 66

4 | Metabolism in the Central Nervous System, 72

5 | Receptors, 86
DEFINITION, 87
ASSAYS, 91
IDENTIFICATION, 93
KINETICS AND THEORIES OF DRUG ACTION, 95

6 | Modulation of Synaptic Transmission, 106
DEFINITIONS, 106
SECOND MESSENGERS, 109

7 | Amino-Acid Transmitters, 124
GABA, 124
PHARMACOLOGY OF GABAERGIC NEURONS, 139
GLYCINE, 155
GLUTAMIC ACID, 161

8 | Acetylcholine, 173
ASSAY PROCEDURES, 173
SYNTHESIS, 175
CHOLINE TRANSPORT, 176
CHOLINE ACETYLTRANSFERASE, 178
ACETYLCHOLINESTERASE, 180
THE GENESIS OF THE CHOLINERGIC TRIAD IN NEURONS, 185
UPTAKE, SYNTHESIS, AND RELEASE OF ACH, 186
CHOLINERGIC PATHWAYS, 190
CHOLINERGIC RECEPTORS, 193
DRUGS THAT AFFECT CENTRAL CHOLINERGIC SYSTEMS, 195
ACH IN DISEASE STATES, 196

9 | Catecholamines I: General Aspects, 203
METHODOLOGY, 203
DISTRIBUTION, 212
LIFE CYCLE OF THE CATECHOLAMINES, 215
AXONAL CATECHOLAMINE TRANSPORT, 247
NEUROTRANSMITTER ROLE, 249
PHARMACOLOGY OF CATECHOLAMINE NEURONS, 252

10 | Catecholamines II: CNS Aspects, 259
SYSTEMS OF CATECHOLAMINE PATHWAYS IN THE CNS, 261
CATECHOLAMINE METABOLISM, 271
BIOCHEMICAL ORGANIZATION, 274
PHARMACOLOGY OF CENTRAL CATECHOLAMINE-CONTAINING NEURONS, 276
PHARMACOLOGY OF DOPAMINERGIC SYSTEMS, 281
SPECIFIC DRUG CLASSES, 288
CATECHOLAMINE THEORY OF AFFECTIVE DISORDER, 304
DOPAMINE HYPOTHESIS OF SCHIZOPHRENIA, 309

11 | Serotonin (5-hydroxytryptamine) and Histamine, 315
SEROTONIN, 315

BIOSYNTHESIS AND METABOLISM OF SEROTONIN, 316
PINEAL BODY, 323
CELLULAR EFFECTS OF 5-HT, 328
HISTAMINE, 340

12 | Neuroactive Peptides, 352
SOME BASIC QUESTIONS, 352
THE GRAND PEPTIDE FAMILIES, 361
INDIVIDUAL PEPTIDES WORTH TRACKING, 378
A READER'S GUIDE TO PEPTIDE POACHING, 387

Index, 395

The Biochemical Basis of Neuropharmacology

1 | Introduction

Neuropharmacology can be defined simply as the study of drugs that affect nervous tissue. This, however, is not a practical definition since a great many drugs whose therapeutic value is extraneural can affect the nervous system. For example, the cardiotonic drug digitalis will not uncommonly produce central nervous system effects ranging from blurred vision to disorientation. For our purposes we must accordingly limit the scope of neuropharmacology to those drugs specifically employed to affect the nervous system. The domain of neuropharmacology would thus include psychotropic drugs that affect mood and behavior, anesthetics, sedatives, hypnotics, narcotics, anticonvulsants, analeptics, analgetics, and a variety of drugs that affect the autonomic nervous system.

Since, with few exceptions, the precise molecular mechanism of action of these drugs is unknown, and since recitations of their absorption, metabolism, therapeutic indications, and toxic liability can be found in most textbooks of pharmacology, we have chosen to take a different approach to the subject. We will concentrate on the biochemistry and physiology of nervous tissue, emphasizing neurotransmitters, and will introduce the neuropharmacologic agents where their action is related to the subject under discussion. Thus a discussion of LSD is included in the chapter on serotonin and a suggested mechanism of action of the antipsychotic drugs in Chapter 10.

It is not difficult to justify this focus on either real or proposed neurotransmitters since they act at junctions rather than on the events that occur with axonal conduction or within the cell body. Except for local anesthetics, which interact with axonal membranes, all neuropharmacological agents whose mechanisms of action are to some extent documented seem to be involved primarily with synaptic events. This finding appears quite logical in view of the regulatory mechanisms in the transmission of nerve impulses. The extent to which a neuron is depolarized or hyperpolarized will depend largely on its excitatory and inhibitory synaptic inputs, and these inputs must obviously involve neurotransmitters or modulators. What is enormously difficult to comprehend is the contrast between the action of a drug on a simple neuron, which causes it either to fire or not to fire, and the wide diversity of central nervous system effects, including subtle changes in mood and behavior, which that same drug will induce. As will become clearer in subsequent chapters, at the molecular level, an explanation of the action of a drug is often possible, at the cellular level, an explanation is sometimes possible, but at a behavioral level, our ignorance is abysmal. There is no reason to assume, for example, that a drug that inhibits the firing of a particular neuron will therefore produce a depressive state in an animal. There may be hundreds of unknown intermediary reactions involving transmitters and modulators between the demonstration of the action of a drug on a neuronal system and the ultimate effect on behavior.

The fact, however, that one can find compounds with a specific chemical structure to control a given pathological condition is an exciting experimental finding, since it suggests an approach that the neuropharmacologist can take to clarify normal as well as abnormal brain chemistry and physiology. The use of drugs that affect the adrenergic nervous system has, for instance, uncovered basic and hitherto unknown neural properties such as the uptake, storage, and release of the biogenic amines. The recognition of the analogy between curare poisoning in animals and myasthenia gravis in humans led to the understanding of the cholinergic neuromuscular transmission problem in myasthenia gravis and to subsequent treatment with anticholinesterases.

We have already referred to neuroactive agents involved in synaptic transmission as neurotransmitters, neuromodulators, and neurohormones so definitions are now in order. Although we can define these terms in a strict, rigid, fashion, it will be apparent—as noted later—that it is an exercise in futility to apply these definitions to a neuroactive agent as a classification unless one both understands its activity and specifies its locus. Briefly, the traditional definition of a *neurotransmitter* states that the compound must be synthesized and released presynaptically, it must mimic the action of the endogenous compound that is released on nerve stimulation, and, where possible, a pharmacological identity is required where drugs that either potentiate or block postsynaptic responses to the endogenously released agent also act identically to the administered suspected neurotransmitter. Conventionally, based on the studies of ACh at the neuromuscular junction, transmitter action was thought to be a brief and highly restricted point-to-point process. If one takes the word *modulation* literally, then a *neuromodulator* has no intrinsic activity but is only active in the face of ongoing synaptic activity where it can modulate transmission either pre- or postsynaptically. In many instances, however, a modulating agent does produce changes in conductance or membrane potential. Typically, modulatory effects involve a second messenger system. A *neurohormone* can be released from both neuronal and nonneuronal cells, and, most important to the definition, travels in some circulation to act at a site distant from its release site. Just how far a neurohormone has to travel before it loses its neurotransmitter status and becomes a neurohormone has never been decided.

We stated earlier that while we could define these terms it would be of little use to pigeonhole known neuroactive compounds until the site of action and the activity of the agent was specified. For example, dopamine is a certified neurotransmitter in the striatum, yet it is released from the hypothalamus and travels through the hypophysial-portal circulation to the pituitary where it inhibits the release of prolactin. Here it obviously fits the definition of a neurohormone. Similarly, serotonin is a neurotransmitter in the raphe nuclei, yet at the facial motor nucleus it acts primarily as a neuromodulator and secondarily as a transmitter. Most peptides with their multiple activities in the brain and gut are generally considered to be neuromodulators, yet Substance P fulfills the criteria of a transmitter at sensory afferents to the dorsal horn of the spinal cord. In sum, the plethora of exceptions to the aforementioned definitions of a transmitter, modulator, or hormone has generated confusion in the literature. Better to describe the activity of a neuroactive agent at a specified site rather than attempt a profitless definition.

The multidisciplinary aspects of pharmacology in general are particularly relevant in the field of neuropharmacology, where a "pure" neurophysiologist or neurochemist would be severely handicapped in elucidating drug action at a molecular level. The neuropharmacologist should be aware of the tools that are available for the total dissection of a biological problem. These would include morphological techniques such as electron microscopy, fluorescence microscopy and freeze-etching, and immunological techniques as a basis for developing radioimmunoassays, immunocytochemistry, and monoclonal antibodies as well as the classical electrophysiological and biochemical procedures. In addition, if the investigator is concerned with certain aspects of the action of psychotropic drugs, he should have some knowledge of the techniques of behavioral testing.

In science, one measures something. One must know what to measure, where to measure it, and how to measure it. This sounds rather obvious, but the student should be aware that, particularly in the neural sciences, these seemingly simple tasks can be enormously difficult. For example, suppose one were interested in elucidating the presumed biochemical aberration in schizophrenia. *What* would one measure? ATP? Glucose? Ascorbic acid? Unfortunately, this problem has been zealously investigated in the last dozen years by people who have measured everything they could think of, generally in the blood, in their search for differences between normal individuals and schizophrenics. As could be predicted, the problem has not been solved. (It may be assumed, however, that these studies have produced a large population of anemic schizophrenics from all the bloodletting.) The situation is the same for a variety of neurological diseases. Even in epilepsy, where there is some evidence that points to a neurochemical lesion, we have no idea what to measure.

Deciding *where* to measure something in neuroscience is complicated by the heterogeneity of nervous tissue: In general, unless one has a particular axon to grind, it is preferable to use peripheral nerve rather than the CNS. Suburban neurochemists have an easier time than their CNS counterparts, since it is not only a question of which region of the brain to use for the test preparation but which of the multitude of cell types within each area to choose. If a project involved a study of amino-acid transport in nervous tissue, for example, would one use isolated nerve-ending particles (synaptosomes), glial cells, neuronal cell bodies in culture, a myelinated axon, a ganglion cell? Up to the present time most investigators have used cortical brain slices, but the obvious disadvantage of this preparation is that one has no idea which cellular organelle takes up the amino acid.

How to measure something is a surprisingly easy question to answer, at least if one is dealing with simple molecules. With the recent advances in microseparation techniques and in fluorometric, radiometric, and immunological assays, there is virtually nothing that cannot be measured with a high degree of both specificity and sensitivity. In this regard one should be careful not to overlook the classical bioassay, which tends to be scorned by young investigators but in fact is largely responsible for the striking progress in our knowledge of both the prostaglandins and the opiate receptor with its peptide agonists. The major problem is with macromolecules. How can neuronal membranes be quantified, for example, if extraneuronal constituents are an invariable contaminant and markers to identify unequivocally a cellular constituent are often lacking? The quantitative and spatial measurement of receptors utilizing autoradiography is also a key problem. Where labeled ligands are employed to map receptors in brain via light microscopy, a mismatch is often encountered. Reasons offered for this problem are (1) except for autoreceptors, neurotransmitters and receptors are located in different neurons, (2) in addition to the synapse, receptors and transmitters are found throughout the neuron, (3) ligands may label only a subunit of a receptor or only one state of the receptor, and (4) autoradiography is subject to quenching. With immunohistochemical peptide mapping, a possible problem is the recognition by the antibody of a prohormone or, alternatively, a fragment of a peptide hormone in addition to the well-recognized problem of cross-reactivity of the antibody with a physiologically different peptide.

This harangue about measurement is meant to point out that what would on the surface appear to be the simplest part of research can in fact be very difficult. It is for this reason that in each section of this book a critical assessment of research techniques is made. It is vital that students learn not to accept data without an appraisal of the procedures that were employed to obtain the results.

Finally, although the theme is not explicitly dealt with in this book, students may find it educational and often entertaining to attempt to define patterns of research design in neuropharmacology as well as current trends in research areas. One common pattern is for someone to observe something in brain tissue, trace its regional distribution in the brain, and then perform a developmental study of the phenomenon in laboratory animals from prenatal through adult life. Another common pattern is for someone to develop a technique and then search (sometimes with what appears to be desperation) for projects that will utilize the technique. Yet another is a somewhat simplistic idea of attempting to relate a behavioral effect to a changing level of a single neurotransmitter, invariably the one that a team has just learned how to measure. Current trends in the neural sciences include isolating ion channels, utilizing molecular genetics to uncover new peptides, neural cartography, that is, the mapping of transmitters and neuroactive peptides in the CNS, searching for toxins with specific effects on conduction or transmission, isolating and characterizing receptors for drugs as well as endogenous neuroactive agents, and identifying trophic factors involved in synaptogenesis. It can also easily be predicted that within the next few years an intensive search will be undertaken to explain the function and integration of the approximately three dozen "classical" neurotransmitters, the neuroactive peptides, and the unclassifiable items such as adenosine, in eliciting behavioral changes. Clearly, in this search neuropharmacologic agents will be invaluable probes.

2 | Cellular Foundations of Neuropharmacology

As we begin to consider the particular problems that underlie the analysis of drug actions in the central nervous system, it may be asked, "Just what is so special about nervous tissue?" Nerve cells have two special properties that distinguish them from all other cells in the body. First, they can conduct bioelectric signals for long distances without any loss of signal strength. Second, they possess specific intercellular connections with other nerve cells and with innervated tissues such as muscles and glands. These connections determine the types of information a neuron can receive and the range of responses it can yield in return.

Cytology of the Nerve Cell

We do not need the high resolution of the electron microscope to identify several of the more characteristic structural features of the nerve cell. The classic studies of Cajal (Ramón y Cajal) with metal impregnation stains demonstrated that nerve cells are heterogeneous with respect to both size and shape. An essential structural feature of the nervous system is that each specific region of the brain and each part of each nerve cell often have several synonymous names. So, for example, we find that the body of the nerve cell is also called the soma and the perikaryon—literally, the part that surrounds the nucleus. A fundamental scheme classifies nerve cells by the number of cytoplasmic processes they possess. In the simplest case, the perikaryon has but one process, called an axon; the best examples of this cell type are the sensory fibers whose perikarya occur in groups in the sensory or dorsal root ganglia. In this case, the axon conducts the signal—which was generated by the sensory receptor in the skin or other viscera—centrally through the dorsal root into the spinal cord or cranial nerve nuclei. At the next step of complexity we find neurons possessing two processes: the bipolar nerve cells. The sensory receptor nerve cells of the retina, the olfactory mucosa, and the auditory nerve are of this form, as is a class of small nerve cells of the brain known as granule cells.

All other nerve cells tend to fall into the class known as multipolar nerve cells. While these cells possess only one axon or efferent-conducting process (which may be short or long, branched or straight, and which may possess a recurrent or collateral branch that feeds back onto the same type of nerve cell from which the axon arises), the main differences are in regard to extent and size of the receptive field of the neuron, termed the dendrites or dendritic tree. In silver-stained preparations for the light microscope, the branches of the dendrites look like trees in winter time, although the branches may be long and smooth, short and complex, or bearing short spines like a cactus. It is on these dendritic branches, as well as on the cell body, where the termination of axons from other neurons makes the specialized interneuronal communication point known as the synapse.

The Synapse

The last specialized structures of the neuron we shall discuss are the contents of the nerve ending and the characteristic specialized contact zone that has been presumptively identified as the site of functional interneuronal communication, that is, the synapse. As the axon approaches the site of its termination, it exhibits structural features not found more proximally. Most striking is the occurrence of dilated regions of the axon (varicosities) within which are clustered large numbers of microvesicles (synaptic vesicles). These synaptic vesicles tend to be spherical in shape, with diameters varying between 400 and 1200 Å. Depending upon the type of fixation used, the shape and staining properties of the vesicles can be related to their neurotransmitter content. The nerve endings also exhibit mitochondria, but do not exhibit microtubules unless the varicosity is

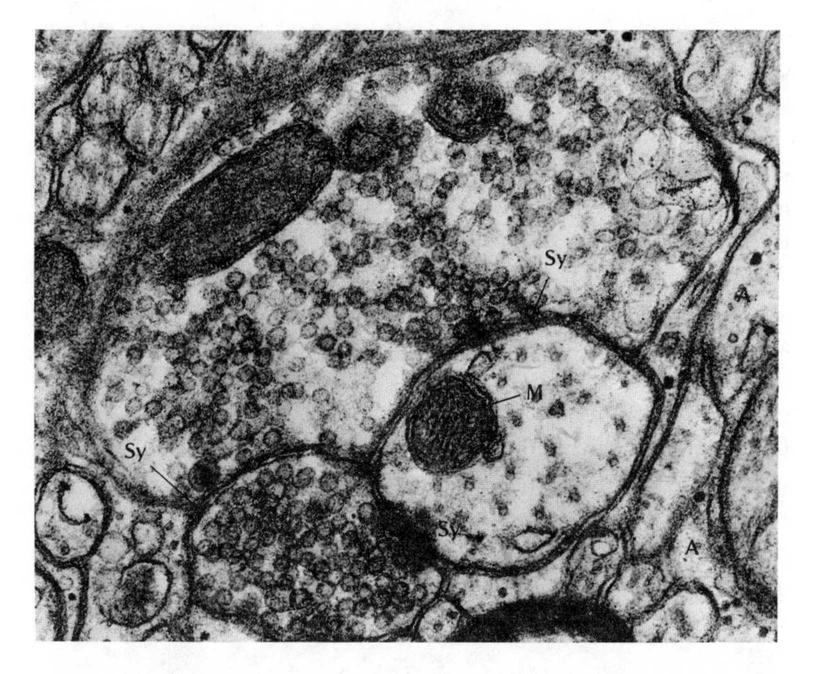

FIGURE 2-1. High-power view of two nerve terminals contacting a small dentritic spine. At this magnification the synaptic vesicles can be seen clearly, as can the zones of specialized contact (Sy). Astrocyte processes containing glycogen (A) can be seen. Note that the larger nerve terminal makes a specialized contact on the small terminal (axo-axonic) as well as on the dendrite (axo-dentritic) × 12000.

one of several such regions along an axon as it extends towards its terminal target. One or more of these varicosities may form a specialized contact with one or more dendritic branches before the ultimate termination. Such endings are known as *en passant* terminals. In this sense, the term "nerve terminal" or "nerve ending" connotes a functional transmitting site rather than the end of the axon.

Electron micrographs of synaptic regions in the central nervous system reveal a specialized contact zone between the axonal nerve ending and the postsynaptic structure (Fig. 2-1). This specialized contact zone is composed of presumed proteinaceous material lining the intracellular portions of the pre- and postsynaptic membranes and filling the synaptic cleft between the apposed cell surfaces. Such types of specialized contacts are a general form of the specialized cell contacts seen between many types of cells derived from the embryonic ectoderm, of which the nerve cell is but one. However, the specialized contact between neurons is polarized; that is, the presynaptic terminal intracellular material is composed of interrupted presynaptic dense projections measuring about 500–700 Å in diameter and separated from each other by distances of 300–400 Å. This material may be present only to bind specific presynaptic nerve endings permanently to specific postsynaptic cell sites. Alternatively, the specialized contact zone could assist in the efficiency of transmission and could constitute one potential method for modulating synaptic transmission in terms of discharge frequency. All aspects of the release and re-uptake of transmitter function quite well, however, in the peripheral nervous system with no apparent specializations. In some neurons, especially these single-process small granule cell types, the "dendrite" may also be structurally specialized to store and release transmitter.

Glia

A second element in the maintenance of the axon's integrity depends on a type of cell known as neuroglia, which we have not yet discussed. There are two main types of neuroglia. The first is called the fibrous astrocyte, a descriptive term based on its starlike shape in the light microscope and on the fibrous nature of its cytoplasmic organelles, which can be seen in both light and electron microscopy. The astrocyte is found mainly in regions of axons and dendrites and tends to surround or contact the adventitial surface of blood vessels. Functions such as insulation (between conducting surfaces) and organization (to surround and separate functional units of nerve endings and dendrites) have been empirically attributed to the astrocyte, mainly on the basis of its structural characteristics.

The second type of neuroglia is known as the oligodendrocyte. It is called the satellite cell when it occurs close to nerve cell bodies, and the Schwann cell when it occurs in the peripheral nervous system. The cytoplasm of the oligodendrocyte is characterized by rough endoplasmic reticulum, but its most prominent characteristic is the enclosure of concentric layers of its own surface membrane around the axon. These concentric layers come together so closely that the oligodendrocyte cytoplasm is completely squeezed out and the original internal surfaces of the membrane become fused, presenting the ringlike appearance of the myelin sheath in cross section. Along the course of an axon, which may be many centimeters in length, many oligodendrocytes are required to constitute its myelin sheath. At the boundary between adjacent portions of the axon covered by separate oligodendrocytes, there is an uncovered axonal portion known as the node of Ranvier.

Many central axons and certain elements of the peripheral autonomic nervous system do not possess myelin sheaths. Even these axons, however, are not bare nor exposed to the extracellular fluid, but rather they are enclosed within single invaginations of the oligodendrocyte surface mambrane. Because of this close relationship between the conducting portions of the nerve cell, its axon, and the oligodendrocyte, it is easy to see the origin of the proposition that the oligodendrocyte may contribute to the nurture of the nerve cell. While this idea may be correct, no evidence is yet available. Clearly, the glia are incapable of supporting the axon when it has been severed from the cell body, for example, by trauma or by surgically inducing lesioning. This incapacity is fortuitous, since one of the primitive methods of defining nerve cell circuits within the brain is the staining of degenerating axons following brain lesions.

Brain Permeability Barriers

While the unique cytological characteristics of neurons and glia are sufficient to establish the complex intercellular relationships of the brain, there is yet another histophysiologic concept to consider. Numerous chemical substances pass from the bloodstream into the brain at rates that are far slower than for entry into all other organs in the body. There are similar slow rates of transport between the cerebrospinal fluid and the brain, although there is no good standard in other organs against which to compare this latter movement.

These permeability barriers appear to be the end result of numerous contributing factors that present diffusional obstacles to chemicals on the basis of molecular size, charge, solubility, and specific carrier systems. The difficulty has not been in establishing the existence of these barriers, but rather in determining their mechanisms. When the relatively small protein (mol wt = 43,000) horseradish peroxidase is injected intravenously into mice, its location within the tissue can be demonstrated histochemically with the electron microscope. As opposed to the easy transvascular movement of this substance across muscle capillaries, in brain the peroxidase molecule is unable to penetrate through the continuous layer of vascular endothelial cells. The endothelial cells of brain capillaries differ from those of other tissue such as muscle and heart in that the intercellular zones of membrane apposition are much more highly developed in the brain, and are virtually continuous along all surfaces of these cells. Furthermore, cerebral vascular endothelial cells show a distinct lack of pinocytotic vesicles, which have been related to transvascular carrier systems of both large and small molecules.

Since the enzyme marker can neither go through nor between the endothelial cells, an operationally defined barrier exists. Whether or not the same barrier is also applicable to highly charged lipophobic small molecules cannot be determined from these observations. As neuropharmacologists, what concerns us more here are the factors that retard the entrance of these smaller molecules, such as norepinephrine and serotonin, their amino-acid precursors, or drugs that affect the metabolism of these and other neurotransmitters. For smaller molecules, the poorly conceived and minimally studied barriers may be tentatively regarded as solvent partitions. Many charged molecules are able to diffuse widely through the extracellular spaces of the brain, when permitted entry via the cerebrospinal fluid. Thus, once a substance can enter the perivascular extracellular spaces of the brain, it is likely to encounter few barriers that prevent it from migrating between cells—presumably along concentration gradients, but possibly moved by specific carriers—to reach those neurons or glia capable of incorporating it, responding to it, or metabolizing it.

Substances that find it difficult to get into the brain also find it exceedingly difficult to leave. Thus, when the monoamines are increased in concentration by the blocking of their catabolic pathways (see Chapter 9), the high levels of amine persist until the inhibiting agents are metabolized or excreted. One such route is the "acid transport" system by which the choroid plexus and/or brain parenchymal cells actively secrete acid catabolites. The step can be blocked by the drug probenecid, resulting in increased brain and CSF catabolites of the amines.

Since the precise nature of these barriers can still not be formulated, the student would be wise to avoid the "great wall of China" concept and lean toward the possibility of a series of variously placed, progressively selective filtration sites that discriminate substances on the basis of several molecular characteristics. With lipid soluble, weak electrolytes, a characteristic of most centrally acting drugs, transport occurs by a process of passive diffusion. Thus, a drug will penetrate the cell only in the undissociated form and at a rate consonant with its lipid solubility and its pKa. Specialized neurons exist within the ventricular system, known as the circumventricular organ systems. These neurons receive and send connections to adjacent brain regions but are on the blood side of the blood–brain barrier, and may allow the brain to monitor bloodborne hormones or metabolites.

Bioelectric Properties of the Nerve Cell

Given these structural details, we can now turn to the second striking feature of nerve cells, namely, the bioelectric property. However, even for this introductory presentation, we must understand certain basic concepts of the physical phenomena of electricity in order to have a working knowledge of the bioelectric characteristics of living cells.

The initial concept to grasp is that of a difference in potential existing within a charged field, as occurs when charged particles are separated and prevented from randomly redistributing themselves. When a potential difference exists, the amount of charge per unit of time that will flow between the two sites (i.e., current flow) depends upon the resistance separating them. If the resistance tends to zero, no net current will flow since no potential difference can exist in the absence of a measurable resistance. If the resistance is extremely high, only a minimal current will flow and that will be proportional to the electromotive force or potential difference between the two sites. The relationship between voltage, current, and resistance is Ohm's law: $V = I \cdot R$.

When we come to the measurement of the electrical properties of living cells, these basic physical laws apply, but with one exception. The pioneer electrobiologists, who did their work before the discovery and definition of the electron, developed a convention for the flow of charges based not on the electron but on the flow of positive charges. Therefore, since in biological systems the flow of charges is not carried by electrons but by ions, the direction of flow is expressed in terms of movement of positive charges. To analyze the electrical potentials of a living system, we use small electrodes (a microprobe for detecting current flow or potential), electronic amplifiers for increasing the size of the current or potential, and oscilloscopes or polygraphs for displaying the potentials observed against a time base.

Membrane Potentials

If we take two electrodes and place them on the outside of a living cell or tissue, we will find little, if any, difference in potential. However, if we injure a cell so as to break its membrane or insert one ultrafine electrode across the membrane, we will find a potential difference such that the inside of the cell is 50 or more millivolts negative with respect to the extracellular electrode (Fig 2-2). This transmembrane potential difference is found in almost all types of living cells in which it has been sought; such a membrane is said to be electrically polarized. By passing negative ions into the cell

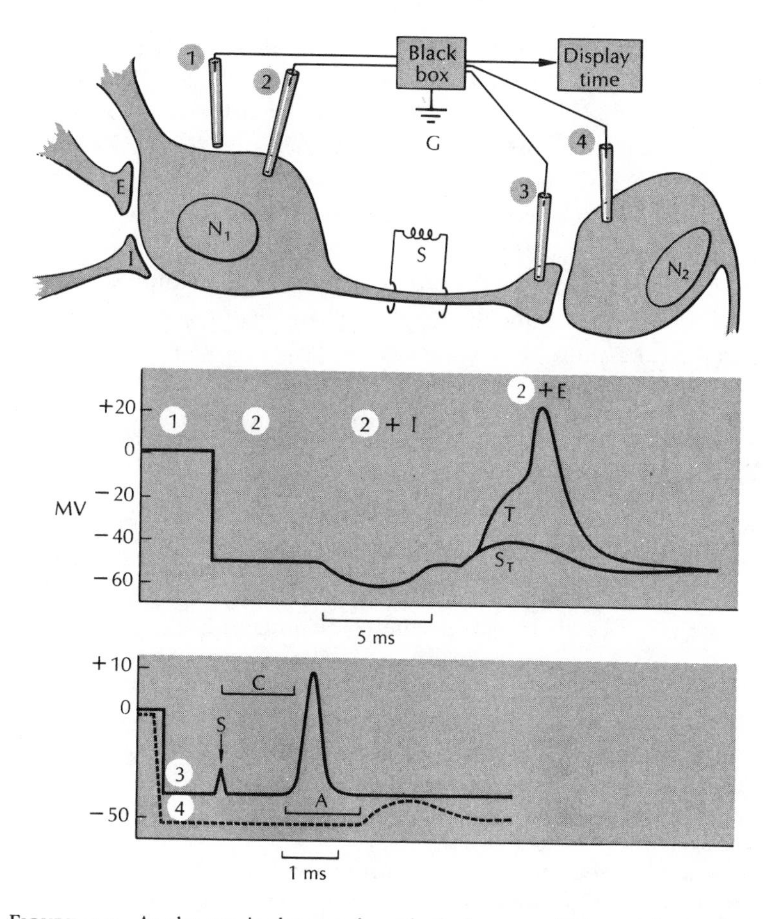

FIGURE 2-2. At the top is shown a hypothetical neuron (N_1) receiving a single excitatory pathway (E) and a single inhibitory pathway (I). A stimulating electrode (S) has been placed on the nerve cell's axon; microelectrode 1 is extracellular to nerve cell 1, while microelectrode 2 is in the cell body, and microelectrode 3 is in its nerve terminal. Microelectrode 4 is recording from within postsynaptic cell 2. The potentials and current, recorded by each of these electrodes, are being compared through a "black box" of electronics with a distant extracellular grounded electrode and displayed on an oscilloscope screen. When the cell is resting and the electrode is on the outside of the cell, no potential difference is observed (1). In the resting state, electrode 2 records a steady potential difference between inside and outside of approximately minus 50 millivolts (2). While recording from electrode 2 and stimulating the inhibitory pathway, the membrane potential is

through the microelectrode (or extracting cations), the inside can be made more negative (hyperpolarized). If positive current is applied to the inside of the cell, the transmembrane potential difference is decreased, and the potential is said to be depolarized. The potential difference across the membrane of most living cells can be accounted for by the relative distribution of the intracellular and extracellular ions.

The extracellular fluid is particularly rich in sodium and relatively low in potassium. Inside the cell, the cytoplasm is relatively high in potassium content and very low in sodium. While the membrane of the cell permits potassium ions (K^+) to flow back and forth with relative freedom, it resists the movement of the sodium ions (Na^+) from the extracellular fluid to the inside of the cell. Since the potassium ions can cross the membrane, they tend to flow along the concentration gradient—which is highest inside the cell. Potassium diffusion out of the cell leaves a relative negative charge behind owing to the negative charges of the macromolecular proteins. As the negative charge inside the cell begins to build up, the further diffusion of potassium from inside to outside is retarded.

---

hyperpolarized during the inhibitory postsynaptic potential (2 + I). When recording from electrode 2 and stimulating the excitatory pathway, a subthreshold stimulus (S^+) produces an excitatory postsynaptic potential indicated by a brief depolarization of the resting membrane potential (2 + E). When the excitatory effects are sufficient to reach threshold (T), an action potential is generated which reverses the inside negativity to inside positivity (2 + E). On the lower scale, potentials recorded by electrodes 3 and 4 are compared on the same time-base following axonal stimulation of nerve cell 1, which is assumed to be excitatory. The point of stimulus is seen as an electrical artifact at point S. The action potential generated at the nerve terminal occurs after a finite lag period due to the conduction time (c) of the axon between the stimulating electrode and the nerve terminal. The action potential in the nerve ending does not directly influence postsynaptic cell 2 until after the transmitter has been liberated and can react with nerve cell 2's membrane, causing the excitatory postsynaptic potential indicated by the dotted line. The time between the beginning of the action potential recorded by microelectrode 3 and the excitatory postsynaptic potential recorded by electrode 4 (A) is the time required for excitation secretion coupling in the nerve terminal and the liberation of sufficient transmitter to produce effects on nerve cell 2.

Eventually an equilibrium point will be reached that is proportional to certain physical constants and to the relative concentrations of intracellular and extracellular potassium and chloride ions. These concepts of ionic diffusion potentials across semipermeable membranes apply generally, not only to nerve and muscle but to blood, glandular, and other cells large enough to have their transmembrane potential measured.

Membrane Ion Pumps

When the nerve cell or muscle fiber can be impaled by electrodes to record transmembrane potential, the relation between the membrane potential and external potassium concentration can be directly tested by exchanging the extracellular fluid for artificial solutions of varying potassium concentration. When this experiment is performed on muscle cells, we find that the membrane potential bears a linear relationship to the external potassium concentration at normal to high potassium concentrations but that it deviates from this linear relationship when the external potassium concentration is less than normal. To account for this discrepancy we must now reexamine an earlier statement. While the plasma membranes of nerve and muscle cells and other types of polarized cells are relatively impermeable to the flow of sodium ions along the high concentration gradient from extracellular to intracellular, they are not completely impermeable. With radioisotope experiments it can be established that a certain amount of sodium "leaks" into the resting cell from outside. The amount of measurable sodium entry into the cell occurs at a rate sufficient to double the intracellular sodium concentration in approximately one hour if there were not some opposing process to maintain the relatively low intracellular sodium concentration. The process that continuously maintains the low intracellular sodium concentration is known as active sodium transport, or colloquially as the "sodium pump." This pump mechanism ejects sodium from the inside of the cell against the high concentration and electrical gradients forcing it in. However, the "pump" does not handle sodium exclusively but requires the presence of extracellular potassium. Thus, when a sodium ion is ejected from the cell, a potassium ion is incorporated into the cell.

When the external potassium concentration is near normal, the transmembrane potential, which is based mainly on potassium concentration differences, behaves as if there were actually more extracellular potassium than really exists. This is because the sodium–potassium exchange mechanism elevates the amount of potassium coming into the cell. Remember that potassium permeability is relatively high and that potassium will tend to diffuse out of the cell because of its concentration gradient but to diffuse into the cell because of charge attraction. Therefore, two factors operate to drive potassium into the cell in the presence of relatively low external potassium concentration: (1) the electrical gradient across the membrane and (2) the sodium–potassium pump mechanism. The latter system could be considered "electrogenic" since at low external K^+ concentrations it modifies the electrical status of the muscle membrane. Other metabolic pumps operate simply to exchange cationic species across the membrane and are "nonelectrogenic." The relative "electrogenicity" of a pump may depend on the ratio of the exchange cations (i.e., 1:1 or 2:2 or 3:2). The pump is immediately dependent upon metabolic energy and can be blocked by several metabolic poisons such as dinitrophenol and the rapid-acting cardiac glycoside, ouabain.

The Uniqueness of Nerve

All that we have said regarding the transmembrane ionic distributions applies equally as well to the red blood cell or glia as to the nerve membrane. Thus, the possession of a transmembrane potential difference is not sufficient to account for the bioelectric properties of the nerve cell. The essential difference between the red blood cell and the nerve cell can be brought out by applying depolarizing currents across the membrane. When the red blood cell membrane is depolarized, the difference in potential across the cell passively follows the imposed polarization. However, when a nerve cell membrane, such as the giant axon of an invertebrate, is depolarized from a resting value of approximately −70 millivolts to approximately −10 to −15 millivolts, an explosive self-limiting process occurs by which the transmembrane potential is reduced not merely to zero but overshoots zero, so that the inside of the membrane now becomes positive with respect to the outside. This overshoot may extend for 10 to 30 millivolts in the positive direction. Because of this explosive response to an electrical depolarization the nerve membrane is said to be electrically excitable, and the resultant explosion is known as the action potential.

Analysis of Action Potentials

In an elegant series of pioneering experiments that are now classical, Hodgkin, Huxley, and Katz were able to analyze the various ionic steps responsible for the action potential. When the cell begins to depolarize from stimulation current, the current flow across the membrane is carried by potassium. As the membrane becomes more depolarized, the resistance to Na decreases (or Na conductance increases) and more sodium enters the cell along its electrical and concentration gradients. As sodium enters, the membrane becomes more and more depolarized, which further increases the conductance to sodium and thus further depolarizes the membrane at a greater rate. Such conductance changes are termed voltage-dependent. This self-perpetuating process continues, driven by the flow of sodium ions tending toward their equilibrium distribution, which should be proportional to the original extracellular and intracellular concentrations of sodium.

However, the peak of the action potential does not attain the equilibrium potential predicted on the basis of transmembrane sodium concentrations because of a second phase of events. The voltage-dependent increase in Na^+ conductance and the consequent depolarization also activates a voltage-dependent K^+ conductance, and K flow then also increases along its concentration gradient, that is, from inside to outside the cell. This process restricts the height of the reversal potential since it tends to maintain the inside negativity of the cell and also begins to reduce the membrane conductance to sodium, thus making the action potential a self-limiting phenomenon. In most nerve axons, the action potential lasts for approximately 0.2 to 0.5 milliseconds (ms), depending on the type of fiber and the temperature in which it is measured.

Once the axon has reached threshold, the action potential will be propagated at a rate that is proportional to the diameter of the axon and that is further accelerated by the presence of the glial myelin sheaths, which restrict the active conducting points to the node of Ranvier. As mentioned above, these are the only sites on the myelinated axons at which the axonal membrane is directly exposed to the extracellular fluid and thus are the only sites at which transmembrane ionic flows can take place. Therefore, instead of the action potential propagating from minutely contiguous sites of the membrane, the action potential in the myelinated axon leaps from node to node. This saltatory conduction is consequently much more rapid.

The threshold level for an all-or-none action potential is also inversely proportional to the diameter of the axon: Large myelinated axons respond to low values of imposed stimulating current, whereas fine and unmyelinated axons require much greater depolarizing currents. Local anesthetics appear to act by blocking activation of the sodium conductance preventing depolarization.

Once threshold has been reached, a complete action potential will be developed unless it occurs too soon after a preceding action potential, during the so-called refractory phase. This phase varies for different types of excitable nerve and muscle cells and appears to be related to the activation process increasing sodium conductance, a phenomenon that has a finite cycling period; that is, the membrane cannot be reactivated before a finite interval of time has occurred. Potassium conductance increases with the action potential and lasts slightly longer than the activation of sodium conductance. This results in a prolonged phase of after-hyperpolarization due to the continued redistribution of potassium from inside to outside the membrane. If the axonal membrane is artificially maintained at a transmembrane potential equal to the potassium equilibrium potential, no after-hyperpolarization can be seen.

Ion Channels

The experiments of Hodgkin and Huxley defined the kinetics of cation movement during nerve membrane excitation without constraint on the mechanisms accounting for the movement of ions through the membrane. The discovery that cation movement can be selectively blocked by drugs, and that Na^+ permeability (blocked by tetrodotoxin) can be separated from K^+ permeability (blocked by tetraethylammonium [TEA]), made more detailed analysis of the ion movement mechanisms feasible. Membrane physiologists now agree that there are several ion-specific pathways that form separate and independent "channels" for passive movement of Na^+, K^+, Ca^{2+}, and Cl^-. According to Hille, the channels are pores that open and close in an all-or-nothing fashion on time scales of 0.1 to 10 ms to provide aqueous channels through the plasma membrane that ions can traverse. In his view, the channels can be conceived as protein macromolecules within the fluid lipid plane of the membrane; these channel macromolecules can exist in several interconvertible conformations, one of which permits ion movement while the others do not. The conformational shifts from one form to another are sensitive to the bioelectric fields operating on the membrane; by facilitating or retarding the conformational shifts, the ion channels are "gated." In this concept, Ca^{2+} acts at the membrane surface to alter permeability only by virtue of the effect its charge has on the electric fields of otherwise fixed (mainly negative) organic charges. The altered fields in turn can gate the channels because a part of the channel protein is able to sense the field, and thus to modulate the conformational shifts that open or close the gate. When ions flow across the membrane, the ionic current changes membrane potential and other membrane properties.

From a variety of experimental methods, including methods that can sample single ion channels in cultured neurons and other excitable cells, a large number of specific ion channels have been described. In current terminology, one recognizes (1) "non-gated," passive ion channels, previously referred to as "leakage channels," that are continuously open; (2) voltage-gated (i.e., voltage sensitive) channels whose opening and closing are affected by the membrane potential inside of the cell; (3) chemically gated channels whose openings and closings are affected by receptors on the external plasma membrane, such as those affected by drugs and other transmitters; and (4) ion-gated channels, whose openings and closings are affected by shifts in intracellular ion concentrations. Ion-gated channels are often also sensitive to membrane potential and to external regulatory receptors, and chemically gated channels are often also voltage sensitive. These various modes of interaction provide an extremely rich spectrum of responses, thus greatly complicating what were once simple rules of excitability and ion conductance regulation.

Junctional Transmission

While these ionic mechanisms appear to account adequately for the phenomena occurring in the propagation of an action potential down an axon, they do not *per se* explain what happens when the action potential reaches the nerve ending. At the nerve ending the membrane of the axon is separated from the membrane of the postjunctional nerve cell, muscle, or gland by an intercellular space of 50–200 Å (Fig. 2-1). When an electrode can be placed in both the terminal axon and the postsynaptic cell, depolarization of the nerve terminal does not result in a direct instantaneous shift in the transmembrane potential of the postsynaptic element, except in those cases in which the connected cells are electronically coupled. With this exception, the junctional site seldom exhibits direct electrical excitability like the axon.

Postsynaptic Potentials

With the advent of microelectrode techniques for recording the transmembrane potential of nerve cells *in vivo*, it was possible to determine the effects of stimulation of nerve pathways that had previously been shown to cause either excitation or inhibition of synaptic transmission. From just such studies Eccles and his colleagues observed that subthreshold excitatory stimuli would produce postsynaptic potentials with time durations of 2 to 20 ms. The excitatory postsynaptic potentials could algebraically summate both with the excitatory and inhibitory postsynaptic potentials. Most importantly, the duration of these postsynaptic potentials was longer than could be accounted for on the basis of electrical activity in the preterminal axon or on the electrotonic conductive properties of the postsynaptic membrane (Fig. 2-2). This latter observation combined with the fact that synaptic sites are not directly electrically excitable provides the conclusive evidence that central synaptic transmission must be chemical: The prolonged time course is compatible with a rapidly released chemical transmitter whose time course of action is terminated by local enzymes, diffusion, and re-uptake by the nerve ending.

By such experiments it was possible to work out the basic ionic mechanisms for inhibitory and excitatory postsynaptic potentials. When an ideal excitatory pathway is stimulated, the presynaptic element liberates an excitatory transmitter that activates an ionic conductance of the postsynaptic membrane. This response leads to an increase in one or more transmembrane ionic conductances, depolarizing the membrane toward the sodium equilibrium potential; in the resting state, as has already been discussed, the membrane resides near the potassium equilibrium potential. If the depolarization reaches the threshold for activating adjacent voltage-dependent conductances, an all-or-none action potential (spike) will be triggered. For many neurons the axon hillock has the lowest spike threshold. If the resultant depolarization is insufficient to reach threshold, the cell can still discharge if additional excitatory postsynaptic potentials summate adequately.

The postsynaptic potential resulting from the stimulation of an ideal inhibitory pathway to the postsynaptic cell has been explained in terms of the fact that inhibitory transmitter selectively activates channels for Cl^- or K^+ resulting in a diffusion of ions and a hyperpolarization of the membrane. This counterbalances the excitatory postsynaptic potentials.

Because the sites of synaptic or junctional transmission are electrically inexcitable, the postsynaptic membrane potential can be maintained at various levels by applying current through intracellular electrodes and changing the intracellular concentrations of various ions. By such maneuvers, it is possible to poise the membrane at or near the so-called equilibrium potentials for each of the ionic species and to determine the ionic species whose equilibrium potential corresponds to the conductance change caused by the synaptic transmitter. This is the most molecular test for the identification of actions of a synaptic transmitter substance. (However, certain objections can be raised to this test in terms of those nerve endings making junctional contacts on distal portions of the dendritic tree. Here, the postsynaptic potentials may be incompletely transmitted to the cell body where the recording electrode is placed.)

On re-reading the above section, note the insidious use of the term "ideal." Present active research in synaptic physiology has come to exploit several *in vitro* experimental preparations such as neurons in long-term culture and "slice" preparations of specific brain regions. Using such *in vitro* preparations, and sophisticated electronic bridge circuits to clamp the transmembrane potential of neurons during intracellular recordings far longer than is possible *in situ*, the variety of ion channels and ion conductance control mechanisms which mammalian neurons appear to exhibit routinely has been greatly expanded. It is generally considered that classically acting neurotransmitters produce their effects on receptor-coupled ion conductances which are voltage independent, that is, the receptor will alter the coupled ion channel regardless of the membrane potential at the moment. Nevertheless, many non-classical transmitters do seem to operate on receptors coupled to voltage-sensitive mechanisms. Some neurons, for example, can be shifted into burst-firing modes by this type of synaptic action: a transmitter turns off an ion conductance, which only becomes activated when the neuron is partially depolarized. Because such transmitters can not be observed to produce an effect on membrane properties in the absence of the depolarizing voltage shift which activates the conductance to be modified, the complex form of this interaction has been considered to be a "modulatory" action rather than a transmitting action. Transmitters whose receptors are associated with activation of cyclic nucleotide synthesis frequently produce these complex forms of interaction. Similarly, many neuropeptides appear to affect certain of their target cells by modifying responses to other transmitters while not showing any direct shifts in membrane potential or conductance when tested for actions on their own.

Although, it may eventually be possible to evolve a set of experimental criteria to discriminate between classical transmitters acting to alter voltage-independent receptor mechanisms and "modulators" acting to modify the effects of the classical transmitters on their targets, this is not now the case. The student is advised to maintain an appreciative awareness of these potentially complex interaction systems without attributing true or false flags to which molecules transmit which signals. In the following section, we examine some of these less-classical synaptic events and their advantageous properties. However, two other aspects of ionic mechanisms bear some passing mention. First, despite our preoccupation with action potentials and their modification, Bullock has pointed out that the most numerous CNS neurons, the small single-process type of granule-like cell, may conduct its neuronal business within its restricted small spatial domain with no need ever to fire a spike. Second, those neurons that do fire spikes may sometimes do so unconventionally by using influx of Ca^{++} ions (a voltage-sensitive Ca conductance) rather than Na^{+}. This Ca spike may represent a mechanism to transmit activity from the cell body out to the dendritic system and may play a functional role in those neurons whose dendrites can also release transmitter—such as the catecholamine cell-body nuclei. Thus, the simplified ideal version of ionic mechanisms may be only one of many regulatory mechanisms between connected cells.

It is important to recognize that while there are only four major permanent ions that exist in significant quantities in the body (Na, K, Ca, and Cl) there are far more varieties of regulatable ion channel. This means that not all Ca channels, for example, are created equally nor are they expected to produce the same effects. The ones on the dendrites of some neurons that can generate action potentials are different channels than those on other synaptic domains of the dendrites and perikaryonal membrane that open when the membrane potential shifts and influences the operation of other (ion-gated) channels. For example, the Ca-dependent K channel is opened when Ca rushes into a neuron during active depolarization by voltage-sensitive Na channels. While all this sounds terribly exciting, especially for enhancing our understanding of how transmitters can act to regulate the activity of neurons, don't assume that only neurons have regulatable ion channels. Many secretory cells have such channels, such as those of the pancreatic islets and the anterior or intermediate lobes of the pituitary. Even lymphocytes, that secrete antibodies and respond to blood-borne messengers referred to as antigens, have regulatable channels. Last, but certainly not least, the most numerous cells of the nervous system, namely the glia, have also been reported to have regulatable channels for K and Ca, responding both to voltage and to chemical transmitters. These properties of glia are, in fact, mainly observed when the glia are cultured from cell lines or neonatal brains and kept *in vitro*, so they may not normally be expressed by glia differentiating *in situ*. If one only knew what glia did, it might be possible to ask whether such channels as they may possess *in vivo* play any part in their actions.

Slow Postsynaptic Potentials

Most of the postsynaptic potentials described by Eccles and his colleagues were relatively short, usually 20 ms or less, and appear to result from passive changes in ionic conductance. Recently postsynaptic potentials of slow onset and several seconds' duration have been described (Fig. 2-2), both of a hyperpolarizing nature and of a depolarizing nature. While such prolonged postsynaptic potentials could be due either to a prolonged release of transmitter or to a persistence of the transmitter at postsynaptic receptor sites, there is substantial support for the possibility that slow postsynaptic potentials could also be caused by other forms of synaptic communication. Many of these slower synaptic potentials are not accompanied by the expected increase in transmembrane ionic conductances, but are instead accompanied by increased transmembrane ionic impedance. The most simple explanations for membrane potential changes accompanied by increased membrane resistance are: (1) that the effect is generated at a "presynaptic" site whereby tonically active excitatory or inhibitory synapses are inactivated by synapses on their terminals; (2) that the action of the transmitter released by a slow synapse is to inactivate a normal finite resting conductance, such as that of sodium or potassium. Since most of the sites where slow synaptic potentials have been observed do not fulfill the anatomical prerequisite of axo-axonic presynaptic synapses, the second explanation may be more likely. A third explanation for such potentials has also been advanced, namely, that the transmitter activates an electrogenic pump mechanism, that is, a metabolically driven ionic "pump" that exchanges unequal numbers of similarly charged ions across the membrane. Studies on invertebrate neurons indicate that such electrogenic mechanisms should have several distinguishable properties: They should be virtually independent of the distribution of the ions to be pumped; they should not exhibit an equilibrium potential; they should be temperature dependent; and they should be sensitive to metabolic poisons and uncouplers of oxidative phosphorylation. Such data are extremely complex to analyze as several membrane properties also include temperature dependency and metabolic poison sensitivity. Thus, at the present time, a true electrogenic synapse in the mammalian central nervous system remains to be described. However, it also seems likely that transmitters such as the catecholamines can activate the synthesis of cyclic nucleotides, which in turn can activate intraneuronal protein kinases that can phosphorylate specific membrane proteins. The phosphorylation of a membrane protein could be expected to alter its ionic permeability, and perhaps such changes lie at the root of the membrane effects of several types of neurotransmitters.

Conditional Actions of Transmitters

Frequently, transmitters produce novel actions unlike those of classically conceived transmitters. These unconventional actions suggest that broader definitions are useful for conceptualizing the range of regulatory signals involved in interneuronal communication, and helpful for examination of transmitter actions. For example, when the beta-adrenergic effects of locus ceruleus stimulation are examined, target cell responses no longer adhere to standard concepts of inhibition. Rather, they appear to fit better the designation of "biasing" or "enabling." The latter term indicates that the enabling transmitter (in this example, norepinephrine) can enhance or amplify the effectiveness of other transmitter actions converging on the common target neurons during the time period of the enabling circuit's activity (see Chapter 7).

Such actions would have been difficult to appreciate in earlier eras. To re-explore the issue of time course on the more complex interactions, it may be useful to speak of "conditional" and "unconditional" actions. Unconditional actions would refer to those which a given transmitter evokes by itself (i.e., in the absence of other transmitters acting on the common target cell). Conditional actions, occurring either pre- or postsynaptically, would include, but would not be limited to, the type of enhancement that is subsumed by "enabling." In such a conditional interaction each transmitter would act at its own pre- or postsynaptic transmitter receptor, and would interact on that target cell when both transmitters occupy their receptors simultaneously.

Conditional transmitter actions may well have been overlooked because early experiments had rightly to focus on unconditional effects. However, consider the rapidly growing bodies of actions ascribed to neuropeptides, and in particular the voltage-dependent and ion-sensitive effects they produce along with monoamine transmitters. The later effects will perhaps become more commonly observed as the requirements of *in vitro* electrophysiological analysis become better defined. The voltage-dependent mechanism might easily appear as "no actions" at all, because they often reveal that the application of a neuropeptide produces no observable shift in either membrane potential or membrane resistance when tested at normal resting membrane potentials levels. However, when the test cell is moved from basal conditions, as by the depolarizing action of a simple amino acid, the resultant depolarization can be reduced or eliminated by the simultaneous application of the same dose of the peptide. This complex conditional interaction is antipodal to the enabling response, and so could be called "disenabling."

The pairs of neurotransmitters which coexist within certain neurons are a major topic (see Chapter 12). Such coexisting pairs, or larger groups of transmitters, broadly defined, may also act in a coordinated manner. Such interactions need not be viewed as simply conditional or unconditional reactions: One transmitter of a pair could simply initiate an action and the coreleased signal would then act, *inter alia*, to terminate that action. In fact, based on actions reported for some peptides, a coexisting peptide could "enable" or "disenable" its partner transmitters in some contexts or act as conventional transmitters in others (see Iversen, Bloom).

Thus, it can be seen that there are abundant circuits, abundant transmitters and, for each, many classes of chemically coupled systems exist to transduce the effects of active transmitter receptors. These receptors can operate either actively or passively, conditionally or unconditionally, over a wide range of time through nonspecific, dependent, or independent metabolic events. Clearly, neurons have a broad but finite and, as yet, incompletely characterized repertoire of molecular responses that messenger molecules (transmitters, hormones, and drugs) can elicit. The power of the chemical vocabulary of such components is their combinatorial capacity to act conditionally and coordinatively and to integrate the temporal and spatial domains within the nervous system.

Transmitter Secretion

We have already seen that the cellular machinery of the neuron suggests it functions as a secretory cell. The secretion of synaptic transmitters is the activity-locked expression of neuronal activity induced by the depolarization of the nerve terminal. Recently, it has been possible to separate the excitation–secretion coupling process of the presynaptic terminal into at least two distinct phases. This has been made possible through an analysis of the action of the puffer fish poison tetrodotoxin, which blocks the electrical excitation of the axon but does not block the release of transmitter substance from the depolarized nerve terminal. The best such experiments have been performed in the giant synaptic junctions of the squid stellate ganglion, in which the nerve terminals are large enough to be impaled by recording and stimulating microelectrodes, and with recording from the postsynaptic and presynaptic neurons. In this case, when tetrodotoxin blocks conduction of action potentials down the axon, electrical depolarization of the presynaptic terminal still results in the appearance of an excitatory postsynaptic potential in the ganglion neuron. Since tetrodotoxin selectively blocks the voltage-dependent Na^+ conductance, the excitation secretion must be coupled more closely to other ions. Present evidence strongly favors the view that a voltage-sensitive Ca^{2+} conductance is required for transmitter secretion. In review, we see that the spike-generating and conducting events rest on voltage-dependent ion conductance changes, while synaptic events rest on voltage-independent or voltage-sensitive conductances.

Biochemical, ultrastructural, and physiological experiments have led to the concept that transmitter molecules are stored within vesicles in the nerve terminal and that the Ca-dependent excitation–secretion coupling within the depolarized nerve terminal requires the transient exchange of vesicular contents into the synaptic cleft. It is unclear whether the vesicle simply undergoes a rapid fusion with the presynaptic specialized membrane to allow the transmitter stored in the vesicle to diffuse out, or whether the process of exocytotic release simultaneously requires the insertion of the vesicle membrane into the synaptic plasma membrane, reappearing later by the reverse process, namely, endocytosis. Information on the lipid and protein components of the two types of membranes once suggested long-term fusion–endocytosis cycles were unlikely, but more recent data are compatible with either fusion release or contact release. In noradrenergic vesicles, for example, the transmitter is stored in very high concentrations in ternary complexes involving ATP, Ca, and possibly additional lipids or lipoproteins. Unfortunately, neurochemically homogeneous vesicles from central synapses have never been completely purified, and therefore all such analyses remain somewhat open to interpretation. For other molecules under active consideration as neurotransmitters, storage within brain synaptic vesicles has been extremely difficult to document chemically. The difficulties arise from the fact that homogenization of the brain to prepare synaptosomes disrupts both structural and functional integrity, and under these conditions the failure to demonstrate that amino-acid transmitters are stored in vesicles is rationalized as uncontrollable leakage. With electron microscopy and autoradiography, however, sites accumulating transmitters for which there is a high-affinity, energy-dependent uptake process can be demonstrated. Under these conditions, authentic "synaptic terminals" are identified but glial processes are also labeled. The vesicle story is further discussed in Chapter 8.

In some cases, release of the transmitter can be modulated "presynaptically" by the neuron's own transmitter (autoreceptors) or by the effects of transmitters released by other neurons in the vicinity of the terminal or the cell body. Autoreceptors are conceived to be receptors that are generally distributed over the surface of a neuron and are sensitive to the transmitter secreted by that neuron. In the case of the central dopamine-secreting neurons, such receptors have been related to the release of the transmitter and to its synthesis. Such effects seem to be achieved through different receptor mechanisms than those by which the same transmitter molecule acts postsynaptically.

An Approach to Neuropharmacological Analysis

You can now see that the business of analyzing bioelectrical potentials can be very complicated, even when restricted to the changes in single neurons or to small portions of contiguous neurons. But if we restrict our examination of centrally active drugs to analyses of effects on single cells, we can ask rather precise questions. For example, does drug X act on resting membrane potential or resistance, on an electrogenic pump, or on the sodium- or potassium-activation phase of the action potential? Or does it act by blocking or modulating the effects of junctional transmission between two specific groups of cells?

Unfortunately, both for us and for the literature, we have not had these precise electrophysiological tools for very long. Earlier neuropharmacologists were thus required to examine effects of drugs on populations of nerve cells. This was usually done in one of two ways. Large macroelectrodes were employed to measure the potential difference between one brain region and another. These electroencephalograms reflect mainly the moment-to-moment algebraic and spatial summations of slow synaptic potentials, and almost none of their electrical activity is due to actual action potentials generated by individual neurons (unit discharges). A second type of analysis was based on evoked changes occurring when a gross sensory stimulation (such as a flash of light or a quick sound) was delivered. Changes in recordings from cortical or subcortical sites along the sensory pathway were then sought during the action of a drug. While we can criticize the technical and interpretative shortcomings of such methods of central drug analysis, these methods were able to reflect the population response of a group of neurons to a drug, something that single unit analysis can do only after many single recordings are collated. The macroelectrode methods are now receiving increased attention again since, as noninvasive methods, they can be used to examine drug actions clinically.

Approaches

If, as modern-day neuropharmacologists, we are chiefly concerned with uncovering the mechanisms of action of drugs in the brain, there are several avenues along which we can organize our attack: We could choose to examine the way in which drugs influence the perception of sensory signals by higher integrative centers of the brain. This is compatible with a single-neuron and ionic conductance type of analysis, directed, say, at how drugs affect inhibitory postsynaptic potentials. Drugs that cause convulsions, such as strychnine, have been analyzed in this respect, but all types of inhibitory postsynaptic potential are not affected by strychnine.

A second basic approach would be to use both macroelectrodes and microelectrodes to compare the drug response of single units and populations of units in the same brain region. This approach is clearly limited, however, unless we understand the intimate functional relations between the multiple types of cells found even within one region of the brain.

A third approach is also possible. We could choose to separate the effects of drugs between those affecting generation of the action potential and its propagation and those acting on junctional transmission. For this type of analysis, we must identify the chemical synaptic transmitter for the junctions to be studied. Many of the interpretative problems already alluded to can be attacked through this approach. Thus, as you might expect, there is likely to be more than one type of excitatory and inhibitory transmitter substance, and a convulsant drug may affect the response to one type of inhibitory transmitter without affecting another. Moreover, a drug may have specific regional effects in the brain if it affects a unique synaptic transmitter there. In fact, by using this approach it may be possible to find drug effects not directly reflected in electrical activity at all but related more to the catabolic or anabolic systems maintaining the required functional levels of transmitter. We shall conclude this chapter by considering the techniques for identifying the synaptic transmitter for particular synaptic connections. The chapters that follow are organized to present in detail our current understanding of putative central neurotransmitter substances.

Identification of Synaptic Transmitters

How then, do we identify the substance released by nerve endings? The entire concept of chemical junctional transmission arose from the classical experiments of Otto Loewi, who demonstrated chemical transmission by transferring the ventricular fluid of a stimulated frog heart onto a nonstimulated frog heart, thereby showing that the effects of the nerve stimulus on the first heart were reproduced by the chemical activity of the solution flowing onto the second heart. Since the phenomenon of chemical transmission originated from studies of peripheral autonomic organs, these peripheral junctions have become convenient model systems for central neuropharmacological analysis.

Certain interdependent criteria have been developed to identify junctional transmitters. By common-sense analysis, one would suspect that the most important criterion would be that a substance suspected of being a junctional transmitter must be demonstrated to be released from the prejunctional nerve endings when the nerve fibers are selectively stimulated. This criterion was relatively easily satisfied for isolated autonomic organs in which only one or at most two nerve trunks enter the tissue and the whole system can be isolated in an organ bath. In the central nervous system, however, satisfaction of this criterion presumes: (1) that the proper nerve trunk or set of nerve axons can be selectively stimulated and (2) that release of the transmitter can be detected in the amounts released by single nerve endings after one action potential. This last subcriterion is necessary since we wish to restrict our analysis to the first set of activated nerve endings and not examine the substances released by the secondary and tertiary interneurons in the chain, some of which might reside quite close to the primary endings. The biggest problem with this criterion in the brain, however, is the lack of a method for detecting release that does not in itself destroy the functional and structural integrity of the region of the brain being analyzed. Such techniques as internally perfused cannulae or surface suction cups are chemically at the same level of resolution as the evoked potential and the cortical electroencephalogram. Each of these methods records the resultant activity of thousands if not millions of nerve endings and synaptic potentials. Release has also been studied in brain slices or homogenate subfractions incubated in "physiological buffer solutions" *in vitro*. These techniques can demonstrate the effects of depolarizing drugs or electric fields to simulate events in the living brain.

Localization

Because it is difficult, if not impossible, to identify the substance released from single nerve endings by selective stimulation, the next-best evidence might be to prove that a suspected synaptic transmitter resides in the presynaptic terminal of our selected nerve pathway. Normally, we would expect the enzymes for synthesizing and catabolizing this substance to also be in the vicinity of this nerve ending, if not actually part of the nerve-ending cellular machinery. In the case of neurons secreting peptides or simple amino-acid transmitters, however, these metabolic requirements may need further consideration. To document the presence of neurotransmitter, several types of specific cytochemical methods for both light microscopy and electron microscopy have been developed. More commonly employed is the biochemical population approach, which analyzes the regional concentrations of suspected synaptic transmitter substances. However, presence *per se* indicates neither releasability nor neuroeffectiveness (e.g., acetylcholine in the nerve-free placenta or serotonin in the enterochromaffin cell). Although it has generally been conceived that a neuron makes only one transmitter and secretes that same substance everywhere synaptic release occurs, neuropeptide exceptions to this rule have become common.

Synaptic Mimicry: Drug Injections

A third criterion arising from the peripheral autonomic nervous system analysis is that the suspected exogenous substance mimics the action of the transmitter released by nerve stimulation. In most pharmacological studies of the nervous system, drugs are administered intravascularly or onto one of the external or internal surfaces of the brain. The substances could also be directly injected into a given region of the brain, although the resultant structural damage would have to be controlled and the target verified histologically. The analysis of the effects of drugs given by each of these various gross routes of administration is quite complex.

We know that diffusional barriers selectively retard the entrance from the bloodstream of many types of molecules into the brain. These barriers have been demonstrated for most of the suspected central synaptic agents. In addition, we suspect that extracellular catabolic enzymes could destroy the transmitter as it diffuses to the postulated site of action. A further complicating aspect of these gross methods of administration is that the interval of time from the administration of the agent to the recording of the response is usually quite long (several seconds to several minutes) in comparison with the millisecond intervals required for junctional transmission. The delay in response further reduces the likelihood of detecting the primary site of action on one of a chain of neurons.

Microelectrophoresis

The student will now realize how important it is to have methods of drug administration equal in sophistication to those with which the electrical phenomena are detected. The most practical micromethod of drug administration yet devised is based upon the principle of electrophoresis. Micropipettes are constructed in which one or several barrels contain an ionized solution of the chemical substance under investigation. The substance is applied by appropriately directing the current flow. The microelectrophoretic technique, when applied with controls to rule out the effects of pH, electrical current, and diffusion of the drug to neighboring neurons, has been able to overcome the major limitations of classical neuropharmacological techniques. Frequently, a multiple-barreled electrode is constructed from which one records the spontaneous extracellular discharges of single neurons while other attached pipettes are utilized to apply drugs. One can also construct an intracellular microelectrode glued to an extracellular drug-containing multielectrode, so that the transmembrane effects of these suspected transmitter agents can be compared with the effects of nerve pathway stimulation. The intracellular electrode can also be used to poise the relative polarization of the membrane and let us detect whether the applied suspected transmitter and that released by nerve stimulation cause the membrane to approach identical ionic equilibrium potentials.

Considerable experimentation with this technique has made possible certain generalizations regarding the actions of each putative neurotransmitter that has been studied. These substances are reviewed in detail in each of the chapters that follow. However, it should be borne in mind at the outset that certain substances have more or less invariable actions, for example, gamma-aminobutyrate and glycine always act to inhibit, while glutamate and aspartate always excite. Insofar as we know, these actions arise from increased membrane conductances to Na, K, Ca, or Cl in every case. Other substances have many kinds of effects, depending upon the nature of the cell whose receptors are being tested. Thus, acetylcholine frequently excites but can also inhibit, and the receptors for either response can be nicotinic or muscarinic. Similarly, dopamine, norepinephrine, and serotonin almost always inhibit, but they have a few excitatory actions that are probably not completely artifactual.

Pharmacology of Synaptic Effects

The fourth criterion for identification of a synaptic transmitter requires identical pharmacological effects of drugs potentiating or blocking postsynaptic responses to both the neurally released and administered samples. Because the pharmacological effects are often not identical (most "classical" blocking agents are extrapolated to brain from effects on peripheral autonomic organs), this fourth criterion is often satisfied indirectly with a series of circumstantial pieces of data. Recently, with the advent of drugs blocking the synthesis of specific transmitter agents, the pharmacology for certain families of transmitters has been improved.

The electrophysiological analysis of drug and transmitter actions in the central nervous system was traditionally accomplished in terms of single cell activity *in vivo*. Four types of physiological responses served as the major indices to compare exogenously applied transmitter candidates and drugs with the effects of endogenous transmitters: (1) spontaneous activity, (2) orthodromic synaptically evoked activity, (3) antidromic activity, and (4) relative responses to independently acting excitatory or inhibitory transmitter released from another experimental source, such as another barrel of a multibarrel pipette. These techniques have been most successful when applied to large neurons, whose selected afferent pathways can be stimulated and the transmembrane effects specifically analyzed.

However, when the unit recording techniques are applied to the mammalian brain, visualization and selection of the neuron under investigation are almost impossible. Although we can generally select a relatively specific brain region in which to insert the microelectrode assembly, we must utilize electrophysiological criteria for cell identification. We must also depend upon geometrical attributes of the nerve cell to encounter its electrical activity as the electrode is navigated by blind mechanical means through the brain. Our ability to encounter cells in the brain depends partly on the size and type of the electrode assembly we use, partly on the size and spontaneous activity of the cell we are approaching, and partly on the surgical or chemical means by which we have prepared the animal (viz., presence or absence of anesthesia, homeostatic levels of blood pressure, and tissue oxygen and carbon dioxide). Even with very fine microelectrodes, the analysis does not take place in a completely undisturbed system since many connections must be broken by the physical maneuvering of the electrode. The leakage of cytoplasmic constituents, including potential transmitter agents and metabolic enzymes, can only be considered as part of the background artifact of the system. In the past few years, additional electrophysiological methods pertinent to transmitter action have been developed that can be applied to neuronal systems analyzed *in vitro* to provide additional insight into ionic channel regulation.

Analysis of Membrane Actions of Drugs and Transmitters in *Vitro*

The development of methods for the nearly complete functional maintenance of central neurons *in vitro* for several hours, such as the tissue slice preparations, and for several days to weeks, in single-cell or in explant culture systems, has led to a proliferation of additional electrophysiological methods to examine transmitter and drug action. In slice preparations, neuronal targets can be readily localized by inspection, and intracellular electrodes can be inserted into suitably large neurons under visual control, while sources of afferent circuitry within the slice may be activated by additional stimulating electrodes. Transmitters and drugs can then be applied to the whole slice by superfusion within oxygenated buffered salt solutions, or more locally by micropressure pulse or iontophoretic application methods. In many cases, excellent intracellular recordings can be obtained for long periods of time because there are no annoying respiratory, cardiac, or other movements to dislodge the electrode.

Unfortunately, the effects of transmitters analyzed *in vitro* in this way often disagree with the effects of the same transmitters when evaluated the old fashioned way *in vivo*. It is not yet clear why this situation prevails, but as in many other cases, the data that are easier to acquire often accumulate at faster rates and submerge the old data without disproving them. Since the solutions used as drug vehicles may lack all the components of arterial blood, and since the cells in slices have undergone what must be at least partial anoxia, one should not necessarily conclude that the *in vitro* actions equate with *in vivo*.

Nevertheless, when long-term intracellular recordings can be obtained, two additional sources of information on transmitter actions can be analyzed. In *voltage clamp* analysis, the experimenter inserts one or two electrodes into the cell and by injecting current holds the membrane potential of the neuron at a constant value. The cell is usually poised at a membrane potential more negative than resting in order to prevent spontaneous spikes. Transmitter action is monitored by the amount of current required to keep the membrane potential constant, and thus measures transmembrane current flow directly. However, since the membrane potential stays constant, any of the nonlinear properties of that neuron's response that could occur when sufficient depolarization has occurred will be prevented.

The clamp can also be quickly changed to a new level of membrane potential, and the neuron's responses to this shift used as the basis for a pharmacological dissection (e.g., with ion channel blockers or ion substitutions) of the degree to which the effects of a transmitter can be explained as ion dependent or voltage dependent. Many of the actions ascribed to neuropeptides and some of those ascribed to monoamines fit the concept of voltage dependent, since they are modest effects at best at resting membrane potential levels, but emerge as more substantive effects when the responding neuron is depolarized or hyperpolarized by other convergent transmitters.

Another of the newer methods, termed *noise analysis*, also examines ion channel activity more directly than the standard *in vivo* methods. This method assumes that ion channels are either open or closed and that they switch instantaneously, and do so independently, between the two conditions. Using intracellular electrodes, the fluctuations in membrane potential (or, if voltage clamping is used, the fluctuations in membrane current flow) are analyzed statistically to infer the conductance of individual types of channels and the mean time they are open in the absence or presence of the transmitter to be analyzed.

When the neurons considered to be the appropriate targets of a specific transmitter can be maintained in long-term tissue culture, the method of *patch clamp analysis* can be applied, as the current superstar of membrane action analysis methods. This method offers the ability to study the behavior of single-ion channels and to do so under conditions of almost unbelievable precision. Special "fire-polished" microelectrodes are placed on the neuron's surface, and a slight vacuum applied to the pipette to attain a very tight junction with the exposed surface of the neuronal membrane, thus requiring near-nude neurons for best application. The resulting cell electrode junction will have such a high electrical resistance (gigohms!) that the patch of enclosed membrane within the microelectrode's tip will be essentially isolated from the rest of the cell. Current flowing within that patch can then be analyzed independently of the responses of the rest of the neuron. With state-of-the-art, low-noise amplifiers, current flow through individual channels can be monitored and transmitter actions evaluated in terms of open time, amplitudes (number of channels opened), and closing times.

In addition, with clever micromanipulations, the patch clamps can be done in three configurations. In the "cell attached" mode, the pipette is sealed to the intact cell and the measurements made with no further physical disruption. However, further application of slight vacuum allows the patch of enclosed membrane to be removed from the cell, but with enclosed ion channels still viable and responsive. In the "inside-out" patch, the previously intracellular surface will be on the exterior of the sealed membrane patch, and the ion channel can be examined for regulation by simulations of changes in intracellular ions or, for example, catalysts of protein phosphorylations. It is also possible to demonstrate, that with clever handling, and with further negative pressure before pulling the membrane patch off the cell, an "outside-out" patch can be obtained. Here the original patch is ruptured, the perimeter remains attached, and then the surrounding external membrane segments reseal once they are excised from the cell surface. By placing the outside surface into solutions with differing doses of transmitter, drug, or ion channel toxin, it is possible to analyze very discrete, single-channel pharmacology.

The Steps of Synaptic Transmission

Let us now conclude this chapter by examining in a summary fashion the mechanisms of presumed synaptic transmission for the mammalian central nervous system. Each step in such transmission constitutes one of the potential sites of central drug action (Fig. 2-3). A stimulus activates an all-or-none action potential in a spiking axon by depolarizing its transmembrane potential above the threshold level. The action potential propagates unattenuated to the nerve terminal where ion fluxes activate a mobilization process leading to transmitter secretion and "transmission" to the postsynaptic cell. From companion biochemical experiments (to be described in the next chapters), the transmitter substance is believed to be stored within the microvesicles or synaptic vesicles seen in nerve endings by electron microscopy. In certain types of nerve junctions, miniature postsynaptic potentials can be seen in the absence of conducted presynaptic action potentials. These miniature potentials have a quantatized effect on the postsynaptic membrane in that occasional potentials are statistical multiples of the smallest measurable potentials. The biophysical quanta have been related to the synap-

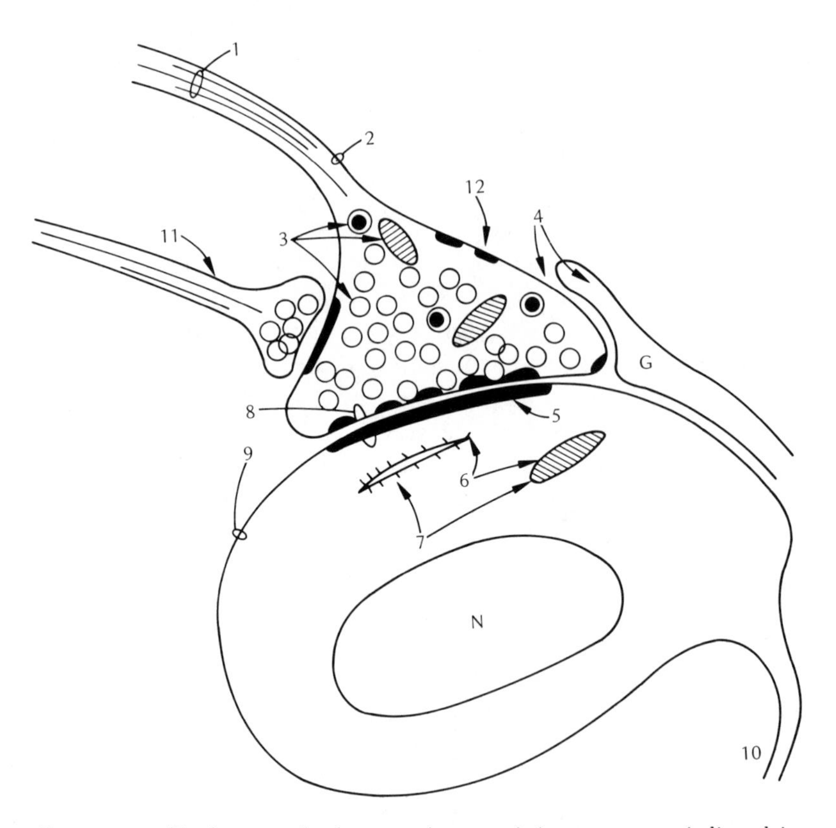

FIGURE 2-3. Twelve steps in the synaptic transmission process are indicated in this idealized synaptic connection. Step 1 is transport down the axon. Step 2 is the electrically excitable membrane of the axon. Step 3 involves the organelles and enzymes present in the nerve terminal for synthesizing, storing, and releasing the transmitter, as well as for the process of active re-uptake. Step 4 includes the enzymes present in the extracellular space and within the glia for catabolizing excess transmitter released from nerve terminals. Step 5 is the postsynaptic receptor that triggers the response of the postsynaptic cell to the transmitter. Step 6 shows the organelles within the postsynaptic cells which respond to the receptor trigger. Step 7 is the interaction between genetic expression of the postsynaptic nerve cell and its influences on the cytoplasmic organelles that respond to transmitter action. Step 8 includes the possible "plastic" steps modifiable by events at the specialized synaptic contact zone. Step 9 includes the electrical portion of the nerve cell membrane which, in response to the various transmitters, is able to integrate the postsynaptic potentials and produce an action potential. Step 10 is the continuation of the information transmission by which the postsynaptic cell sends an

tic vesicles, although the proof for this relationship is still circumstantial.

When the transmitter is released from its storage site by the presynaptic action potential, the effects on the postsynaptic cells cause either excitatory or inhibitory postsynaptic potentials, depending upon the nature of the postsynaptic cell's receptor for the particular transmitter agent. If sufficient excitatory postsynaptic potentials summate temporally from various inputs onto the cell, the postsynaptic cell will integrate these potentials and give off its own all-or-nothing action potential, which is then transmitted to each of its own axon terminals, and the process continues.

Transmitters, Modulators, and Neurohormones

So far, we have only considered the chemical controls normally operating at receptors sensitive to transmitters secreted at synapses, such as those diagrammed in Figure 2-3. These actions may not be sufficient, however, to account for the sorts of effects ascribed to "neuromodulators," "neuroregulators," or "neurohormones." These latter ambiguous terms are difficult to define, and this may be a virtue; some overly enthusiastic investigators use the terms to cover new candidates as transmitter molecules before sufficient data are available to assure their transmitter status (see Chapter 1). Actually, the term "neurohormone" was coined by Scharrers in the early 1950s to cover what was then an almost unbelievable situation: neurosecretory cells that secreted their "hormones" broadly via the bloodstream in response to conventional neuronal signals relayed through synapses. Neurons such as the supraoptic and paraventricular hypothalamic cells whose axons constitute the neuro-

---

action potential down its axon. Step 11, release of transmitter, is subjected to modification by a presynaptic (axoaxonic) synapse; in some cases an analogous control can be achieved between dendritic elements. Step 12, release of the transmitter from a nerve terminal or secreting dendritic site, may be further subject to modulation through autoreceptors that respond to the transmitter which the same secreting structure has released.

hypophysis clearly fall into this category, and so would the innervated cells of the adrenal medulla. However, in the CNS, peptide-secreting hypothalamic neurons also make synaptic contact with each other and with other neurons, and present data suggest that they secrete the same peptide at these sites. Other peptide-secreting cells in the CNS do not propagate their secretion via the bloodstream at all, and thus the fact that some peptidergic neurons do secrete into the blood should not be taken to imply that neurally released peptides act differently than conventionally conceived transmitter agents.

In trying to solve the problem of interneuronal chemical communication, it may be useful, nevertheless, to maintain an open mind with regard to three dimensions by which neuronal circuits can be characterized: (1) the spatial domain (those areas of the brain or peripheral receptive fields that feed onto a given cell and those areas into which that cell sends its efferent signals); (2) the temporal domain (the time spans over which the spatial signals are active); and (3) the functional domain (the mechanism by which the secreted transmitter substance operates on the receptive cell). When the receptive cell is closely coupled in time, space, and function to the secreting cell, almost everyone would agree that "real" synaptic actions have occurred. When the effects are long lasting and widely distributed, however, many would prefer to call this action something else, even though the molecular agonist is stored in and released presynaptically from neurons onto the nerve cells they contact.

Selected References

Bloom, F. E. (1984). The functional significance of neurotransmitter diversity. *Am. J. Physiol. 246*, C184–C194.

Bloom, F. E. (1985). Neurohumoral transmission and the central nervous system. In *The Pharmacological Basis of Therapeutics, 7th ed.* (A. G. Gilman, L. Goodman, and A. Gilman, eds.), pp. 235–257. Macmillan, New York.

Bullock, T. H. (1980). Spikeless neurones: Where do we go from here? *Soc. Exp. Biol. Seminar Series 6*, 269–284.

Brodie, B. B., H. Hurz, and L. S. Schanker (1960). The importance of dissociation constant and lipid solubility in influencing the passage of drugs into the cerebrospinal fluid. *J. Pharmacol. 130*, 20.

Cowan, W. M., and M. Cuenod (1975). *The Use of Axonal Transport for Studies of Neuronal Connectivity*. Elsevier, Amsterdam.

Eccles, J. C. (1964). *The Physiology of Synapses.* Academic Press, New York.

Grafstein, B., and D. S. Forman (1980). Intracellular transport in neurons. *Physiol. Rev. 60*, 1167–1283.

Hille, B. (1984). *Ionic Channels of Excitable Membranes.* Sinauer Associates, Inc. Publishers, Sunderland, MA, 426 pp.

Hodgkin, A. L., and A. F. Huxley (1952). Currents carried by sodium and potassium ions through the membrane of the giant axon of Loligo. *J. Physiol. 116*, 449.

Iversen, L. L. (1984). The Ferrier Lecture, 1983. Amino acids and peptides: Fast and slow chemical signals in the nervous system? *Proc. Roy. Soc. Lond. B221*, 245–260.

Katz, B. (1966). *Nerve, Muscle and Synapse.* McGraw-Hill, New York.

Kupferman, I. (1979). Modulatory actions of neurotransmitters. *Ann. Rev. Neurosci. 2*, 447–465.

Loewi, O. (1921). Über humorale Übertragbarkeit der Herznervenwirkung. *Pfluegers Arch. 189*, 239.

Reuter, H. (1983). Calcium channel modulation by neurotransmitters, enzymes and drugs. *Nature 301*, 869.

Schmitt, F. O. (1984). Molecular regulators of brain function: A new view. *Neuroscience 13*, 994–1004.

Siggins, G. R. and D. L. Gruol (1986). Synaptic mechanisms in the vertebrate central nervous system. In *Handbook of Physiology, Intrinsic Regulatory Systems of the Brain* (F. E. Bloom, ed.), Vol. IV. The American Physiological Society, Bethesda, MD.

3 | Molecular Foundations of Neuropharmacology

The ultimate resolution of a biochemical basis for any drug's actions on the nervous system depends on understanding the molecular interactions involved: A drug acts selectively when it elicits responses from discrete populations of cells that possess "drug recognizing" macromolecules. Such recognition sites are known more colloquially as receptors. Drug receptors with important regulatory actions on the nervous system are generally on the external surface of neurons if they involve sites where neurotransmitters act. Some drugs can act on the nervous system because they resemble, at the molecular level, endogenous intercellular signals like those of a neurotransmitter. However, drugs may also act by regulating intracellular enzymes critical for normal transmitter synthesis or breakdown. Receptors recognize drugs for a variety of reasons that this book explores in subsequent chapters. Once having made that recognition, the activated receptor usually interacts with other molecules to alter membrane properties or intracellular metabolism. These cellular changes in turn regulate the interactions between cells in circuits. These circuit changes regulate the performance of functional systems (like the sensory, motor, or vegetative control systems) and eventually the behavior of the whole organism.

Thus, understanding the actions of drugs on the function of the brain, whether it be on single cells or behavior, is a multilevel, multifaceted process that begins with and builds upon the concept of molecular interactions. Even beginning students of drug action on the nervous system will probably accept this statement as a reasonable hypothetical principle. In practice, however, this principle is severely compromised because most molecules in a very complex organ like the brain are unknown. Furthermore, until the recent explosion in molecular biology, there was little hope that the necessary details would be obtained.

Molecular biological research blossomed in the late 1970s with the recognition that genetic material from mammals and bacteria could be combined experimentally to create new bits of genetic information. The application of these methods to the most complex cellular system in the body—the brain—represented a natural continuation of the exploitation by neuroscientists of the other biophysical and biochemical methods that had propelled earlier waves of molecular and cellular understanding.

The power of molecular biological methods is gained from several related but independent developments: (1) the ability to *clone* genetic information (i.e., to isolate a segment, purify it, and create large amounts of the purified selected sample), (2) the ability to obtain the molecular *sequence* of genetic information (i.e., to determine rapidly the complete molecular structure of genes, the basic pieces of genetic instruction that allows cells to produce the specific molecules they require), and (3) the ability to practice *genetic engineering* (i.e., to perturb and control gene expression and alter the structure of gene products by chemically modifying precise sites in the molecular structure of the genes).

All of these developments have contributed to a very rapid advance towards a truly molecular basis for the understanding of the nervous system and the way it can be altered by drug actions. In this chapter we explore these molecular foundations.

Cellular Variation

Before we can deal effectively with the critical details of molecules that regulate the function of the nervous system and mediate the responses to drugs that act there, we must begin by briefly considering how the cells of the brain differ from the other cells of the body, and what it is that allows for differences between types of cells. Except for erythrocytes, all mammalian cells have a nucleus that separates the basic units of genetic information from the cytoplasm. The cytoplasm and its organelles allow the cell to generate energy that the cell uses to synthesize the structural and enzymatic molecules that give it and its enveloping plasma membrane the functional properties by which it contributes to the overall operation of the organism. Liver cells, kidney cells, skin cells, white blood cells, and cells of the nervous system all possess these basic similarities and these basic individual properties. In a single individual organism, all of these somatic cells (with only one exception), including the cells of the central and peripheral nervous system, possess entirely the same basic set of genetic information. The exception to this basic rule is the antibody-producing lymphocytes, and they have their own books.

The total set of potential genetic instructions of an individual, its *genotype*, is composed of basic instructional units—the *genes*—each having a specific location on a specific chromosome. Each type of specialized cell expresses a subset of genes that encode the special structural and enzymatic proteins that endow the cell with its size, shape, location, and other functional characteristics. This set of characteristic features that are expressed by a cell is termed its *phenotype*.

The bridge between the genotype of a cell and its phenotype is a series of elemental biological processes that are critical for our understanding of the molecular basis of drugs acting in the brain. In the nervous system we traditionally subdivide the cells into two major cell type classes, the neurons and the glia. We examine their individual cellular properties in more detail in Chapter 2. However, for the present discussion we can assume that these cellular archetypes, and the almost innumerable variants within each of those two large classes, are the outward reflection of their corresponding specific expression of subsets of their genes. The next section provides a brief overview of this process, with additional complexities in the expanded coverage that follows. Neuropharmacology students recently exposed to biology or biochemistry courses may find the next few sections "old hat." However, we include this material so that everyone reaching the end of this chapter will have the fundamentals necessary for a full appreciation of the material to follow (see Fig. 3-1).

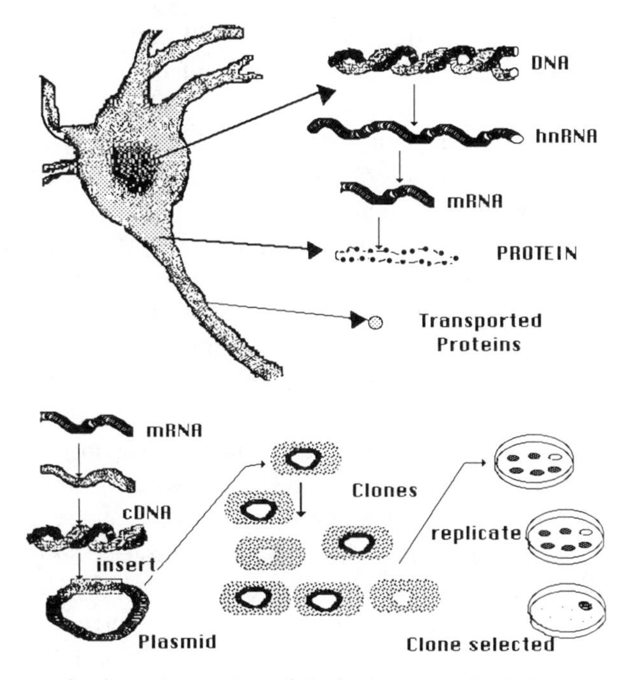

FIGURE 3-1. A schematic overview of the basic steps and cellular compartments involved in determining the specific phenotype of a neuron (above) and in cloning the mRNAs that allow the neuron to translate its genetic information into specific proteins. An mRNA is converted to a single-stranded DNA that is further converted into a double-stranded segment that is then inserted into a plasmid. Bacteria are infected with individual plasmids, and individual, plasmid-infected bacteria are grown into colonies. Replicates of the culture plate are screened by nucleic acid or antibody probes to select clones of interest.

1. The genetic information, stored in the form of long strands of deoxyribonucleic acid chains (DNA), is selectively expressed. That is, each specialized cell attains its specialized functional and structural status by expressing a subset of all of its genetic instructions. To express selective segments of the genome, the DNA-encoded information is converted, or *transcribed*, into a second similar molecular form as strands of ribonucleic acid (RNA) under the control of special proteins and RNA synthesizing enzymes (polymerases)

that perform the transcription steps. We will deal in slightly more detail with the actual molecular properties of DNA and RNA below.

2. The primary transcript form of the selected genetic information is then edited, by several rapid steps and exported from the nucleus to the cytoplasm. The edited RNA transcript, or *messenger RNA* (mRNA), is then *translated* by special cytoplasmic organelles, also made of RNA and proteins, called *ribosomes.* The translation is a chemical language shift from the nucleic acid code of the RNA into the amino acid sequence of the protein that is to be expressed.

3. In some brain cells, such as the glia, that make their proteins largely for use within the cell, the ribosomes occur free within the cytoplasm. In neurons, and other dedicated secretory cells, the translated protein undergoes *post-translational processing*, in which the protein's structure may be modified to attain the folded, globular, or linear structural properties that allow it to become associated with the proper intracellular compartments (e.g., within the plasma membrane or within the cytoplasm) where it is intended to function.

4. In cells, like neurons, that transport large amounts of protein products within the cell's interior for purposes of secretion, more extensive post-translational processing occurs. In such cells, the ribosomes are physically associated with a special set of endoplasmic reticular membranes, giving the membranes a "rough" appearance, for which they are known as the "rough ER." Within the channels of this inside-the-cell network, the newly synthesized proteins are lead to a set of smooth endoplasmic reticular membranes, the *Golgi apparatus*, where they are packaged into secretory organelles for transport to the secretory, or releasing, segments of the cell.

All of these organelle systems of the cell, essential for the selective transcription, translation, and packaging or compartmentalization of the specific proteins by which a given class of cells attains its specific phenotype, were well known to classical cytologists long before the molecular mechanisms underlying these events were understandable. Although the details of these molecular revelations are well beyond the capacities of this book, an exploration of the basic features will help the interested reader begin to comprehend the special analytic advantages that arise from the methods of molecular biology.

Fundamental Molecular Interactions

The cornerstone discovery of molecular biology was the formulation by Watson and Crick in 1953 of the *double-stranded helix* model of DNA structure. The insightful model they developed provided a coherent integration of the regular x-ray crystallographic structure of partially purified DNA with the previously known quantitative chemical data on the relative frequency within DNA of its four nucleotide bases, thus explaining why the purine–pyrimidine pairs adenine (A)–thymidine (T) and guanine (G)–cytosine (C) occur in precisely equal frequency. The Watson-Crick molecular model for DNA also accurately predicted the basic mechanism of DNA replication.

Nucleic Acid Base Pairing Complementarity

In the Watson–Crick double helix, two right-handed helical polynucleotide chains coil around the same central axis, making a complete helical turn every 10 nucleotides. (Fig. 3-2) In the interior of the helix, the purine and pyrimidine bases (A with T and G with C) are paired through hydrogen bonding of their complementary structures, placing the phosphate groups around the outside of the helix. The crucial structural feature that is the focal point for gene expression is the precise molecular complementarity between the primary sequences of nucelotide bases in one strand of the DNA helix with the antiparallel sequence of the second strand. The strand that encodes the genetic information is termed the sense strand. Wherever a particular base occurs in the sense strand, there will be a complementary base, and only that base in the antisense strand, such that A always pairs with T, and vice versa, and G always pairs with C, and vice versa.

The base pair complementarity allows for duplication of the ge-

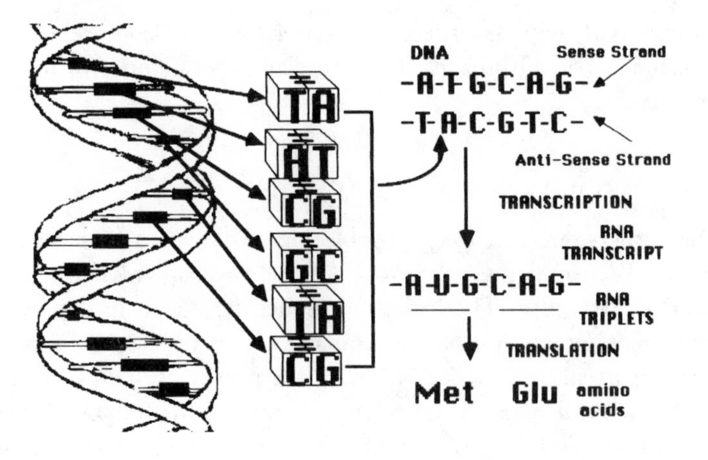

FIGURE 3-2. The arrangement of bases within the DNA double helix and the relationships between nucleic acid base pairs, RNA, and amino acids during the process of transcription and translation.

netic information in dividing cells. This is accomplished by enzymes known as DNA polymerases that open the helix and replicate each single strand back into double strands according to the single strand's template. The double-stranded complementarity also provides a means to repair the DNA should it be damaged, since whichever single strand survives the damage can act as a template for the repair.

In a similar manner, the information-bearing, or sense strand of the helical DNA chain is copied into a complementary single-stranded RNA during the process of transcription. RNA is thus a single-stranded complementary copy of the DNA antisense strand (so that its sequence resembles closely that of the DNA sense strand; see Fig. 3-2). RNA differs chemically with the substitution of uridine for thymidine, and ribose phosphates for deoxyribose phosphates. The process of transcription is accomplished by enzymes known as RNA polymerases. The affinity of the base pairs along sequences of a single DNA strand for their complementary base pair sequences in DNA or RNA are so precise that small segments can be used as *probes* for the detection of homologous sequences between large domains of DNA and RNA because the molecular complementarity will allow the probe to bind to its complementary structure only when there is a long sequence of consistent match. The ability of a single-stranded nucleic acid to bind, or hybridize, to its complementary sequence is an essential component of many molecular biological techniques.

The Genetic Code

To translate genetic information from sequences of RNA into linear sequences of amino acids in proteins requires a strict coding, in order that the 20 or so amino acids commonly found in proteins can be specified by various combinations of the four nucleotides. Through an ingenious series of important experiments, that space does not permit us to describe here, it was demonstrated that sets of three RNA bases (triplets) provide the code words that specify the order of amino acids to be incorporated into protein (Fig. 3-2), and that other triplet sequences mark the point at which synthesis would begin or end or be modified in other ways.

Often, mRNAs encode far more protein sequence than is represented in the final form of the processed gene product. One feature that was only detected through recent cellular biological insight was the *signal peptide:* a 15 to 30 amino acid sequence at the *N*-terminal end of the encoded gene product in which the amino acids are highly hydrophobic. The signal peptide is a near-constant feature of proteins intended for secretion, such as neuropeptides; its function seems to be to guide the nascent protein chain through the endoplasmic reticulum membrane for susequent packaging by the Golgi membrane apparatus. (A corollary that might be anticipated is that proteins that are not intended for secretion would lack signal peptides and thus represent a means to predict from the structure of novel proteins whether they are or are not secretory products; regrettably, this corollary does not hold.)

DNA Segments for Genes Are Interrupted

Almost as essential to the progress in application of this base pairing complementarity and nucleic acid to protein translation was another totally unanticipated wrinkle in the molecular basis for gene regulation in eukaryotic cells that stemmed from even earlier observations on the adenoviruses. In the late 1970s, with the analysis of the genes encoding immunoglubulins and the hemoglobins, it was recognized that the basic organization of genes in eukaryotic cells (cells with nuclei, as in all multicellular organisms) did not follow the principles that had been uncovered from the study of prokaryotes (cells, like bacteria, that lack a separable genetic compartment). Instead, higher organisms (and some viruses) have their gene segments split up into coding regions that are expressed *(exons)* separated by intervening regions *(introns)* that are not found in the mRNA.

This sort of interrupted DNA coding segmental structure leads to two outcomes: (1) the primary gene transcript, formally termed the heterogeneous nuclear RNA, or hnRNA, will contain extra RNA sequences, and the introns must be edited out before the mRNA can successfully direct protein synthesis by ribosomes (see Fig. 3-3). The editing process consists of opening up the hnRNA, removing intronic segments and resplicing the cut ends. (2) In some cells, including neurons, the composition of the transcribed mRNA can also be edited by splicing out certain exon segments. The editing-splicing process provides a means by which a gene containing several exons can give rise to several different gene product proteins, in which certain protein domains may be shared and others will be unique. When gene products share similar nucleotide and protein sequences, we often speak of them as a "structural family."

This editing and splicing may seem to be an unnecessarily complicated route to follow for a process that is intended to translate important genes with great fidelity into equally important enzymatic and structural proteins. However, the reader should be cautioned that such biological complexity almost always implies important and unanticipated regulatory control and enrichment. In the

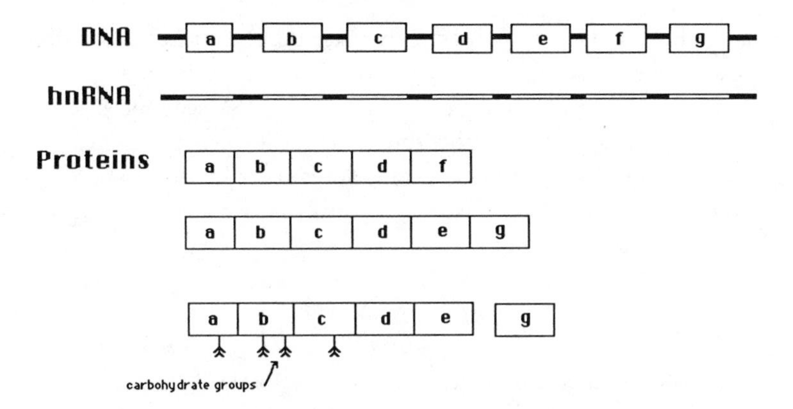

FIGURE 3-3. The relationship between DNA base sequences in introns and exons (a–g), the resulting primary RNA transcript, the subsequently edited forms of mRNA (two different forms of hypothetical editing and splicing are depicted), and the resulting proteins, which can then undergo posttranslational processing to yield small peptides, or to add carbohydrates or other chemical modifications to the molecule.

case of gene regulation and expression, the added complexities offer the means by which new life forms can evolve.

Nucleic Acid Sequence Determinations

The final fundamental procedural development that accelerated discoveries made with molecular biological methods was the ability to determine the sequence of DNA molecules, even those several thousand base pairs in length. (The methods for doing this sequencing are clearly outside the province of this book. The reader interested in the two different approaches (that of Maxam and Gilbert, and that of Sanger), each based on some very clever chemistry, with profound enough importance to merit Nobel recognition would be well served to invest in outside reading). From these DNA sequences, it is possible to deduce the nucleotide sequences of the RNA, and thereby the amino-acid sequence of the protein product. The structures of the DNA and of the gene product can then be analyzed, often via computer, to compare their sequences with data banks of previously characterized proteins or nucleic acids. In addition to structural comparisons of gene or RNA sequences, additional important clues to the functional features of the encoded molecule may be inferred from the domains of hydrophobic, hydrophilic, or other consensus structural sites for possible post-translational modification.

Once-Over Quickly Cloning

When mRNAs are converted into DNA by the enzyme *reverse transcriptase*, the copied double-stranded DNA form, or *cDNA*, can be incorporated (inserted) into specific sites within an infectious vector, or plasmid. The sites for insertion are selected by identifying DNA sequences that can be "cut" by the actions of *restriction endonucleases*, enzymes from purified bacterial sources that cleave DNA sequences at specific palindromic repeated sequence sites. By using plasmids of known DNA sequence that can be tailored to include the proper restriction cleavage sites to allow for the DNA insertion, the same enzyme can then later be used to cleave out the insert. The restriction sites for insertion are typically chosen within plasmid genes that code for some discernible functional property (such as antibiotic resistance). Thus, interruption of its coding and expression (when insertion has been successful) leads to the loss of that functional property and a means to identify which plasmids have successful inserts. Each plasmid can generally only incorporate one cDNA insert, and with a great excess of host bacteria, each insert-bearing plasmid will infect only a single host (almost always *Escherichia coli*).

By growing up these infected bacteria in such a way that each individual bacterium gives rise to a *colony* of identical bacteria bearing identical replicates of the plasmid and insert, the DNA has been *cloned*. The *cDNA* can then be recovered from the plasmid through another exposure to the restriction enzyme selected for the original opening of the plasmid insertion site. Thus, in a relatively few steps, one can start with a mixture of mRNAs in widely differing proportions from common to very rare, and purify them individually, as well as preparing virtually unlimited pure samples of the DNA insert.

This highly superficial survey should indicate the facility of cloning DNA segments (taken directly from genomic digests, or from mRNA copies) and then sequencing the cloned segments and deducing the structure of their product (see Fig. 3-4). However, we have glossed over the one potentially sour note in this rhapsody of high-tech molecular music making: How do you identify which clone is carrying the insert that encodes the gene product you want to identify? Suppose you don't even know what it is you want to look for and clone? As it turns out, through more cleverness and a few good breaks from Mother Nature, biologists have found multiple methods to do these feats of molecular magic, although some of the screening methods have given new meaning to the term "a search for needles in haystacks."

Molecular Strategies in Neuropharmacology

The immediate applications of molecular biological strategies within neurpharmacology are shortcuts in molecular isolation and sequence determination, for instance, to uncover new peptides or to provide more complete understanding of enzymes, receptors, channels, or other integral proteins of the cell. A rather likely premise holds that the phenotype of a cell within the nervous system depends upon the structural, metabolic, and regulatory proteins by which it establishes its recognizable structural and functional properties. If this is valid, then complex, multifaceted neurons will probably rely upon hundreds, if not thousands, of special purpose proteins, many of which may exist in rather limited amounts. To purify such rare proteins by the methods that existed before molecular cloning, especially in the absence of a functional assay to guide the purification process, is an overwhelming task, requiring exceptional patience, resources, and a very large supply of the proper starting tissue material. For some of the very rare hypothalamic hypophysiotrophic releasing factors (see Chapter 12), hundreds of

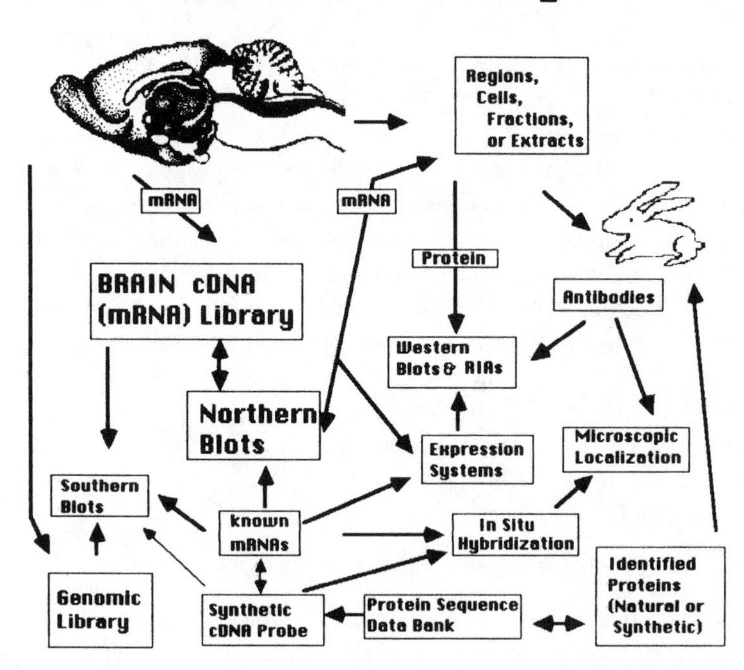

FIGURE 3-4. Applied cloning. Schematic overview of the major currently employed methods for applying molecular cloning methods to resolving the molecular structure of brain proteins. Brain mRNA, taken from whole brain or subfractions, is the starting material for a brain cDNA library that can be further selected for clones of interest by the several indicated strategic protocols. Alternatively (lower left) DNA can be prepared as a total genomic library, or (lower right) selected nucleic acid probes can be tailored to features of known proteins. Immunological methods are also critical in evaluating the gene product's location and possible functional properties.

thousands of hypothalami were required, as well as the development of unique purification schemes for each subsequent factor to be pursued.

Converting the quest for the structure of specific proteins into a molecular biological quest for the mRNA or gene segment that encodes this protein greatly facilitates the experimental analysis, simply because of the powers of cloning, complementarity, and rapid sequencing. Several methods have been developed that increase the chances of finding whatever the researcher is seeking (Fig. 3-4). These depend in part on the nature of the cDNA being pursued and how the investigator is able to probe either for the insert or for the translation of the fusion gene product in a cell system capable of processing the primary translation product into a structural form that will resemble its natural configuration and sometimes even its natural function.

General Strategies for Clone Screening and Selection

Enrichment by Tissue Selection and Preparation

The basic starting point is to select the brain cells or regions that are presumed to express the molecule to be studied and then to enrich sources of mRNA to favor the detection of the one being pursued. Cell lines and even tumors that produce large amounts of a hormone (such as pheochromocytomas or VIP-omas) or bear large numbers of the desired receptors or channels (like electroplax or striated muscle) have proved to be excellent starting materials. Once the cell source is selected, the desired mRNAs can be further enriched (e.g., by sucrose gradient centrifugation or electrophoresis) provided that some characteristics of the mRNA being sought are known.

Recognizing the Wanted Clone

A general strategy for detecting desired colonies of cloned bacteria is known as *colony hybridization.* In this approach, the bacteria are grown on a special culture plate from which their colonies can be copied (transferred as a group by lightly pressing them to another supporting surface, called "replica plating," and thereby sampling and preserving the spatial identity of all the colonies on the plate.)

The bacteria on the replicate supports are treated to expose their DNA and are screened with radioactive nucleic acid probes. Colonies that hybridize with the probe can then be identified by autoradiography. Alternatively, if the plasmid-carrying inserts were tailored to allow for expression of the protein encoded by the transferred genetic material, it might also be possible to identify the desired clones by immunological reactivity. When a reactive colony has been identified, its original bacterial colony is recovered from the original culture plate and the living bacteria are then grown up in large amounts to provide the starting material for DNA sequence analysis.

Building Your Own DNA

If partial protein (or peptide) sequences are known, it is possible to make predictions of what the mRNA sequence should be (by back-translating the genetic code for amino acids) and including enough alternatives to overcome the ambiguous cases where a specific amino acid may be encoded by several variant triplets. From this predicted RNA structure, it is then possible to design and synthesize the hypothetical cDNA. This approach has been used, for example, to create hypothetical cDNAs for the hypophysiotrophic hormones whose amino-acid sequences had been accurately determined "the old-fashioned way," by earning it one amino acid at a time from highly purified brain extracts. The main reason to do this with a biologically active protein or peptide whose structure is already known would be to get at the complete structure of its prohormone, or to obtain its complete genomic structure and analyze its regulatory control and expression mechanisms. As will be seen in Chapter 12, when this is done, more often than not, the prohormone of the known peptide is found to encode more than one active product. However, because of the redundancy of triplet RNA codons for some amino acids (there are six different codes for leucine alone) it is generally difficult to acquire a functional full-length mRNA by predictive synthesis.

An alternative approach is to synthesize a shorter complementary single-stranded DNA (a so-called "oligodeoxynucleotide probe") and use it as a *probe* to screen libraries of clones. Such libraries may be prepared from mRNA extracts or from special digests of the whole genome. In the former case, the starting material would be brain, while in the latter case, in theory, any somatic cells could be used to prepare the library. With the availability of automated "gene machines" it is now possible to synthesize a proper probe or two overnight, and use them to screen an awaiting genomic or cDNA library and thereby determine the complete coding sequences for a partially purified protein within a few weeks.

On such a screening expedition, likely candidate colonies can be cross-screened by a second synthetic oligodeoxynucleotide probe, based upon another separate domain of the full protein. Clones positive for both probes would then have to contain the gene sequence that encodes the two sequences against which the probes were made as well as the sequence between them. This strategy has been used with many neuropeptide mRNAs (see Chapter 12).

When You Haven't Got a Clue

It is also possible to penetrate into the large treasure trove of cellular proteins that have not yet been identified by conventional strategies. Given the length of the mammalian genome, and the relatively short list of identified specific molecules in cells of all classes, we must conclude that there is an awful lot left to be identified, and few clues as to what it is we don't know about. The methods of molecular biology can help here too.

In nonneural tissues, for example, it has been possible to identify the unique proteins expressed in a male versus a female liver, or those that are unique to thymus-derived versus bone marrow-derived lymphocytes by exploiting the fact that a very high proportion of the proteins expressed in these pairs of tissues are, for the most part, very similar. Depending on how the pairs of tissues or cell types are defined and how similar or dissimilar they actually turn out to be, this method can be made highly sensitive and can reveal unique differences in cell-specific gene expression.

The discovery strategy can also be broadened to look for large sets of tissue specific genes. For example, Sutcliffe, Milner, and Bloom have employed molecular cloning methods to ask what proteins are made by brain generally that are not found in other major tissues, and to determine the degree to which neurons differ in specific proteins underlying their phenotypes that at present are only defined empirically. In their general approach, a cDNA library was prepared from mRNAs extracted from whole rat brain, and the individual cDNAs were characterized for their ability to hybridize to mRNAs extracted from brain, liver, or kidney. Those expressed in brain, but not detectable in extracts of liver or kidney, were found to represent well over half of the total brain mRNA population (approximately 30,000 out of an estimated total of 50,000) suggesting that much of the genome contains information pertinent to the generation of neuronal function. Individual brain-specific clones are then analyzed further by determining their nucleotide sequence, and deducing the amino acid sequence of its encoded protein. Proteins that are unique to the data base of known sequences can then be identified further by raising antisera against synthetic peptides that mimic selected regions of the deduced protein structure. One such clone appears to resemble a possible precursor for neuropeptides.

Yet another strategic approach involves injecting mRNAs from enriched or prepared sources into frog oocytes where complex eukaryotic genes may be expressed more efficiently. Because the oocytes are relatively large, their expression of a novel functional protein can be assessed physiologically or biochemically to identify species of mRNA for isolation and cloning. By injecting groups of mRNAs and evaluating the oocytes for the response one seeks, it is possible through trial and error to identify the mRNA for a specific functional protein, like an ion channel or a cell surface receptor.

An All-points Search

The would-be molecular neuropharmacologist attending a seminar involving these approaches may find his or her initial exposure to the jargon confusing without some orientation. (Orientation is an appropriate term, for such presentations are sprinkled with what may sound like references to compass points; see Fig. 3-4.) When DNA is fractionated by restriction enzymes, and the resulting fragments are separated by gel electrophoresis, it is possible to transfer, or "blot," the resulting fragments, separated mainly on the basis of their lengths, from the acrylamide gel to a nitrocellulose or nylon support, and there to analyze them for the ability to hybridize with cDNA or RNA probes. This method is referred to as a "Southern" blot analysis, named for the scientist (Dr. E. M. Southern) who started the evolution of this method. Later, a similar approach was devised in which the starting material was RNA. Here, the separated RNAs were blotted for probing with radioactive, single-stranded cDNA probes. Because the starting material is, nucleic acidly speaking, the opposite of the Southern, this RNA blot is referred to as a "Northern blot."

More recently, immunological methods have been used to probe protein extracts that were separated by acrylamide gel electrophoresis and then "blotted" (by an electrical transfer) for identification by peroxidase or radioisotope-labeled antibodies to specific protein antigens. The result is termed a "Western blot." If RNA or protein samples are simply dried directly onto nitrocellulose for probing analysis without first separating them for size, the resulting blots are termed "slot blots" or "dot blots." These blots are useful when screening a large number of clone extracts for insert or expressed products quickly. As of this writing, no method has yet earned the accolade of being an "Eastern blot."

Given the resourcefulness with which molecular biologic methods are being applied to all aspects of specialized cell biology, including the actions of drugs on the nervous system, one senses that we are about to experience a logarithmic increase in the number of specific molecules that will be fully characterized, and that will enable pharmacological engineers to shape drug molecules precisely to fit the pocket of receptors or enzymes for the ultimate in specificity. Any biological phenomenon in the brain that is mediated by proteins, which hardly seems to exclude any brain function, is therefore amenable to the ultimate in molecular analysis: The range of such specific molecules would then include (but is definitely not limited to) the enzymes needed for transmitter synthesis, storage, release and catabolism, the receptors and related macromolecules needed for response, and response mediation to broader-scale events such as establishing the shape and orientation of a cell by the proper anchoring of its microtubules and filaments to the plasma membrane and the repair and maintainence of synaptic connectivity.

Beyond the Clones

Although we may rightly marvel at the advances that have been achieved through the use of molecular biological techniques, all that we have in essence discussed so far in this chapter is a set of methods that provides a novel, powerful, and accurate way to identify, isolate, and characterize the amino-acid sequences of a host of intracellular proteins and their possible subsequent metabolic products. While this is unquestionably a very major advance in the research armamentarium, it still does not really begin to deal with a wide range of other important questions that are also approachable through the molecular tools that recombinant methodologies provide.

For example, once a cDNA has been proven to represent the mRNA for a specific molecule, the deduced protein's sequence can be inspected to infer potential functional properties. Thus, the acetylcholine receptor molecule and the myelin proteolipid protein exhibit several stretches of 20 to 24 hydrophobic amino acids in a row, which are strong presumptive evidence of membrane-crossing domains, and thus suggestive of plasma membrane constitutive proteins. In any case, when the protein structure can be deduced, the entire molecule or selected fragments of it can be synthesized and used to raise antisera. These antisera can then be used to develop radioimmunoassays for the protein. The antisera can also be used for immunocytochemical analysis of the nervous system to determine which cell and which compartments of those cells exhibit the protein that has been identified. The synthetic fragments can be used to determine whether the protein's domains may be substrates for posttranslational modification, being processed by further proteolytic cleavage or by structural modification with glycosylation, phosphorylation, sulfation, or acylation. Subcellular fractions and ultrastructural cytochemistry may suggest organelle specialization or cell surface associations.

The products of cDNA cloning can also be taken back to the genome, to probe the regions around the location of the exons to search for their molecular mechanisms for control of expression. Once the location of the surrounding elements in the genome have been located, the cDNA probes can be used to determine the degree to which the mRNA or the underlying gene exons have been conserved across eukaryotic species and to determine the position on the chromosomes to which the gene can be mapped. The chromosomal location of many proteins has been determined in this or a similar manner. However, the human genome, and that of most mammals, is estimated to be on the order of 3×10^9 base pairs long, of which fewer than 1000 genes have been mapped, most of them being on the relatively small X chromosome. What this means is that there are vast expanses of the genome with no known markers of any kind. For students of genetic diseases of the nervous system, the situation is even worse, as much less than 1 percent of the genome is associated with known nervous system markers even though a significant fraction of the genome may be expressed selectively in brain.

These considerations of length and complexity of neuronal gene expression suggest that efforts to link specific patterns of DNA polymorphism may be a critical future development. The genetic linkages are defined on the basis of Southern blot analysis of family members whose genomic DNA has been treated with various different restriction enzymes to provide *restriction fragment length polymorphisms.* Because there can be considerable individual variation in nucleotide sequences without disturbing the function of the encoded proteins, the degree to which the patterns of restriction endonuclease-digested fragments differ reflects this individual variation. The ability to link fragments of DNA with inheritance of genetic disorders and to specific markers within the digested fragments helps to establish approximate locations of genetic mutations, such as the localization of the gene for Huntington's disease to human chromosome 4.

Although new modifications on these basic strategies are reported continuously, and although there is a steady growth of molecular information on the elements of neuronal function, most of the details that follow in this book deal with those still relatively rare molecules, (e.g., the major known neurotransmitters, their synthetic and catabolic enzymes, their receptors and response mediators), whose nature is already partly established. While this list may well be lengthened ever more rapidly by the shortcuts made possible by molecular biological methods, the student should recognize that identifying new molecules *per se* is merely a first step towards an important pharmacological end, but is nowhere near the true strength of what molecular biology may have to offer our field.

We have already spoken of some important tools as though they were only to be used for molecular identification, and it would be somewhat misleading to let those impressions stand without a pedagogical challenge. The awareness of how to "tailor" purified mRNAs can be taken far beyond simply sticking them into a plasmid: More complex tailoring, leading to real "designer genes," can be used to prepare synthetic genes that carry with them the basis for being incorporated and being expressed by novel cellular hosts, bacterial or mammalian. Directly injecting mRNAs into frog oocytes, and allowing them to be translated and incorporated into the oocyte's cellular machinery, is useful for detecting specific mRNAs for specific functional proteins, but again there is more: The cell bearing a known receptor, whose complete amino-acid sequence is known, could be used as an ideal template for drug design, by combining x-ray crystallographic information on the gene product protein, amino-acid sequences and functional responses in a responsive cell. Such methods have already been employed in part to derive potent enzyme inhibitors for membrane lipolytic enzymes. Furthermore, once that mRNA has been identified, synthetic versions of it lacking certain nucleotide segments can be evaluated to determine the functional consequences of tinkering around with the molecule in specifically designated sites. This strategy would be capable of indicating the active sites of a receptor, or its membrane anchoring, or its sites of interactions with the ion channels or enzymes that the receptor can regulate. Such mutagenesis can also be achieved by other treatments for the same objective.

The ability to inject mRNAs into unfertilized oocytes can be taken yet another step by injecting a gene, together with adjacent genetic segments that may help to achieve incorporation or expression, into a single cell fertilized mamalian ovum. When the injected genetic material is successfully incorporated into the embryonic genome, all of the embryo's cells will contain the novel gene, producing "transgenic" embryos. In certain cells of the transgenic animal, the gene will be expressed. In this way, "supermice" have been made by giving them extra genes for growth hormone production, and attempts are in progress to make supergoats and sheep. Furthermore, once the genes have been successfully incorporated into the genome, they are also present in the germ line and can be passed on to ensuing generations.

If you are still with this trip into a soon-to-be-here genetic future pharmacology, ponder two more highly likely extensions: It should be possible to eliminate undesirable gene products by injecting the cell with an "antisense strand" mRNA that will bind to the undesirable mRNA and prevent its translation. Similarly, it should be possible to replace bad or missing genes by targeting good copies of them for the cells that need them. Right now, certain types of viruses, that survive in nature by finding the right host cells and making them synthesize the virus's genome, seem like possible carriers for such molecular transplantations. These examples are by no means exhausting the idea banks now being formulated, and if some of these ideas don't seem as old hat to you as you thought they might, perhaps it's time you paid a trip to your intellectual haberdashery and check out your skullcap for the reading material necessary to work in the gene library world.

Selected References

General Sources

Cold Spring Harbor Symposium on Quantitative Biology (1984). Vol. 48, parts 1 and 2.

Darnell, J. E. (1983). The processing of RNA. *Sci. Am.*, 90–100.

Schmitt, F. O., S. J. Bird, and F. E. Bloom (1982). *Molecular Genetic Neuroscience.* Raven Press, New York, 492 pp.

Stryer, L. (1981). *Biochemistry*, 2nd ed. W.H. Freeman and Co., New York.

Watson, J. D., J. Tooze, and D. T. Kurtz (1983). *Recombinant DNA. A Short Course.* W.H. Freeman and Co., New York.

Specific References

Crick, F. H. C., L. Barnett, S. Brenner, and R. J. Watts-Tobin (1961). General nature of the genetic code for proteins. *Nature 192*, 1227–1232.

Cohen, S. A., A. Chang, S. Boyer, and R. Helling (1973). Construction of biologically functional bacterial plasmids in vitro. *Proc. Nat. Acad. Sci. U.S.A.* 70) 3240–3244.

Gilbert, W. (1985). Genes-in-pieces revisited. *Science 228*, 823.

Gusella, J. F., R. E. Tanzi, M. A. Anderson, W. Hobbs, K. Gibbons, R. Raschtchian, T. C. Gilliam, M. R. Wallace, N. S. Wexler, and P. M. Conneally (1984). DNA markers for nervous systems diseases. *Science 225*, 1320–1325.

Maxam, A. M., and W. Gilbert (1977). A new method of sequencing DNA. *Proc. Nat. Acad. Sci. U.S.A. 74)* 5463–5467.

Rosenfeld, M. G., S. G. Amara, and R. M. Evans (1984). Alternative RNA processing: Determining neuronal phenotype. *Science 225*, 1315–1319.

Ruddle, F. H. (1981). A new era in mammalian gene mapping. Somatic cell genetics and recombinant DNA. *Nature 294*, 115–120.

Sanger, F., and A. R. Coulson (1975). A rapid method for determining sequences in DNA by primed synthesis with DNA polymerase. *J. Mol. Biol. 94*, 444–448.

Southern, E. M. (1975). Detection of specific sequences among DNA fragments separated by gel electrophoresis. *J. Mol. Biol. 98*, 503–517.

Sutcliffe, J. G., R. J. Milner, J. M. Gottesfeld, and W. Reynolds (1984). Control of neuronal gene expression. *Science 225*, 1308–1314.

Sutcliffe, J. G., R. J. Milner, T. M. Shinnick, and F. E. Bloom (1983).

Identifying the protein products of brain-specific genes with antibodies to chemically synthesized peptides. *Cell 33*, 671–682.
Watson, J. D. and F. H. C. Crick (1953). Molecular structure of nucleic acids: A structure for deoxyribose nucleic acid. *Nature 171*, 737–738.

4 | Metabolism in the Central Nervous System

One way to approach the biochemistry of the brain is to go through the litany of the Embden-Meyerhof glycolytic pathway, the tricarboxylic acid cycle, the pentose phosphate shunt, and the biosynthetic and metabolic pathways of carbohydrates, lipids, nucleic acids, and proteins. All of these enzyme systems occur in brain and have been studied in more or less detail, but qualitatively nothing of striking import distinguishes these enzyme activities in nervous tissue from their activity in extraneuronal tissue. For example, the synthesis of ATP from glycolysis and oxidative phosphorylation appears not to be fundamentally different in brain than in other tissues and does nothing to illuminate the brain as a specialized organ. In particular, if one recalls the diversity as well as the specificity of effects of neurotropic agents, it is difficult to see the relevance of such a universal and basic mechanism as energy production in a discussion of neuropharmacology. Unless neuropharmacological agents had a peculiar affinity for nervous tissue, one would not expect to encounter such a specificity among these agents if they directly inhibited glycolysis or oxidative phosphorylation. More logically, the mechanism of action of these agents might be anticipated to lie in their ability ultimately to affect neurotransmission. This could be primarily by affecting the synthesis, release, catabolism, storage, or re-uptake of transmitters, by altering ion movement in axons, or by altering receptor sensitivity.

To state the obvious, then, since the brain is an organ that transmits and stores information, it seems most profitable in discussing the biochemistry of the brain to look at enzyme systems and substrates that may contribute to this specialization. In other words, if one were asked to prove biochemically that a piece of tissue was from brain, what would one measure? Accepting this challenge, we can now proceed to list the types of compounds and enzyme systems which can be used to elucidate the biochemical uniqueness of brain as well as to point out profitable future lines of experimental investigations:

(A) *Compounds with demonstrated electrophysiological activity.*
(B) *Compounds that either occur only in the CNS or whose concentration in the brain is extraordinarily high compared with that of other tissues.* A concentration of >3mM has been arbitrarily chosen as a cutoff point.
(C) *Compounds in which a change in concentration in brain has been shown to correlate with a change in the behavioral state of the animal.*

The composite flow sheet of Figure 4-1 covers to a large extent the listings above. The arrows are not intended to imply direct conversions but merely precursor relationships. For convenience, compounds in Group A are in capital letters; compounds in Group B are screened and compounds in Group C are underlined.

The figure properly begins with glucose because in brain, in contrast to other tissues, virtually everything except essential fatty acids and amino acids is derived from this substrate. For example, when labeled glucose is incubated with brain cortical slices, labeled aspartate and glutamate are synthesized within a few minutes after the addition of the glucose to the flasks. The same occurs *in vivo* as well as in brain perfusion experiments. The brain accounts for about 20 percent of the total oxygen consumption of the body; the energy associated with this oxygen consumption comes from the utilization of glucose.

The knowledge that glucose is under normal conditions the primary if not the sole substrate of the brain *in vivo* derives from two experimental findings. First, in measuring A-V differences in O_2 and CO_2 in the blood going through the brain, a respiratory quotient of 1 was obtained, indicating that the oxidation of carbohydrates rather than amino acids or fatty acids is the primary source of energy. Further, when the A-V differences of glucose, gluta-

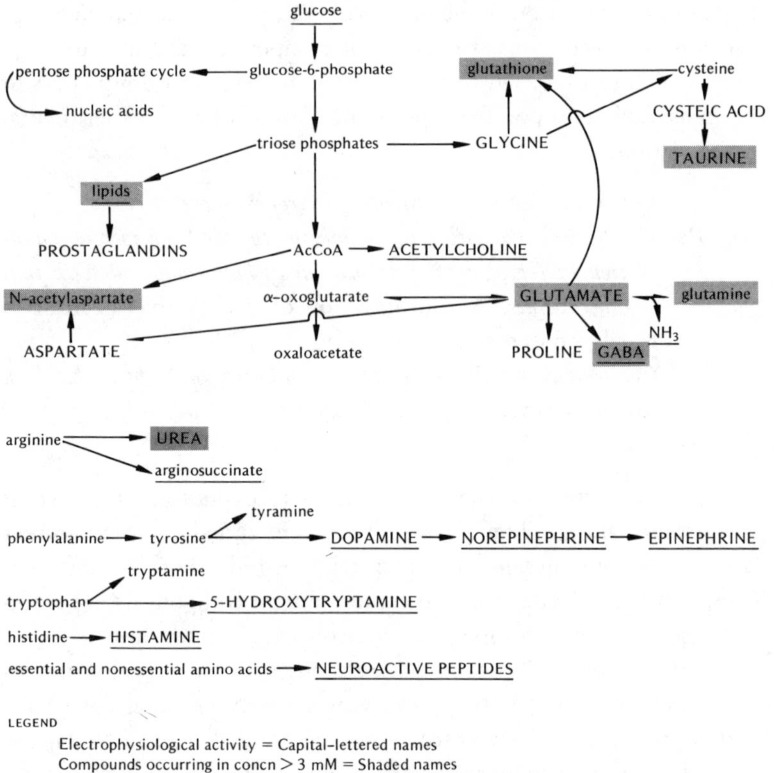

FIGURE 4-1. Various interrelationships among compounds of neurochemical interest.

mate, lactate, and pyruvate were determined, the only compound removed in large amounts from the arterial blood was glucose, and the only product that was released in large quantities was CO_2. The second line of evidence implicating glucose derives from experiments on the hypoglycemia induced either by hepatectomy or by the administration of insulin. In these experiments the fall in blood glucose correlated with a cerebral dysfunction that ranged from mild behavioral impairment to coma. This CNS effect can be completely reversed by the administration of glucose, maltose, or mannose: Maltose is hydrolyzed to glucose; mannose is phosphorylated and converted to fructose-6-phosphate. The CNS effect is not reversed by lactate, pyruvate, or glutamate (perhaps because these agents do not penetrate the blood–brain barrier at a fast enough rate). In insulin coma, with its concomitant low cerebral metabolic rate, there is enough glucose and glycogen to last about ninety minutes; then other cerebral metabolites are utilized. This calculation correlates with the observation that in insulin coma irreversible damage occurs after ninety minutes despite the subsequent administration of glucose. This irreversibility may be due to the fact that vital amino acids and lipids are utilized in an attempt to maintain cerebral metabolism.

Although a diversion, it is worthwhile in discussing glucose metabolism to mention the imaginative 2-deoxyglucose technique developed by Sokoloff and co-workers to correlate functional brain activity and energy metabolism. By injecting labeled 2-deoxyglucose whose metabolism essentially stops at the 2-deoxyglucose-6-phosphate level, and ultimately measuring radioactivity in brain sections by autoradiography, these researchers made possible the pictorial representation of relative rates of glucose utilization in all brain areas and the mapping of functional neural pathways under pharmacological and pathological conditions on the basis of the evoked metabolic response. Recently, this metabolic encephalographic technique has been developed for *in vivo* human use by employing the derivative [^{18}] fluorodeoxyglucose and scanning with positron-emission tomography.

The first step in glucose metabolism involves phosphorylation by ATP to form glucose-6-phosphate. About 90 percent of this phosphate ester is metabolized through the well-known Embden-Meyerhof glycolytic pathway and about 10 percent is oxidized via the pentose phosphate cycle. This latter system, sometimes referred to as the hexose monophosphate shunt, does not lead directly to ATP synthesis, as is the case with glycolysis or oxidative phosphorylation. Instead, it appears to function primarily in providing reduced nicotinamide adenine dinucleotide phosphate (NADPH), vital in lipid biosynthesis, and the pentose phosphates that are required in the *de novo* synthesis of nucleotides.

The Embden-Meyerhof pathway leads to triose phosphates, which are involved in the synthesis of lipids. The high concentration of lipids is a unique feature of brain since about 50 percent of the dry weight of brain is lipid compared with 6 to 20 percent in other organs of the body. In further contrast to other tissues, cholesterol occurs almost exclusively in the free form, with only a minute amount esterified; the concentration of triglycerides and free fatty acids is minimal compared to other organs, and, whereas other organs contain glucocerebroside, brain contains only galactocerebroside.

In addition to cholesterol, brain lipids include glycerides and sphingolipids of very complex composition. The composition of some of these compounds is shown in Table 4-1. Protein-lipid complexes are also found in nervous tissue and are referred to as proteolipids or lipoproteins, depending on the physical characteristics of the molecule. Phosphatidopeptides have also been reported in the brain. Although the chemistry of brain lipids is fairly well known and much is being learned currently of the synthesis and catabolism of lipids, the function of these compounds is still unclear. What is clear is that most lipids serve as structural components, either in myelin or in membranes of the various cellular organelles, and exhibit a slow turnover. However, a metabolic function for phosphoinositides has received considerable attention ("the P. I. effect") as a modulatory mechanism and is discussed in Chapter 6. Another recent development in lipid metabolism, deriving from the imaginative investigations of Hirata and Axelrod, indicates that methylation of membrane-bound phospholipids is initiated by a variety of agents acting on the receptors of cells. This transduction of receptor-mediated signals is also discussed in Chapter 6.

Although the function of the complex lipids is unknown, their implication in certain genetic types of mental retardation known as the cerebral lipidoses is well documented. Thus, for example, in Tay-Sachs disease there is an accumulation of gangliosides, in

TABLE 4-1. Lipid structures

Basic structure	R_1	R_2	X	*Compound*
Glycerophosphate: $H_2C{-}O{-}C({=}O){-}R_1$ / $R_2{-}C({=}O){-}O{-}CH$ / $H_2C{-}O{-}P({=}O)(OH){-}O{-}X$	Fatty acid	Fatty acid	$CH_2CH_2\overset{+}{N}H_2$	Phosphatidylethanolamine
	Fatty acid	Fatty acid	$CH_2CH(\overset{+}{N}H_3)$-COOH	Phosphatidyl serine
	Fatty acid	Fatty acid	$CH_2CH_2\overset{+}{N}(CH_3)_3$	Phosphatidylcholine
	Fatty acid	Fatty acid	Inositol ring: positions (1)–(6), OH	Phosphatidylinositol (di- and triphosphoinositides have phosphate on position 4 and 5)
	OPO_3H	H		Cardiolipin (diphosphatidyl glycerol)

Basic structure	*Addition*	*Compound*
$CH_3(CH_2)_{12}CH=CH(OH)CH(NH_2)CH_2OH$ Sphingosine	Fatty acid	Ceramide
	Fatty acid, phosphorylcholine	Sphingomyelin
	Galactose	Psychosine
	Galactose, fatty acid	Cerebroside
	Galactose, fatty acid, sulfate	Sulfatide
	Glucose, galactose, fatty acid, *N*-Acetylneuraminic acid, *N*-Acetylgalactosamine	Ganglioside

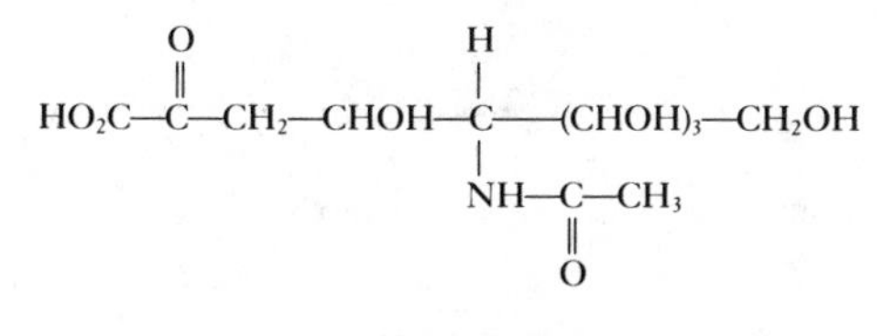

N-Acetylneuraminic acid (sialic acid)

Gaucher's disease there is a rise in the cerebroside content of the brain, and in Niemann-Pick disease there is an accumulation of both sphingomyelin and gangliosides. It is also known that in demyelinating diseases the cerebroside and sphingomyelin content of the brain declines and that there is a formation of cholesterol esters.

Gangliosides are of particular interest in that they bind black-widow spider venom as well as cholera, tetanus, and botulinum toxins; recently a cholinergic terminal antigen, referred to as chol-1, has been identified as a ganglioside. In addition, these glycolipids have been shown repeatedly to induce neurite extension in a variety of neuronal cultures. Finally, antibodies to gangliosides when injected intracerebrally produce a variety of CNS effects that are relatively selective. Thus, with this range of effects, among the complex lipids the gangliosides deserve particular attention.

Continuing with the scheme in Figure 4-1, the offshoot giving rise to ACh, glycine, and prostaglandins will be dealt with in subsequent chapters.

Glutamate is the next compound we encounter in Figure 4-1 to fulfill our criteria for neuropharmacological interest. Besides its obvious utility in the brain as an energy source in coupled phosphorylation and as an amino acid for protein synthesis, glutamate has five other functions. It serves as:

1. a "detoxifying" substrate trapping NH_3 to form glutamine;
2. the precursor of γ-aminobutyric acid (GABA);
3. the precursor of proline;
4. one of the amino acids in glutathione;
5. a "universal" excitatory amino acid.

The possible neurotransmitter roles of glutamate and GABA will be discussed in Chapter 7. Also to be discussed in that chapter is the "GABA shunt," a sequence of reactions unique to brain, in which glutamate functions catalytically to produce succinate from α-oxoglutarate without invoking α-oxoglutarate dehydrogenase.

It is well established that glutamate exists in brain in at least two pools: a small one thought to be glial and a larger one that is neuronal. It is not known with any certainty how the pools relate to the various activities of this amino acid. (Because of these pools, and since glutamate and the other amino-acid transmitters all derive from glucose, it would be wise to use labeled glucose as the precursor for these amino acids rather than employing the labeled amino acid in experiments involving the release of these agents.)

The relationship between ammonia levels in the brain and convulsions has been demonstrated in a variety of experimental procedures. Although the cerebral origin of endogenous ammonia production is still unclear, the role of glutamate and glutamine synthetase in ammonia disposal is well established. Thus the infusion of ammonium salts into dogs results in an increased level of cerebral glutamine. Glutamine levels are presumably regulated by diffusion of the compound into the general circulation to the kidney where glutaminase would remove ammonia. The ammonia would be eliminated either as a cation or via the urea cycle. Under normal conditions glutamine in glial cells appears to serve as a storehouse for neuronal glutamate and may regulate the level of this amino acid and, consequently, of GABA. The observation that glutaminase is an allosteric enzyme that is activated by phosphate, ammonia, and nucleoside triphosphates and is inhibited by glutamate suggests that the conversion of glutamine to glutamate is under fine metabolic control.

Glutathione, a tripeptide consisting of glutamic acid, cysteine, and glycine, has been known for over forty years, but its function still remains elusive. It does serve as the coenzyme for formaldehyde dehydrogenase, glyoxylase, and for the isomerization of maleylacetoacetate to fumarylacetoacetate, but these reactions are not considered important enough to explain its high concentration in brain. Rather, speculation on the role of the tripeptide has always revolved around its oxidation and reduction. Its primary function may be to destroy free radicals and to maintain a proper ratio of sulfhydryl and disulfide groups in proteins or other cerebral constituents. Alternatively, according to Meister, it may function as the substrate for γ-glutamyl transpeptidase in the γ-glutamyl cycle that operates in amino-acid transport. The oxidized or disulfide form of glutathione is rapidly reduced by a specific NADPH-dependent glutathione reductase in brain; the reaction is essentially irreversible.

The wordly student should also be aware of a final neuropharmacological implication of glutamate in the Chinese-restaurant syndrome. Some individuals experience painful reactions after eating Chinese food. The syndrome has been traced to the large amount of monosodium glutamate used in some recipes.

The transamination of glutamate with oxaloacetate leads to aspartate. Although in some invertebrates this amino acid serves as an organic anion reservoir, in mammalian tissue its role appears to be metabolic. Like glutamate, aspartate has central excitatory activity on iontophoretic application, can serve as a readily available energy source, and can also trap ammonia to form asparagine. Aspartic acid is a requisite in *de novo* nucleotide synthesis and in the arginosuccinic acid synthetase reaction in brain.

In the presence of acetyl CoA and a membrane-bound enzyme from brain or spinal cord, aspartate forms *N*-acetylaspartate. This compound is of interest in that (1) next to glutamic acid it is the amino acid or derivative in highest concentration in the brain, (2) it appears to have a primarily neuronal localization, and (3) the hydrolysis of *N*-acetylaspartate is catalyzed by a relatively specific enzyme designated acylase II that resides primarily in glial tissue. Although in weanling animals the acetyl group can contribute to lipid biosynthesis in the brain, in adults the compound is believed to be somewhat inert metabolically.

The fact that endogenous *N*-acetylaspartate has been shown to be localized in cell cytoplasm, that it has an apparently long turnover time, and that as a dicarboxylic acid it contributes approximately 11 mEq of anion/kg of brain, raises the possibility that acetylaspartate functions solely as an anion to help defray the so-called "anion deficit" that exists. Another possible function relates to the suggestion that *N*-acetylaspartyl peptides may be involved in behavioral changes. Specifically, *N*-acetylaspartyl glutamate, which when applied iontophoretically excites the pryiform cortex, has been implicated as the neurotransmitter in the lateral olfactory tract.

Arising from the metabolism of the essential amino-acid methionine are cysteic acid and taurine. This pair is reminiscent electrophysiologically of the glutamate-GABA duo in that the first member of the pair is excitatory and the second is inhibitory.

A rather puzzling situation exists with respect to urea in the brain. Arginosuccinate synthetase, the arginosuccinate-splitting enzyme, and arginase have been found in the brain. However, attempts to find carbamyl phosphate synthetase and ornithine transcarbamylase have been negative. It is conceivable that citrulline is not formed from ornithine in brain but is transported there, giving rise to a modified urea cycle that is in point of fact acyclic.

It is interesting to look at the three essential aromatic amino acids—phenylalanine, tryptophan, and histidine—as a group since (1) they all are associated with congenital errors of metabolism that produce mental retardation and (2) they all give rise to neuroactive amines. In the case of tyrosine (derived from phenylalanine in the liver or from the diet) and tryptophan, the metabolic pathways are strikingly similar since decarboxylation produces pressor agents (tyramine, tryptamine), whereas oxidation prior to decarboxylation yields neurotransmitters (dopamine, norepinephrine, epinephrine, and serotonin). With histidine, decarboxylation also gives rise to a vasoactive substance (histamine). Carnosine (β-alanylhistidine) is found in high concentrations in olfactory bulb and disappears after deafferentation; also found in the olfactory bulb are carnosine synthetase, carnosinase, and a specific olfactory marker protein. The function of carnosine and of the marker protein is at present obscure.

In this chapter on the biochemistry of the central nervous system, no attempt has been made to discuss the two dozen or more genetic abnormalities that are associated with mental retardation; those that are noted are mentioned only because they were associated with compounds that were discussed for other reasons. Aberrant amino acid, carbohydrate, and lipid metabolism may all produce mental deficiencies. Rather than illuminating a specific reaction that is vital to normal mental development, these genetic blocks

TABLE 4-2. Hereditary diseases associated with cerebral impairment

Disorder	*Defect*
(A) AMINO ACID METABOLISM	
Arginosuccinic aciduria	Arginosuccinase
Citrullinemia	Arginosuccinic acid synthetase
Cystathionuria	Cystathionine-cleaving enzyme
Hartnup disease	Tryptophan transport
Histidinemia	Histidase
Homocystinuria	Cystathionine synthetase
Hydroxyprolinemia	Hydroxyproline oxidase
Hyperammonemia	Ornithine transcarbamylase
Hyperprolinemia	Proline oxidase
Maple syrup urine disease	Valine, leucine, and isoleucine decarboxylation
Phenylketonuria	Phenylalanine hydroxylase
(B) LIPID METABOLISM	
Abetalipoproteinemia (acanthocytosis)	β-lipoproteins
Cerebrotendonous xanthomatosis	Cholesterol
Gaucher's disease	Cerebrosides
Juvenile amaurotic idiocy	Gangliosides (?)
Krabbe's globoid dystrophy	Cerebrosides
Kuf's disease	Gangliosides (?)
Metachromatic leukodystrophy	Sulfatides
Niemann-Pick disease	Gangliosides and sphingomyelins
Refsum's disease	3, 7, 11, 15-Tetramethylhexadecanoic acid
Tay-Sachs disease	Gangliosides
(C) CARBOHYDRATE METABOLISM	
Galactosemia	Galactose-1-phosphate uridyl transferase

TABLE 4-2. *continued*

Disorder	*Defect*
Glycogen storage disease (Type 2)	α-Glucosidase
Hurler's disease	Chondroitin sulfuric acid β, gangliosides
Unnvericht myoclonus epilepsy	Polysaccharides (?)
(D) MISCELLANEOUS	
Cretinism	Thyroid hormone
Hallevorden-Spaatz disease	Iron deposition in basal ganglia
Intermittent acute porphyria	δ-Aminolevulinic acid
Subacute necrotizing encephalomyelopathy (Leigh's disease)	Thiamin triphosphate
Lesch-Nyhan syndrome	Hypoxanthine-guanine Phosphoribosyl transferase
Methylmalonic acidemia	Methylmalonyl CoA mutase
Wilson's disease	Ceruloplasmin

dramatically point out the necessity of totally integrated metabolism. Table 4-2 lists some genetic abnormalities that are associated with cerebral impairment.

No discussion of nucleic acids and proteins has been presented in this chapter since the synthetic and catabolic pathways of these macromolecules in brain do not appear to be basically different from those in other tissues so that they do not fit into the scheme as shown in Figure 4-1. This omission is, of course, not meant to imply that either nucleic acids or proteins of brain are identical with their extraneural counterparts. Several soluble proteins have been isolated which immunologically represent brain-specific proteins. The two most prominently studied proteins are the S-100 and the 14-3-2, the former with a glial localization (function unknown); the latter, with an isozyme that is found in neurons only, has been shown to be enolase.

The attempt in this chapter has been to explain the functional uniqueness of brain in terms of biochemistry. To this end, although we have answered the question posed at the beginning of this chapter, (What biochemical substrates or reactions distinguishes brain from other tissue?), the major items remain to be discussed. These are the neurotransmitters and the enzymes, receptors, and other proteins involved in their function.

The fundamental specialization of the brain, as is true of other organs, is morphological with its basis in genetic transcription. Given the current intense activity in recombinant DNA techniques, it is perhaps not too farfetched to imagine gene replacement therapy not only as a cure of the myriad genetic deficiency diseases such as the cerebral lipidoses, but even as the ultimate therapy for diseases such as convulsive disorders, which are now merely controlled by drugs.

Selected References

Hirata, F., and J. Axelrod (1980). Phospholipid methylation and biological signal transmission. *Science 209*, 1082.

Ledeen, R. W., R. K. Yu, M. W. Rappaport, and K. Suzuki (1984). *Ganglioside Structure, Function, and Biomedical Potential*, AEMB Vol. 174. Plenum, New York.

Margolis, F. (1980). Carnosine: An olfactory neuropeptide. In *The Role of Peptides in Nuronal Function* (J. L. Barker and T. G. Smith, Jr., eds.), p. 546. Marcel Dekker, New York.

McIlwain, H., and H. S. Bachelard (1985). *Biochemistry and the Central Nervous System*. Churchill, London.

Meister, A., and M. E. Anderson (1983). Glutathione. *Ann. Rev. Biochem. 52*, 711.

Nadler, J. V., and J. R. Cooper (1972). Metabolism of the aspartyl moiety of *N*-acetyl-L-aspartic acid in the rat brain. *J. Neurochem. 19*, 2091.

Siegel, G. J., R. W. Albers, R. Katzman, and B. W. Agranoff (1986). Basic Neurochemistry, 4th ed. Little, Brown, Boston.

Sokoloff, L. (1984). Localization of functional activity in the central nervous system by metabolic probes. In IUPHAR Proceeding, London, p. 9.

Stanbury, J. B., J. B. Wyngaarden, and D. S. Fredrickson (1983). *The Metabolic Basis of Inherited Disease*. McGraw-Hill, New York.
Zomzely-Neurath, C., and A. Keller (1977). Nervous system-specific proteins of vertabrates. *Neurochem. Res.* 2, 353.

5 | Receptors

The concept that most drugs, hormones, and neurotransmitters produce their biological effect by interacting with receptor substances in cells was introduced by Langley in 1905. It was based on his observations of the extraordinary potency and specificity with which some drugs mimicked a biological response (agonists) while others prevented it (antagonists). Later, Hill, Gaddum, and Clark independently described the quantitative characteristics of competitive antagonism between agonists and antagonists in combining with specific receptors in intact preparations. This receptor concept has been substantiated in the past few years by the actual isolation of macromolecular substances that fit all the criteria of being receptors. To date, receptors (or, at least, binding sites) have been identified for all the proven neurotransmitters as well as for histamine, opioid peptides, neurotensin, VIP, bradykinin, CCK, somatostatin, Substance P, insulin, angiotensin II, gonadotropin, glucagon, prolactin, and TSH. In addition, as noted in Table 5-1, multiple receptors have been shown to exist for all the biogenic amines, ACh, histamine, opiates and the amino-acid transmitters. These multiple receptors, however, should be viewed skeptically until a physiological response to the ligand has been shown. Many of the receptor subtypes have only been identified by binding techniques (see below) that can lead to erroneous conclusions. This problem is particularly evident with muscarinic receptors where a literature search to determine the reality of the different subtypes, affinities, and classes of agonists that have been proposed can lead to a cholinergic activation of the reader with profuse sweating. All the receptors for neurotransmitters and peptide hormones that have been studied, regardless of whether they have been isolated, are localized on the surface of the cell; among other receptors, only those for steroid and thyroid hormones are apparently intracellular.

In earlier days with the discovery that the action of physostigmine (eserine) was due to its anticholinesterase activity, it was assumed that most drugs probably acted by inhibiting an enzyme. It now appears, however, that with few exceptions, the mechanism of action of most neuroactive drugs is referable to their effect on specific receptors. Predictably, the current search for receptors is among the most intensively investigated areas in the neurosciences. It should be appreciated that this interest is not purely academic. The recent identification of adrenergic, dopaminergic, and histaminergic receptor subtypes has resulted in the synthesis of highly selective drugs that are considerably more specific than their prototypes which were developed after general screening for activity. Further, as technological advances occur in the production of monoclonal antibodies and in recombinant DNA, a rapid means is at hand for determining receptor structure and isolating messenger RNA to produce specific receptors.

Definition

In this rapidly developing field, considerable confusion has arisen as to what functional characteristics are required of an isolated, ligand-binding molecule to qualify as a receptor. This confusion, a semantic problem, developed after the successful isolation of macromolecules that exhibit selective binding properties (see below), which made it mandatory to determine whether this material comprised both the binding element and the element that initiated a biological response, or merely the former. Some investigators use the term "receptor" only when both the binding and signal generation occur and the term "acceptor" if no biological response has been demonstrated. Others are content to ignore the bifunctional aspect and use "receptor" without specifications. In this chapter, we will define a *receptor* as the binding or recognition component and refer to the element involved in the biological response as the *effector*, without specifying whether the receptor and effector reside in the same or separate units. The criteria for receptors will be dealt with shortly; the biological response that is generated by the effector ob-

TABLE 5-1. Classification of multiple receptors

Subtype	*Characteristics*
ADRENERGIC	
α_1	Postsynaptic; contracts smooth muscle, localized in heart, vas deferens, and brain; phenylephrine is moderately selective agonist, prazosin is selective antagonist; agonist potency = EPI > NE > ISO.
α_2	Primarily presynaptic but also postsynaptic, linked to adenylate cyclase; in pancreas, duodenum, and brain; clonidine is selective agonist; idazoxan is a selective antagonist, yohimbine is moderately selective antagonist; agonist potency = EPI ≥ NE > ISO.
β_1	Linked to adenylate cyclase; causes cardiac stimulation and fatty acid mobilization; metoprolol is selective antagonist; agonist potency = ISO > EPI = NE.
β_2	Linked to adenylate cyclase; produces bronchodilation, vasodilation; salbutamol is selective agonist; IPS-339[a] is selective antagonist; agonist potency = ISO > EPI >> NE.
DOPAMINERGIC	
D_1	Stimulates adenylate cyclase; SKF38393[b] is a selective agonist, SCH2339, the 3-methyl 7-chloroanalog of SKF38393 is an antagonist, present in retina, parathyroid gland and pyramidal cells of olfactory tubercle.
D_2	Inhibits adenylate cyclase; quinperole and N0434[c] are agonists, (−) sulpiride and domperidone are antagonists; on terminals and cell bodies of nigrostriatal and mesolimbic dopamine neurons (auto-

Subtype	Characteristics
DOPAMINERGIC	
	receptors) and possibly on terminals of cortico-striate pathway.
SEROTONERGIC	
5-HT_1	Selectively labeled by 5-HT, regulated by guanine nucleotides.
5-HT_2	Localized in prefrontal cortex, preferentially antagonized by ketanserin.
CHOLINERGIC	
MUSCARINIC	Quinuclidinyl benzilate (QNB) is selective antagonist; oxotremorine is selective agonist; M_1 and M_2 subtypes have been postulated where QNB blocks both while pirenzepine is a selective antagonist for M_1 sites (sympathetic ganglia, hippocampus).
NICOTINIC	α-Bungarotoxin and venom from *Naja naja siamensis* are specific antagonists at the neuromuscular junction. Here α-bungarotoxin blocks transmission, but at autonomic ganglia the toxin binds but may not affect transmission. In the CNS, the properties of the nicotinic cholinergic receptor are unclear (see Chapter 8).
OPIATE	
Mu (μ)	Morphine and opiate alkaloids are agonists.
Delta (δ)	Receptor for enkephalins.
Kappa (κ)	Recently certified, ketocyclazocine and dynorphin are preferential agonists.

TABLE 5-1. Classification of multiple receptors (*cont.*)

Subtype	*Characteristics*
Sigma (σ)	Agonist is *N*-allylnormetazocine; phencyclidine (PCP) also interacts with this receptor. Since agonist activity is not reversed by naloxone, the legitimacy of this opiate receptor is unsettled. An epsilon (ϵ) receptor has also been described but awaits certification.
GABAERGIC	
$GABA_A$	Linked to change in chloride conductance, antagonized by bicuculline.
$GABA_B$	Mechanism unclear, baclofen is selective agonist.

EXCITATORY AMINO ACIDS

The status of the receptors for glutamate and aspartate is still evolving with the possibility that other endogenous agents and receptors are probably present. Based on agonist specificity, however, three distinct classes of receptors have been identified.

N-methyl-D-aspartate (NMDA, also called A1).	
	This compound is the distinguishing agonist. D-2-amino-5-phosphonovalerate (D-AP5) is an antagonist. High concentration in cerebral cortex and hippocampus. Implicated in long-term potentiation and convulsive disorders.
Quisqualate (A2).	
	This compound and α-amino-3-hydroxy-5-methylisoxazole-4-propionic acid (AMPA) are selective agonists. No specific antagonists available but γ-D-glutamyltaurine and γ-D-glutamylaminomethylsulfonate will block both quisqualate and kainate receptors but not NMDA receptors. Glutamate has higher affinity than aspartate.
Kainate (A3).	
	This compound is the selective agonist. No selective antagonists available (but see quisqualate above).

In high concentration in limbic system. Implicated in convulsive disorders. Neither glutamate nor aspartate may be the endogenous ligand.

---

[a](*t*-Butyl-amino-3-ol-2-propyl) oximino-9-fluorene.
[b]2,3,4,5-tetrahydro-7,8-dihydroxy-1-phenyl-1H-3-benzazepine.
[c]5-Hydroxy-2(*N*-*n*-propyl-*N*-2-phenylethyl)-aminotetraline.

SOURCE: mainly personal communications (G. K. Aghajanian, S. K. Fisher, J. Kebabian, H. Loh, J. V. Nadler, E. J. Simon, and J. H. Tallman).

viously has a wider range of complexity, from a simple one-step coupling to an unknown number of steps (Fig. 5-1).

Assays

Basically, there are two ways to study the interaction of neurotransmitters, hormones, or drugs with cells. The first procedure, and until recently the only one, is to determine the biological response of an intact isolated organ, such as the guinea pig ileum, to applied agonists or antagonists. The disadvantage of this procedure is that one is obviously enmeshed in a cascade of events beginning with transport, distribution, and metabolism of the agent before it even interacts with a receptor and ranging through an unknown multiplicity of steps before the final biological response of the tissue is measured. Thus, although studies with agonists may be interpretable, it is not difficult to envisage problems when antagonists are employed since these compounds may be competing at a different level than receptor binding. Despite the not unusual problem of a nonlinear relationship between receptor occupancy and biological response, this approach has yielded a considerable amount of information.

The second approach to studying receptors is by measuring ligand binding to a homogenate or slice preparation. This technique has only recently become feasible with the development of ligands of a high specific radioactivity and a high affinity for the receptor. Here the direct method is to incubate labeled agonist or antagonist with the receptor preparation and then separate the receptor–ligand

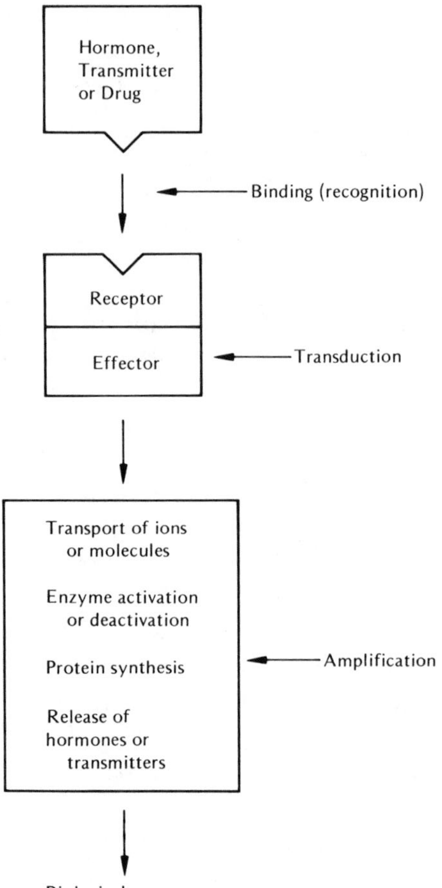

FIGURE 5-1. Schematic model of ligand–receptor interaction.

complex from free ligand by centrifugation, filtration, or precipitation. The indirect technique is to use equilibrum dialysis where the receptor–ligand complex is determined by subtracting the ligand concentration in the bath from that in the dialysis sac.

Though not always appreciated, the advantage of using isolated tissues is that both the efficacy and the functional activity of an agonist is assessed, in contrast to binding procedures using broken cell preparations where only affinity or a biochemical sequella of binding can be appraised. Ideally, both techniques should be used but, alas, this is rarely done. It should be noted in passing that efficacy and affinity are independent. To date it appears that the drugs that exhibit a high affinity but low efficacy have a more efficient coupling to the effector than the reverse situation and thus are the most potent and selective agents. In contrast, an endogenous ligand that exhibited high affinity binding would not be physiologically useful since high affinity generally implies a slow off-rate. Thus it is the low-affinity binding of transmitters to receptors that is thought to be relevant to synaptic transmission. For a detailed discussion of the above, the review by Kenakin is recommended.

Identification

In the midst of an intensive drive to isolate and characterize receptors, some zealous investigators have lost sight of basic tenets that must be satisified before it is certain that a receptor has indeed been isolated. Thus, on occasion enzymes, transport proteins, and merely extraneous lipoproteins or proteolipids that exhibit binding properties have been mistakenly identified as receptors.

Authentic receptors should have the following properties:

1. *Saturability.* The great majority of receptors are on the surface of a cell. Since there a finite number of receptors per cell, it follows that a dose-response curve for the binding of a ligand should reveal saturability. In general, specific receptor binding is characterized by a high affinity and low capacity whereas nonspecific binding usually exhibits high-capacity and low-affinity binding that is virtually nonsaturable.
2. *Specificity.* This is one of the most difficult and important criteria to fulfill because of the enormous mass of nonspecific binding sites compared to receptor sites in tissue. For this reason, in binding assays it is necessary to explore the

displacement of the labeled ligand with a series of agonists and antagonists that represent both the same and different chemical structures and pharmacological properties as the binding ligand. One should also be aware of the avidity with which inert surfaces bind ligands. For example, Substance P binds tenaciously to glass and insulin can bind to talcum powder in the nanomolar range. With agents that exist as optical isomers, it is of obvious importance to show that the binding of the ligand is stereospecific. Even here problems arise. With opiates it is the levorotatory enantiomorph that exerts the dominant pharmacological effect; Synder, for example, has found glass fiber filters that selectively bind the levorotatory isomer.

Specificity obviously means that that one should find receptors only in cells known to respond to the particular transmitter or hormone under examination. Further, it is a truism that a correlation should be evident between the binding affinity of a series of ligands and the biological response produced by this series. This correlation, the *sine qua non* for receptor identification, is unfortunately a criterion that is not often investigated.

3. *Reversibility.* Since transmitters, hormones, and most drugs act in a reversible manner, it follows that the binding of these agents to receptors should be reversible. It is also to be expected that the ligand of a reversible receptor should be not only dissociable but recoverable in its natural (i.e., nonmetabolized) form. This last dictum distinguishes receptor–agonist interactions but not receptor–antagonist binding from enzyme–substrate reactions.
4. *Restoration of function on reconstitution.* Following the isolation and identification of the components of the receptor system, to put them back together again is the ultimate goal of all receptorologists. Although a difficult task, this has been accomplished with the nicotinic ACh receptor and the peripheral β-adrenergic receptor.

Kinetics and Theories of Drug Action

From the law of mass action, the binding of a ligand (L) to its receptor (R) leads to the equation

$$L + R \underset{k_2}{\overset{k_1}{\rightleftharpoons}} LR$$

$$\text{thus } K_D = \frac{k_2}{k_1} = \frac{[L][R]}{[LR]}$$

where k_1 = association rate constant
k_2 = dissociation rate constant
$[L]$ = concentration of free ligand
$[R]$ = concentration of free receptors
$[LR]$ = concentration of occupied binding sites
K_d = dissociation equilibrium constant

Since the total number of receptors $= [R_t] = [R] + [LR]$

$$[R] = [R_t] - [LR]$$

$$\text{therefore } K_d = \frac{[L]([R_t] - [LR])}{[LR]}$$

$$\text{and } [LR] = \frac{[L][R_t]}{[L] + K_d}$$

Since the fraction of receptors occupied $(r) = \frac{\text{bound}}{\text{total}} = \frac{[LR]}{[R_t]}$

$$\text{therefore } r = \frac{[L]}{[L] + K_d}$$

If experiments are performed in which the receptor concentration is kept constant and the ligand concentration is varied, then a plot of r versus [L] will produce a rectangular hyperbole, the usual Langmuir adsorption isotherm. Here r approaches the saturation

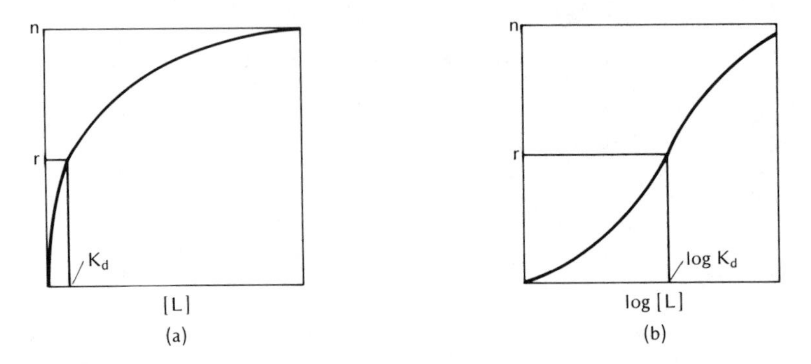

FIGURE 5-2. Ligand–receptor interactions plotted in two ways where {L} is the concentration of ligand (drug, hormone, or neurotransmitter) and r is the biological response, proportional to the moles ligand bound per mole of protein. The number of binding sites per molecule of protein is designated by n.

value equal to the number of binding sites per molecule of ligand. If r is plotted against log [L], a sigmoid curve will result; log [L] at half saturation will give log K_d on the horizontal axis (Fig. 5-2).

This equation can be rearranged as follows:

$$K_d r + r[L] = [L]$$

$$\frac{r}{[L]} = \frac{-r}{K_d} + \frac{1}{K_d}$$

Now if $\frac{r}{[L]}$ is plotted against r, a straight line will result (assuming only one set of binding sites) with two intercepts, the one on the x axis giving the number of binding sites per molecule and the y intercept yielding $1/K_d$. This type of plot is the Scatchard plot (more correctly, Rosenthal plot) widely used in studying receptor-ligand interactions (Fig. 5-3). Among the pitfalls that are encountered in a Scatchard analysis is the problem that the system is not in true equilibrium.

Another useful representation that can be derived from the general equation is the Hill plot (Fig. 5-4).

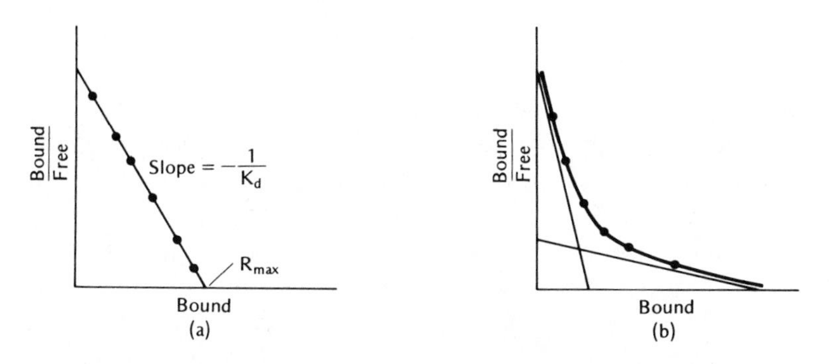

FIGURE 5-3. Scatchard plots of the binding of a ligand to a receptor. In (a) only one binding affinity occurs, but in (b) both a high- and low-affinity binding sites are suggested.

If $\log \frac{E}{E_{max} - E}$ is plotted versus log [L] when E is the effect produced and E_{max} is the maximum effect, then the slope, indicative of the nature of the binding, gives a Hill coefficient of unity in cases where E/E_{max} is proportional to the fraction of total number of sites occupied (r). In many situations, the slope turns out to be a non-integer number different from unity. This finding indicates that

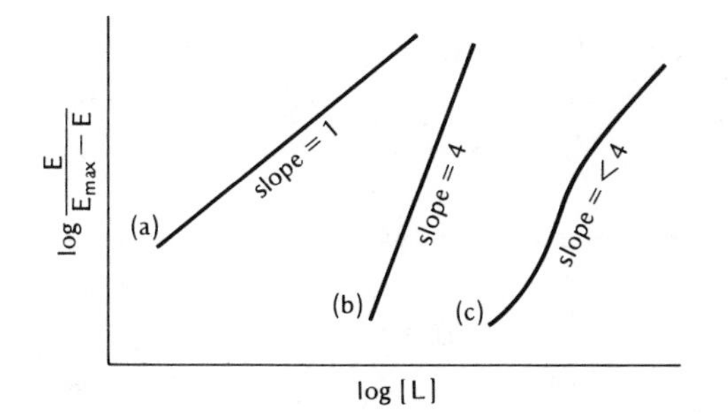

FIGURE 5-4. Hill plots for receptor–ligand binding: (a) noncooperative binding; (b) idealized plot of cooperative binding with four sites; (c) a typical Hill plot of multiple binding sites but less than four. (Modified after Van Holde, *Physical Biochemistry,* Englewood Cliffs, N.J.: Prentice-Hall, Inc., 1971, p. 63)

cooperativity may be involved in the binding of the ligand to the receptor. Cooperativity is the phenomenon whereby the ligand binding at one site influences, either positively or negatively, the binding of the ligand at sites on other subunits of the oligomeric protein. This idea, originally suggested by Monod, Wyman, and Changeux in 1965 to explain allosteric enzyme properties, currently offers the most attractive hypothesis for studying reactions of receptors with hormones, transmitters, or drugs. We will utilize this hypothesis later in an attempt to explain drug action, including the problem of efficacy (intrinsic activity), spare receptors, and desensitization; for a full discussion the review by Rang is recommended.

Another useful anaylsis when competitive inhibitors of the receptor binding ligand are studied is the equation derived by Cheng and Prusoff in their kinetic analysis of enzyme inhibitors:

$$K_i = \frac{I_{50}}{1 + (L)/K_d}$$

where K_i is the equilibrium dissociation constant of the competitive inhibitor and I_{50} is the concentration of the inhibitor producing 50 percent inhibition at the concentration of the labeled ligand that is used in the study.

Clark produced the first model of drug–receptor action, known as the occupation theory, in which the response of a drug was held to be directly proportional to the fraction of receptors that were occupied by the drug. Here, as mentioned earlier, one should find the usual Langmuir absorption isotherm. But instead of the expected rectangular hyperbole when drug concentration was plotted against drug bound, in most cases a sigmoid curve resulted. A second problem with the occupancy theory is that in many instances only a small fraction of the total receptors available are occupied, and yet a maximum response is obtained. These additional receptors, which may represent as much as 95 to 99 percent of the total, are referred to as spare receptors. The occupancy theory was further complicated when the activity of a series of related agents was explored and the biological response varied from maximum to zero, even though the agents all occupied the receptor. In other words, these agents could be full agonists, partial agonists that gave less than a maximal response, and antagonists whose occupancy produced no response. This phenomenon is referred to as efficacy or intrinsic activity of a drug and is obviously not directly related to the binding affinity of the drug. A fourth characteristic of drug–receptor interaction sometimes observed is desensitization, which is defined as the lack or decline of a response in a previously activated receptor.

These problems—the sigmoidal dose-response curve, some anomalous effects with antagonists that spare receptors might account for, efficacy, and desensitization—can all be comfortably fitted into a two-state model of a receptor analogous to the allosteric model of Monod, Wyman, and Changeux, proposed independently by Changeux and by Karlin in 1967.

According to the two-state model, receptors exist in an active (R) and an inactive (T) state, and each is capable of combining with the drug (A):

$$\begin{array}{ccc} A+R & \overset{K_{AR}}{\longleftrightarrow} & AR \\ \updownarrow & & \updownarrow \\ A+T & \overset{K_{AT}}{\longleftrightarrow} & AT \end{array}$$

Here an agonist prefers the R (active configuration of the receptor and the efficacy (i.e., intrinsic action) of the drug will be determined by the ratio of its affinity for the two states. In contrast, competitive antagonists prefer the T form of the receptor and will shift the equilibrium to AT. The sigmoid relationship between the fraction of receptors activated and the drug concentration (i.e., cooperativity) can be explained by this model with its equilibrium between R and T if one designates T as a subunit of R that binds A and thereby influences the further binding of A to R. Cooperativity can also explain anomalous effects of antagonists whenever the effect of an antagonist persists even in the presence of a high concentration of agonist. Here it could be postulated that by tightly binding to one conformational state of the receptor, the antagonist inhibits the binding of the agonist. One might even use this two-state model to account for desensitization where T would be a receptor that has been desensitized perhaps by a local change in the ionic environment. The ionic environment, in fact, may be one of the factors that dictate the two conformational states of a receptor; other possibilities include polymerization (clustering) of the subunits, depolymerization, or phosphorylation. It should be emphasized that this model is conjectural, subject to modifications as the need arises. A ternary complex model involving an unspecified membrane component has recently been suggested.

Table 5-2 lists some characteristics of receptors that have been defined to date. When one considers the finite number of receptors per cell and the fact that virtually all receptors are found on the plasma membrane, it is not surprising that progress in this area has been slow and laborious. It has been calculated that, assuming one binding site and an average molecular weight of 200,000 for a receptor, complete purification of a receptor protein would require about a 25,000-fold enrichment. The extraordinary density of ACh receptors in electric tissue has made this preparation so popular a choice that considerable information is now available on this cholinergic receptor. Two snake neurotoxins, *Naja naja siamensis*, and α-bungarotoxin, which specifically bind nicotinic cholinergic receptors, have been the key agents that have helped in isolating this receptor.

Finally to be considered is the relationship between receptors and effectors as they interact in the fluid mosaic membrane. As envisaged by Singer and Nicolson, membranes are composed of a fluid lipid bilayer that contains globular protein. Some of these proteins extend through the membrane and others are partially embedded in or on the surface. The hydrophilic ends of the protein protrude from the membrane while the hydrophobic ends are localized in the lipid bilayer. Some of the proteins are immobilized but others,

TABLE 5-2. Characteristics of isolated receptors

Receptor	Tissue	Chemical nature	Mol wt.	Mol wt. of subunits	No. of binding sites per cell	Equilibrium constant (M)
β-Adrenergic	Frog erythrocyte	Lipoprotein	140,000–160,000	—	1300–1800	10^{-9}
Cholinergic (nicotinic)	Electric tissue of *Torpedo*	Glycoprotein	250,000–500,000	40,000, 50,000 60,000, 65,000		2×10^{-8}
	Electric tissue of *Electrophorus*	Glycoprotein	500,000–600,000	41,000, 50,000 55,000, 64,000	10^{11}	10^{-7}
Glucagon	Liver	Lipoprotein	190,000	90,000	110,000	1.5×10^{-9}
Insulin	Liver	Glycoprotein	300,000	90,000–130,000	100,000–250,000	1.2×10^{-8}
TSH	Thyroid	Glycoprotein	280,000	24,000, 75,000, 166,000	500	1.9×10^{-9}

SOURCE: Modified after Kahn (1976).

floating in an oily sea of lipids, are capable of free movement. In this concept of membrane fluidity, the receptor protein would be on the surface of the membrane and the effector within the membrane. Although the ratio of receptors to effectors may in some cases be unity (thus explaining instances in which the occupancy theory is satisfied), it may also be greater than one. Consequently, in a situation where multiple hormones activate a response (e.g., there is only one form of adenylate cyclase in a fat cell, but it may be activated by epinephrine, glucagon, ACTH, or histamine), it would be concluded that an excess of receptors over effectors is present. This model would also explain spare receptors and is exemplified by the fact that only 3 percent of insulin receptors need to be occupied in order to catalyze glucose oxidation in adipocytes. With the possibility of easy lateral movement of effectors in the membrane, it is also understandable why one receptor may activate several types of effectors. Membrane fluidity will account for the observation that hormones can influence the state of aggregration of the receptor, thus giving rise to either positive or negative cooperativity as determined in kinetic studies of binding. Although direct evidence of the interaction and migration of receptors and effectors is difficult to obtain (but see the review article by Poo), current information is easily accommodated by the fluid mosaic membrane model.

Many neuroactive agents act on receptors that are coupled to the adenylate cyclase system. The components of this complex effector are the receptor, the catalytic portion of adenylate cyclase that converts ATP to cAMP, and two guanine nucleotide-binding regulatory components referred to as N_s and N_i that are coupled to the catalytic unit of the enzyme. When the receptor is stimulated (e.g., a β_2-adrenergic receptor), the coupling protein is N_s. Conversely, when the receptor is inhibited (e.g., an α_2-adrenergic receptor), N_i is the coupling protein. These regulatory proteins are activated by the hydrolysis of GTP to GDP. Examples of receptors whose activity is linked to the adenylate cyclase complex are the adrenergic receptors α_2, β_1, and β_2 (but not α_1, which may be coupled to

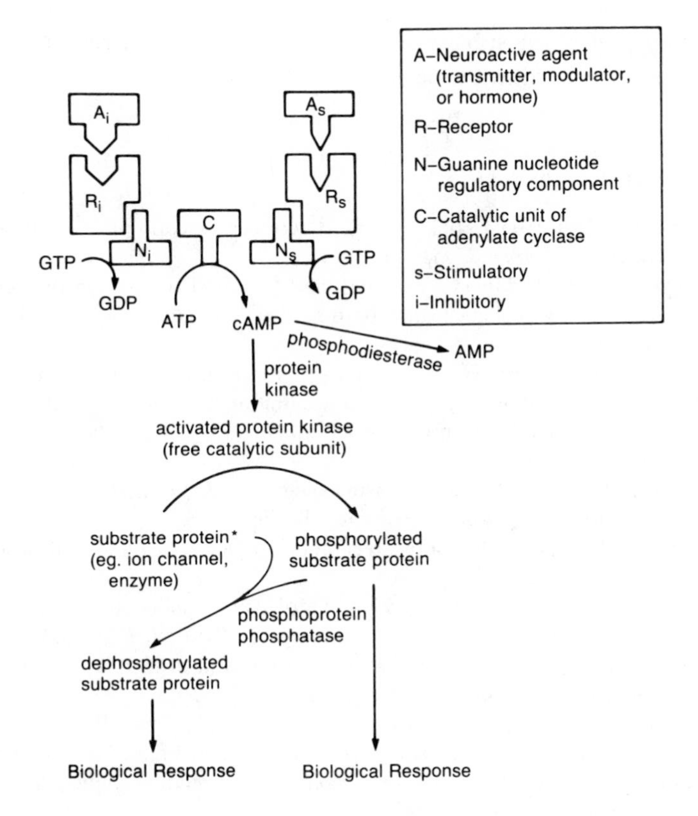

FIGURE 5-5. Components of a receptor-activated, cyclic nucleotide-linked system (after Lefkowitz *et al.*, 1984).

phosphatidyl inositol hydrolysis), adenosine, opiates, and possibly, some muscarinic receptors. Figure 5-5 illustrates this receptor-activated cascade.

Because of space limitations, this abbreviated account has not covered such topics as factors regulating the synthesis and degradation of receptors, complex kinetics, isolation techniques, or the chemical composition of these macromolecules. The student who is interested in these subjects may consult a number of recent books and reviews listed in the selected references.

Selected References

Ariens, E. J., and A. J. Beld (1977). The receptor concept in evolution. *Biochem. Pharmacol. 26*, 913.

Changeux, J. P., J. Thiery, Y. Tung, and C. Kittel (1976). On the cooperativity of biological membranes. *Proc. Nat. Acad. Sci. U.S.A. 57*, 335.

Cheng, Y.-C., and W. H. Prusoff (1973). Relationship between the inhibition constation (K_i) and the concentration of inhibitor which causes 50 percent inhibition (I_{50}) of an enzymatic reaction. *Biochem. Pharmacol. 22*, 3099.

Clark, A. J. (1937). General pharmacology. In *Heffter's Handbuch d-exp. Pharmacol. Erg. band 4*. Springer, Berlin.

Kahn, C. R. (1976). Membrane receptors for hormones and neurotransmitters. *J. Cell Biol. 70*, 261.

Karlin, A. (1967). On the application of a plausible model of allosteric proteins to the receptor for acetylcholine. *J. Theor. Biol. 16*, 306.

Kenakin, K. (1984). The classification of drugs and drug receptors in isolated tissues. *Pharmacol. Rev. 36*, 165.

Kito, S., T. Segawa, K. Kuriyama, H. I. Yamamura, and R. W. Olsen, eds. (1984). *Neurotransmitter Receptors*. Plenum Press, New York.

Laduron, P. M. (1984). Criteria for receptor sites in binding studies. *Biochem. Pharmacol. 33*, 833.

Lefkowitz, R. J., M. G. Caron, and G. L. Stiles (1984). Mechanisms of membrane-receptor regulation. *New Engl. J. Med. 310*, 1570.

McKinney, M., and E. Richelson (1984). The coupling of the neuronal muscarinic receptor to responses. *Ann. Rev. Pharmacol. Toxicol. 24*, 121.

Poo, M. (1985). Mobility and localization of proteins in excitable membranes. *Ann. Rev. Neurosci. 8*, 369.

Rang, H. P. (1975). Acetylcholine receptors. *Q. Rev. Biophys. 7*, 283.

Rosenthal, H. E. (1967). Graphic method for the determination and presentation of binding parameters in a complex system. *Anal. Biochem. 20*, 525.

Ross, E. M. and A. G. Gilman (1980). Biochemical properties of hormone-sensitive adenylate cyclase. *Ann. Rev. Biochem. 49*, 533.

Singer, S. J., and G. L. Nicolson (1972). The fluid mosaic model of the structure of cell membranes. *Science 175*, 720.

Snyder, S. H., and R. R. Goodman (1980). Multiple neurotransmitter receptors. *J. Neurochem.* 35, 5.
Stoof, J. C., and J. W. Kebabian (1984). Two dopamine receptors: biochemistry, physiology and pharmacology. *Life Sci.* 35, 2281.

6 | Modulation of Synaptic Transmission

Contrary to what one might assume, the more we learn about intercellular communication in the nervous system, the more complicated the situation appears. Up to about the mid-1970s, synaptologists smugly focused on a straight point-to-point transmission where a presynaptically released transmitter impinged on a postsynaptic cell. Gradually, situations emerged where previously identified neurotransmitters were observed as not acting in this fashion but rather "modulating" transmission. These departures from the previous norm and the continuing discovery of peptides and small molecules such as adenosine, which exhibited neuroactivity but which did not appear to be transmitters in the classical sense, supported the broader view of modulation of synaptic transmission. In retrospect, it is a concept that should have been apparent early on since it imparts to neural circuitry an extraordinary degree of flexibility that is necessary in considering mechanisms to account for behavioral changes. It should also be noted that modulation is not a neurophysiological property that is seen only in higher forms: Birds do it, bees do it, even educated fleas do it.

Definitions

Via the activation of a receptor (including channels), synaptic transmission may be modulated either presynaptically or postsynaptically. With presynaptic modulation, regardless of the mechanism, the ultimate effect is a change in the amount of transmitter that is released. With postsynaptic modulation, the ultimate effect is a change in the firing pattern of the postsynaptic neuron or in the activity of a postsynaptic tissue (e.g., blood vessel, gland, muscle).

Some confusion has arisen on the correct nomenclature of pre- and postsynaptic receptors so an explanation is in order. It should be recognized that what may be classified as a presynaptic receptor on the terminal of neuron B may be a postsynaptic receptor of neuron A, which is making an axoaxonic contact with neuron B (see Fig. 2-3). Thus depending on which neuron you are investigating, the receptor will be denoted as either pre- or postsynaptic.

Procedures that have been employed to fix activity at the presynaptic receptor level in a terminal are (1) to use synaptosomes, (2) to add tetrodotoxin to the preparation to block action potentials in neighboring interneurons, (3) to use neurons from the peripheral nervous system in cell culture where no postsynaptic cells exist, or (4) where feasible, either to chemically destroy terminals or lesion the neuron and demonstrate by ligand binding the loss of the receptor. To complete the nomenclature on presynaptic receptors, an *autoreceptor* is located on the terminal or somatic-dendritic area of a neuron that is activated by the transmitter(s) released from that neuron. A *heteroreceptor* is a presynaptic receptor that is activated by a modulating agent that originates from a different neuron or cell. As we shall now detail, there exists an extraordinary number of possibilities that are available to alter the point to point synaptic transmission that was mentioned above. Let us count the ways.

Presynaptic modulation can be affected by:

1. Receptor activation of a presynaptic neuron causing:
   a. A change in the firing frequency in the presynaptic neuron: This is probably the most common type of modulation, particularly in the CNS, where it can be assumed that the firing rate of virtually every neuron is governed by inputs on dendrites, soma, or axons. The firing rate determines the frequency of impulse conduction, thus the invasion of action potentials into terminals

or varicosities and consequently the amount of transmitter that is liberated.

b. A change in the transport or re-uptake of a transmitter or precursor, or in the synthesis, storage, release, or catabolism of a transmitter. All of these possibilities will result in a change in the concentration of a transmitter at the terminal. In practice, it has been shown that presynaptic activation of biogenic amine neurons promotes the phosphorylation of both tyrosine and tryphophan hydroxylase, which increases the synthesis of norepinephrine, dopamine, and serotonin. Curiously, phosphorylation of the pyruvate dehydrogenase complex causes a decrease in enzyme activity and in theory would decrease the levels of ACh and the amino-acid transmitters. To date, however, modulation of this enzyme activity by presynaptic receptor activation has not been observed.

c. An effect on ion conductances at the terminal: The three ions and their respective channels that one might focus on would be K^+, Ca^{2+}, or Cl^-. Endogenous neuroactive agents or drugs could alter transmitter release by opening or blocking the channels or changing the kinetics of channel open time via the possible mediation of protein phosphorylation or other second messengers.

2. A direct effect of neuropharmacological agents on some element of the release process. This could be an effect of the modulating agent on vesicular apposition to a terminal, fusion, or fission if an exocytotic mechanism is operating, or, in a nonexocytotic release process, an effect possibly on Na^+, K^+-ATPase.

3. A drug-induced direct effect without the mediation of second messenger systems (see below) on ionic conductance. Like the presynaptic changes described above, Ca^{2+}, Na^+, K^+, or Cl^- channels could be directly altered in their kinetics by the administration of a drug.

Postsynaptic modulation can be effected by:

1. A long-term change in the number of receptors: This is a situation that is not observed under normal physiological conditions. It is, however, commonly noted pharmacologically where the administration of a receptor agonist for a period of time will result in the down regulation of the receptor and conversely, treatment with a receptor antagonist produces an increase in receptor density.
2. A change in the affinity of a ligand for a receptor. The now classical example is the salivary nerve of the cat where both ACh and VIP are colocalized. When VIP is released on electrical stimulation, it increases the affinity of ACh for the muscarinic receptor on the salivary gland up to 10,000-fold with a consequent increase in salivation.
3. An effect on ionic conductances. As discussed in 1-a above, this is frequency modulation. This is a postsynaptic effect on the first neuron in a relay but would be classified as a presynaptic effect on the second neuron.

A listing with references to all the neurons and all the modulating agents that have been investigated is given in the reviews by Starke and by Chesselet and a book by Vizi. These tables can be summarized by stating that virtually every neuronal pathway is modulated and virtually every endogenous neuroactive agent has been shown in one preparation or another to be capable of affecting synaptic transmission. All this information is descriptive: The question of the second messenger systems that may be ultimately involved in regulating activity should now be addressed.

Second Messengers

Currently, five biochemical mechanisms have been proffered:

1. *Protein phosphorylation.* Following the identification of cyclic nucleotides by Sutherland and associates, and an implication as a second messenger system that was preceded by an initial nerve im-

pulse or hormonal signal, E. Krebs and colleagues demonstrated a multistep sequence of events that linked cyclic AMP generation in muscle to regulation of carbohydrate metabolism. Since that time, the cyclic nucleotides have been shown to regulate an enormous diversity of processes ranging from axoplasmic transport to neuronal differentiation and including transmitter synthesis and release and generation of postsynaptic potentials. All these effects are referable to the cAMP or cGMP activating protein kinases and thus protein phosphorylation. Second messenger activity, achieved via protein phosphorylation, can be mediated by cAMP and cGMP-dependent protein kinases as well as calcium–calmodulin-dependent protein kinase and calcium–phosphatidyl serine-dependent protein kinase. In some instances, however, Ca acts as a second messenger without the participation of protein phosphorylation. The diversity of signals that are coupled to protein phosphorylation is depicted in Figure 6-1. For a molecular illustration of receptor coupling, the reader is referred to Figure 5-5 in Chapter 5.

Although as illustrated in Figure 6-1, there is an overwhelming number of systems in which protein phosphorylation is implicated, there are currently less than a dozen situations where direct evidence has been obtained in which this process has been linked to modulation of synaptic transmission, either pre- or postsynaptically. This information is presented in Table 6-1. It is not yet known if the proteins comprising the channels are phosphorylated or if the phosphorylated proteins are somehow associated with the channels. At any rate, with over six dozen proteins in brain that have been shown to be phosphorylated, the story is far from complete. Questions that remain to be answered include (1) the regulation of the almost infinite number of steps in the cascade and the specificity of the phosphodiesterases, protein kinases, substrate proteins, and phosphoprotein phosphatases and (2) the reasons for the difficulties in ascribing neuronal functions to cGMP. Although primarily due to the lack of specific activated catalytic entities and inhibitors of cGMP dependent protein kinase, there is also a lack of known substrate proteins, and several tissues (e.g., retina) contain a high concentration of cGMP but virtually no cGMP-dependent protein ki-

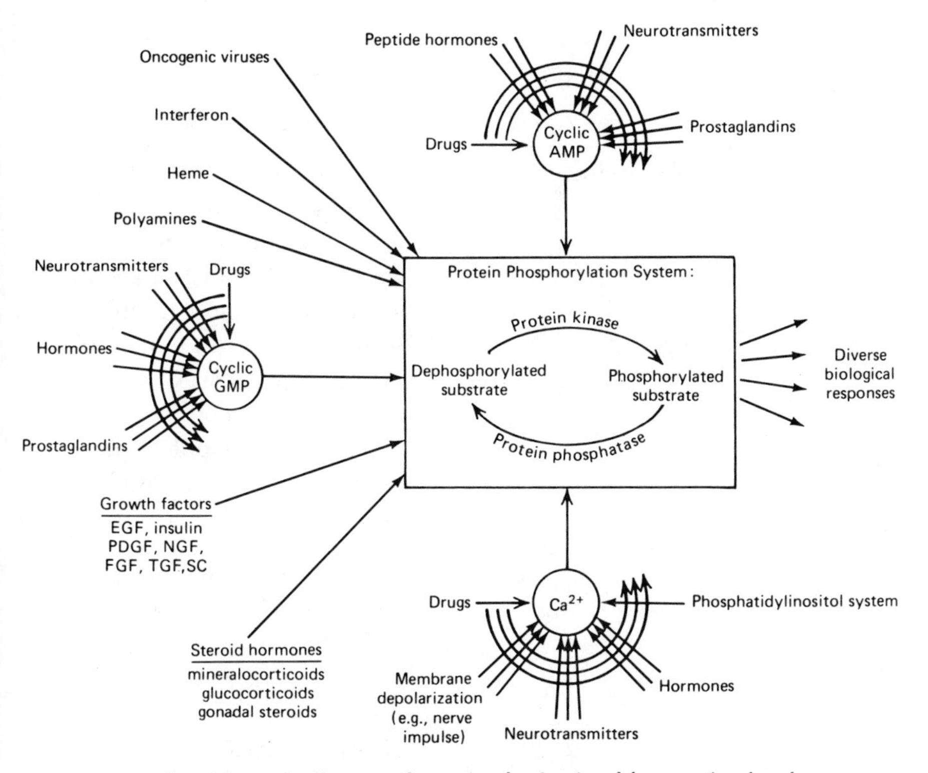

FIGURE 6-1. Schematic diagram of postulated role played by protein phosphorylation in mediating some of the biological effects of a variety of regulatory agents. Many of these agents regulate protein phosphorylation through altering interacellular levels of the second messengers cyclic AMP, cyclic GMP, or calcium. Other agents appear to regulate protein phosphorylation through mechanisms that do not involve these second messengers. Most drugs regulate protein phosphorylation by affecting the ability of other first messengers to alter second messenger levels (curved arms). A small number of drugs (e.g., phosphodiesterase inhibitors, calcium channel blockers) regulate protein phosphorylation by directly altering second messenger levels (straight arrow). Abbreviations: EGF, epidermal growth factors; PDGF, platelet-derived growth factor; NGF, nerve growth factor; FGF, fibroblast growth factor; TGF, transforming growth factor; SC, somatomedin C. (From Nestler and Greengard, 1984)

TABLE 6-1. Modulation of ionic conductances mediated by protein phosphorylation

Genus	*Cell*	*Experimental finding*
Aplysia	Bag cell neurons	Application of cAMP or injection of a cAMP-dependent protein kinase mediates synaptic activation that produces an afterdischarge. The effect is due to decreases in multiple voltage-dependent K^+ channels. Activation of the calcium/phospholipid-dependent protein kinase enhances Ca^{2+} current.
	Sensory neurons	Facilitation of neurotransmitter release mediated by a decreased conductance of a serotonin-dependent K^+ channel via cAMP-dependent protein kinase activity.
	R 15 neuron	Increase in conductance of a serotonin–regulated anomalously rectifying K^+ channel. Blocked by injection of a specific inhibitor of cAMP-dependent protein kinase.
Helix	Unidentified neurons	Injection of catalytic subunit of cAMP-dependent protein kinase increases conductance of calcium-dependent K^+ channels.
Hermissenda	Photoreceptor cells	Kinase injection decreases conductance of both early and

Genus	*Cell*	*Experimental finding*
		late voltage-dependent K^+ channels.
Guinea pig	Ventricular myocytes	Injection of catalytic subunit of cAMP-dependent protein kinase increases amplitude of slow inward calcium current. In contrast, injection of the regulatory subunit decreases the height and width of action potentials by decreasing calcium channels.
Rat	Cerebellar Purkinje cells	Injection of norepinephrine or cAMP produces hyperpolarization and decreased ionic conductance.
	Hippocampal pyamidal cells	As with Purkinje cells.

SOURCE: Modified from Nestler and Greengard (1984).

nase. Conceivably, the major function of this cyclic nucleotide is referable to the finding of a cGMP-dependent phosphodiesterase which regulates the level of cAMP.

2. *Phosphoinositide hydrolysis.* In 1953 Hokin and Hokin showed that the incorporation of inorganic phosphate (Pi) into phosphatidyl inositol (PI) and phosphatidic acid (PA) in pancreas slices was stimulated by ACh and ultimately resulted in the release of amylase. This receptor-activated hydrolysis of phosphoinositides is referred to as the phosphatidyl inositol effect. For nearly thirty years after the Hokin report, the literature on this effect was replete with the traditional scientific jargon "it is tempting to speculate that . . .", with no one having solid evidence as to whether the phosphoinosi-

tides or the inositol phosphates were the message and, if so, exactly what was the medium for the exchange. That this situation has now dramatically improved is shown in Figure 6-2.

The signals that initiate this transduction process in neuronal systems include ACh, norepinephrine, serotonin, histamine, bradykinins, Substance P, vasopressin, TRH, and angiotensin acting on brain, sympathetic ganglia, salivary glands, iris smooth muscle, adrenal cortex, and neuronal tumor cells. Specific receptors that have been implicated are muscarinic cholinergic receptors, α_1-adrenergic receptors, the H_1-histaminergic receptor, and the V_1-vasopressin receptor. In each case, Ca appears to be the intracellular second messenger that activates phosphoinositide hydrolysis. Like the specific GTP binding proteins that link receptors to the adenylate cyclase system discussed earlier, evidence is accumulating to suggest that a specific GTP binding protein is also coupled to the phospho-

FIGURE 6-2. Receptor-activated phosphoinositide metabolism. The binding of an agonist to a receptor on the plasma membrane stimulates the hydrolysis of phosphatidylinositol, 4,5-bisphosphate [PtdIns $(4,5)P_2$] by a phosphodiesterase (PDE), a specific phospholipase whose activity is controlled by a guanine nucleotide regulatory protein to form inositol 1,4,5-trisphosphate ($InsP_3$) and diacylglycerol (DG). The former binds to a receptor (R_2) on the endoplasmic reticulum to release calcium, which can directly produce a biological response or can activate calmodulin kinase to promote protein phosphorylation and a subsequent biological response. In some cells (e.g., mouse atria, neuroblastoma, and glioma hybrid NG108-15), receptor-activated production of $InsP_3$ requires extracellular Ca^{2+}.

The latter parallel arm of the cycle, diacylglycerol, can also promote a biological response via the production of prostaglandins, thromboxanes, and leukotrienes from released arachidonic acid (arachidonic acid has also been reported to stimulate guanylate cyclase to generate cGMP) or via a stimulation of protein kinase C (C-kinase) and subsequent protein phosphorylation. Diacylglycerol has also been reported to promote fusion of synaptic vesicles to terminal membrane.

The phosphoinositides are synthesized from inositol with cytidine diphosphate: diacyglycerol (CDP.DG) as intermediary and the stepwise phosphorylation by kinases (a and b). As shown in the figure, lithium blocks the cycle by inhibiting inositol-1-phosphatase. Although the antimanic activity of lithium has been ascribed to this inhibitory effect, the evidence is not compelling. (Modified from Berridge and Irvine, 1984)

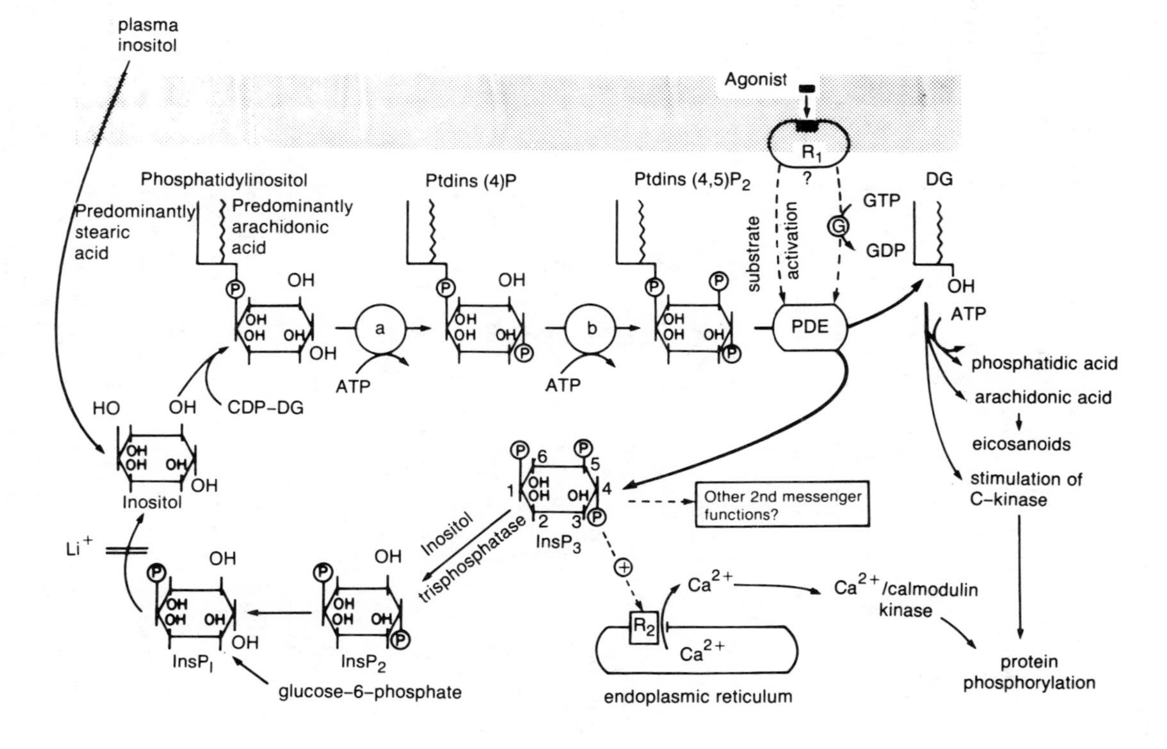

plasma inositol
Agonist
R1
?
Phosphatidylinositol
Predominantly stearic acid
Predominantly arachidonic acid
Ptdins (4)P
Ptdins (4,5)P2
substrate
activation
GTP
G
GDP
DG
OH
a
b
ATP
PDE
phosphatidic acid
arachidonic acid
eicosanoids
stimulation of C-kinase
HO
CDP-DG
Inositol
InsP3
Other 2nd messenger functions?
Inositol trisphosphatase
Li+
InsP2
InsP1
glucose-6-phosphate
R2
Ca2+
Ca2+/calmodulin kinase
endoplasmic reticulum
protein phosphorylation

diesterase that catalyzes the hydrolysis of phosphatidylinositol, 4,5'-bisphosphate.

As noted in Figure 6-2, the key event in the transduction process is the activity of this phosphodiesterase, a specific phospholipase C yielding two separate second messengers—the water-soluble $InsP_3$ and the lipid-soluble diacylglycerol. The former mobilizes calcium, which can act through calmodulin to phosphorylate specific proteins and the latter by activating C-kinase, a calcium–phosphatidylserine-dependent protein kinase, also phosphorylates specific proteins. In addition, since diacylglycerol can activate guanylate cyclase to produce cGMP, an inference of a cGMP-dependent activity (protein kinase or otherwise), must be considered. With an assumed ambidexterity of neuronal cells, these two arms could function singly, cooperatively, or antagonistically depending on the situation thus providing subtle variations on the modulatory mechanism. In addition, as noted in Figure 6-2, calcium may produce a physiological response directly without invoking an activation of calmodulin, so yet another control is indicated. On the subject of control, the activities of the various kinases, esterases, and phospholipases in the PI cycle would be expected to be vital control points. Free inositol in the brain must be derived from glycolysis since plasma inositol cannot pass the blood–brain barrier to any significant degree. Glycolysis, therefore, would be another regulatory factor in the response mechanism.

Finally, although the origin of the mobilized calcium is now clear (it is released from endoplasmic reticulum and not mitochondria), some controversy still exists as to whether phosphoinositide hydrolysis releases only internal calcium or whether external calcium transport is also invoked. Current evidence suggests that for neuronal modulation both or either may be involved, depending on the preparation. This same answer can also be given to answer the other controversy on whether the PI effect is presynaptic or postsynaptic. A complication in the PI cycle that has recently surfaced is the finding that the inositol trisphosphate that is produced is not always or only inositol 1,4,5-triphosphate but may be inositol 1,3,4-trisphosphate. The physiological role of this isomer is currently under investigation.

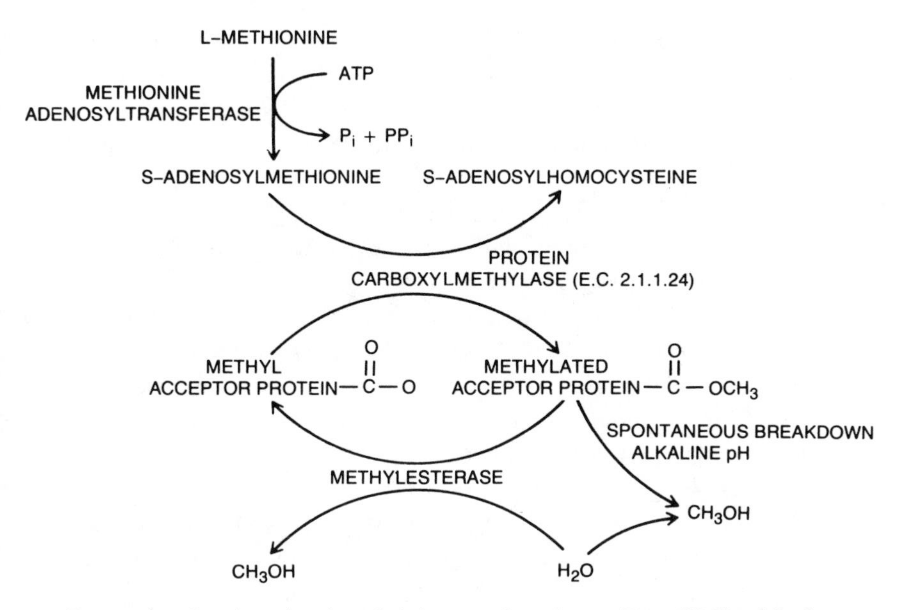

FIGURE 6-3. Protein carboxyl methylation: reaction scheme. (From Wolf and Roth, 1985)

3. *Protein carboxyl methylation.* As pioneered by Axelrod and colleagues and shown in Figure 6-3, protein carboxymethylase (now known as protein-*O*-carboxymethyltransferase) posttranslationally catalyzes the transfer of methyl groups from *S*-adenosylmethionine to protein carboxyl groups. The methyl esters are hydrolyzed by a methyl esterase; methanol can also be released by spontaneous hydrolysis in neutral or alkaline media. Protein carboxymethylase (PCM) is a soluble enzyme, found in highest amounts in brain and testes. Substrates for the enzyme may be either soluble or particulate and to date include calmodulin, phosphodiesterase, protein kinase, calcineurin, and ACTH. The methyltransferase has been postulated to have a regulatory function in leucocyte and bacterial chemotaxis, ion transport, stimulus secretion coupling in hormonal and neuronal systems, and modulation of the nicotinic ACh receptor and dopamine receptor. Recently, Clarke has suggested that carboxyl methylation of abnormal D-aspartyl and L-isoaspartyl res-

idues may mark proteins for repair or degradation. Currently, the most complete evidence linking protein carboxyl methylation to synaptic modulation involves the coupling of dopamine autoreceptor activation to inhibition of dopamine release from striatal nerve terminals.

Using both striatal synaptosomes and slices, the Roth laboratory has shown that dopamine and dopamine agonists stimulate methyl ester formation and this activity can be blocked by specific dopamine antagonists or by pretreatment of rats with 6-hydroxydopamine, the neurotoxin that selectively destroys monoaminergic neurons and their associated autoreceptors. Further, the methyltranserase inhibitor, *S*-adenosylhomocysteine mimics the enhancement of K^+-stimulated dopamine release induced by dopamine antagonists in striatal slices.

It has been proposed that the functional role of PCM in brain is to regulate the activity of calmodulin-stimulated enzymes. Studies using purified PCM have shown that the calmodulin-binding proteins calcineurin, phosphodiesterase, and protein kinase may represent important substrates for PCM. For all of these enzymes, carboxyl methylation results in a significant inhibition of Ca-calmodulin-stimulated activity. Calmodulin itself is also a substrate for PCM, although a greater effect on calmodulin-stimulated phosphodiesterase, calcineurin, or protein kinase activity is observed when the target enzymes themselves, rather than calmodulin, are carboxylmethylated. These findings suggest that protein carboxyl methylation may play multiple and complex roles in the regulation of Ca–calmodulin-dependent synaptic events such as neurotransmitter release and synthesis.

4. *Phospholipid methylation*. As with protein carboxyl methylation described above, *S*-adenosylmethionine methylates not only acidic amino-acid residues but also lipids. As first described by Axelrod and colleagues using adrenal medullary homogenates, the addition of *S*-adenosylmethionine resulted in the stepwise methylation of phosphatidyl ethanolamine to phosphatidyl choline. Two methyltransferase activities (but perhaps not separate enzymes) are involved in these three phospholipid methylations with the first

methyltransfer being rate-limiting and converting phosphatidyl ethanolamine to phosphatidyl monomethylethanolamine and the second activity catalyzing two successive methylations to form phosphatidyl dimethylethanolamine and then phosphatidyl choline. As the lipids are methylated they are translocated from the cytoplasmic side to the outside of the membrane and this translocation apparently decreases membrane viscosity. This can result in an increased evoked release of neuroactive agents, an increase in receptor density, and activation of phospholipase A_2 that would liberate arachidonic acid with a subsequent production of prostaglandins, thromboxanes, and leukotrienes. With respect to a role of lipid methylation in modulation of synaptic transmission however, a causal relationship between this process and any effects observed in decreasing membrane viscosity has not yet been demonstrated. Specific inhibitors are not yet available that would distinguish between protein carboxyl methylation and lipid methylation.

5. *Eicosanoids (arachidonic acid metabolites).* Arachidonic acid (5,8,11,14-eicosatetraenoic acid), derived on demand by either phospholipase A_2 or diglyceride lipase activation, yields a bewildering array of bioactive metabolites as shown in Figure 6-4. The three major groups are prostaglandins, thromboxanes, and leukotrienes.

Everybody knows that the eicosanoids, particularly the prostaglandin series, play an important modulatory role in nervous tissue but it is difficult to write a lucid account of specifically how and where they act. This situation is primarily due to the fact that they are not stored in tissue but are synthesized on demand, particularly in pathophysiological conditions, they act briefly and at extremely low concentrations (10^{-10} M), and although indomethacin is a good inhibitor of cyclooxygenase blocking the conversion of arachidonic acid to prostaglandins, there are no specific inhibitors available to block production of subsequent metabolites. Thus, although it has been postulated that the E series of prostaglandins modulates noradrenergic release, blocks the convulsant activity of pentylenetetrazol, strychnine, and picrotoxin possibly by increasing the level of GABA in the brain, and increases the level of cAMP in cortical

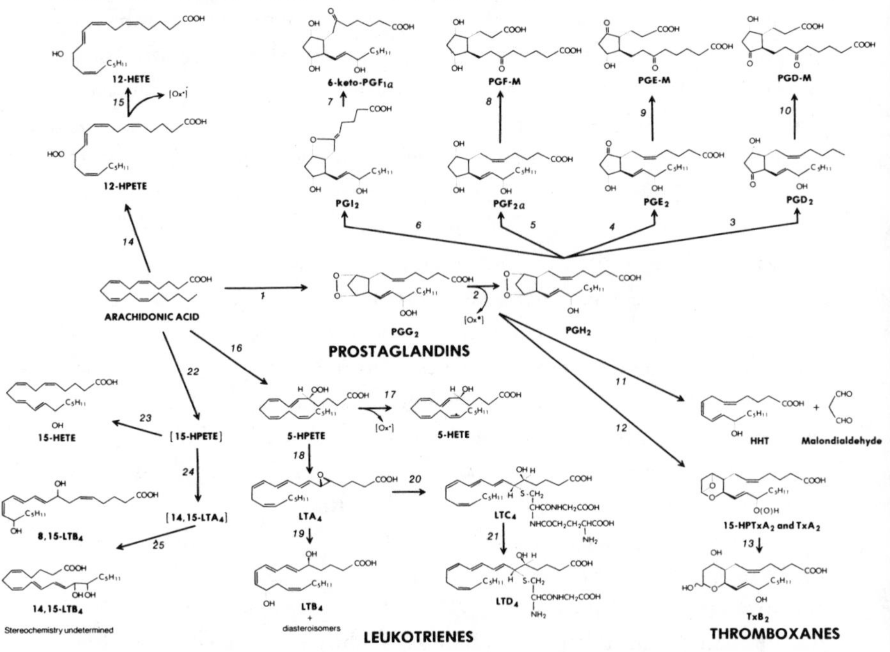

FIGURE 6-4. Pathways of arachidonic acid metabolism. Recent additions to be added to this flow are the identification of two further leukotriene metabolites LTE4 and LTF4, a new class of tetraenes (lipoxin A and lipoxin B), and ω-oxidase metabolites of LTB4. (1) Fatty acid cyclooxygenase, prostaglandin endoperoxide synthetase: nonenzymatic); (6) prostacyclin synthetase; (7) nonenzymatic hydrolysis; (8–10) prostagiandin 15-dehydrogenase, 13-14 reductase followed by β and ω oxidations; (11) nonenzymatic endoperoxide clevage; (12) thromboxane synthetase; (13) nonenzymatic hydrolysis; (14) 12-lipoxygenase; (15) hydroperoxidase; (16) 5-lipoxygenase; (17) hydroperoxidase; (18) leukotriene A synthetase; (19) enzymatic hydrolysis; (20) gluathione *S*-transferase; (21) γ-glutamyltranspeptidase; (22) 15-lipoxygenase; (23) hydroperoxidase; (24) 14, 15-leukotriene A_4 synthetase; (25) enzymatic hydrolysis. (from Wolfe, 1982)

and hypothalamic slices, these effects are noted *in vitro* with the addition of substantial amounts of the prostaglandins. There is very little evidence in intact animals to support these neuronal findings and since we all believe in *In Vivo Veritas*, the physiological relevance of the effect is still in doubt.

Reflecting on the mechanisms available to nervous tissue to modulate synaptic transmission, one cannot fail to be overwhelmed by the almost infinite possibilities that provide the fine tuning for behavioral changes. The key question of course is this: What predicates the cell to decide which process is appropriate? To this mystery, one can add the problem of the trigger and the site of release of endogenous modulators such as the neuroactive peptides. These agents can be released presynaptically, postsynaptically in response to presynaptic receptor activation, from glial cells, or from blood vessels. Currently, the most mysterious modulator is adenosine, which is found at virtually every synapse that has been examined, electrophysiologically tends to inhibit the evoked release of transmitters but also acts postsynaptically, exhibits a variety of behavioral effects ranging from evoking premature arousal in hibernating ground squirrels to anticonvulsant activity, increases cerebral blood flow, and interacts with the benzodiazepine receptor. At least two membrane-localized adenosine receptors have been identified with one stimulating and one inhibiting adenylate cyclase but it is questionable if this is a legitimate classification that explains all the activity of this purine.

It is because of the myriad possibilities for regulation outlined in this chapter that the note of despair was sounded in the introduction. The happy side of this state of affairs, however, is that neuropharmacologists have the luxury of choosing from an almost infinite number of channels, receptors, or enzymes as the site to develop new drugs that selectively interfere with one of the modulatory processes.

Selected References

Berridge M. J., and R. F. Irvine (1984). Inositol trisphosphate, a novel second messenger in cellular signal transduction. *Nature 312*, 315.

Berridge M. J. (1984). Inositol trisphosphate and diacylglycerol as second messengers. *Biochem J. 220*, 345.

Billingsley, M. L. and W. Lovenberg (1985). Protein carboxylmethylation and nervous system function. *Neurochem. Int. 7*, 575.

Clarke S. (1985). Protein carboxyl methyltransferases: Two distinct classes of enzymes. *Ann. Rev. Biochem. 54*, 479.

Chesselet M.-F. (1984). Presynaptic regulation of neurotransmitter release in the brain. *Neurosci. 12*, 347.

Cooper J. R., and E. M. Meyer (1984). Possible mechanisms involved in the release and modulation of release of neuroactive agents. *Neurochem. Int. 6*, 419.

Crews F. T. (1984). Phospholipid methylation in brain and other tissues. In *Brain Receptor Methodologies Part A, (P. J. Marangos, I. C. Campbell, and R. N. Cohen, eds.)* p. 217. Academic Press, New York.

Hirasawa K., and Y. Nishizuka (1985). Phosphatidylinositol turnover in receptor mechanism and signal transduction. *Ann. Rev. Pharmacol. Toxicol. 25*, 147.

Hokin, L. E. (1985). Receptors and phosphoinositide-generated second messengers. *Ann. Rev. Biochem. 54*, 205.

Laduron P. M. (1985). Presynaptic heteroreceptors in regulation of neuronal transmission. *Biochem. Pharmacol. 34*, 467.

Lundberg, J. M., B. Hedlund, and T. Bartfai (1982). Vasoactive intestinal polypeptide enhances muscarinic ligand binding in cat submandibular salivary gland. *Nature 295*, 147.

McGiff, J. C. (1981). Prostaglandins, prostacyclin, and thromboxanes. *Ann Rev. Pharmacol. Toxicol. 21*, 478.

Nairn A. C., H. C. Hemmings, Jr., and P. Greengard (1985). Protein kinases in the brain. *Ann. Rev. Biochem. 54*, 931.

Nestler E. J., and P. Greengard (1984). *Protein Phosphorylation in the Nervous System.* Wiley, New York.

Ogorochi T., S. Narumiya, N. Mizuno, K. Yamashita, H. Miyazaki, and O. Hayaishi (1984). Regional distribution of prostaglandins D_2, E_2, and F_2 and related enzymes in postmortem human brain. *J. Neurochem. 43*, 71.

Reuter H. (1983). Calcium channel modulation by neurotransmitters, enzymes, and drugs. *Nature 301*, 569.

Starke K. (1981). Presynaptic receptors. *Ann. Rev. Pharmacol. Toxicol. 21*, 7.

Vizi, E. S. (1984). *Non-synaptic Interactions Between Neurons: Modulation of Neurochemical Transmission*. Wiley, New York.

William M. (1984). Adenosine—A selective neurotransmitter in the mammalian CNS. *Trends in Neurosci. 7*, 164.

Wolf M. E., and R. H. Roth (1985). Dopamine autoreceptor stimulation increases protein carboxyl methylation in striatal slices. *J. Neurochem. 44*, 291.

Wolfe L. S. (1982). Eicosanoids: Prostaglandins, thromboxanes, leukotoienes, and other derivatives of carbon-20 unsaturated fatty acids. *J. Neurochem. 38*, 1.

Yamada S., M. Isogai, H. Okudaira, and E. Hayashi (1983). Regional adaptation of muscarinic receptors and choline uptake in brain following repeated administration of diisopropylfluorophosphate and atropine. *Brain Res. 268*, 315.

7 | Amino-Acid Transmitters

Over the years several amino acids have gained recognition as major neurotransmitter candidates in the mammalian CNS. Since these substances are also involved in intermediary metabolism, it has been difficult to fulfill all the required criteria that would give these substances legitimate status as neurotransmitters in the mammalian CNS. On the basis of neurophysiological studies, amino acids have been separated into two general classes: excitatory amino acids (glutamic acid, aspartic acid, cysteic acid, and homocysteic acid), which depolarize neurons in the mammalian CNS; and inhibitory amino acids (GABA, glycine, taurine, and β-alanine), which hyperpolarize mammalian neurons. Strictly from a quantitative standpoint, the amino acids are probably the major transmitters in the mammalian CNS, while the better-known transmitters discussed in other chapters (acetylcholine, norepinephrine, dopamine, histamine and 5-hydroxytryptamine) probably account for transmission at only a small percentage of central synaptic sites.

GABA

Synthesized in 1883, γ-aminobutyric acid (GABA) was known for many years as a product of microbial and plant metabolism. Not until 1950, however, did investigators identify GABA as a normal constituent of the mammalian central nervous system and find that no other mammalian tissue, with the exception of the retina, contains more than a mere trace of this material. Obviously, it was thought, a substance with such an unusual distribution must have some characteristic and specific physiological effects that would make it important for the function of the central nervous system.

More than thirty years later we still have no conclusive proof as to the precise role this compound plays in the mammalian central nervous system. However, much evidence has accumulated supporting the hypothesis that the major share of GABA found in brain functions as an inhibitory transmitter. The knowledge that GABA probably functions as an inhibitory transmitter in brain has spurred a prodigious research effort to implicate GABA in the etiology of a host of neurological and psychiatric disorders. Although the present evidence is not overwhelming, GABA has been implicated, both directly and indirectly, in the pathogenesis of Huntington's disease, Parkinsonism, epilepsy, schizophrenia, tardive dyskinesias and senile dementia as well as several other behavioral disorders.

Distribution

In mammals, GABA is found in high concentrations in brain and spinal cord, but it is absent or present only in trace amounts in peripheral nerve tissue such as sciatic nerve, splenic nerve, sympathetic ganglia, or in any other peripheral tissue such as liver, spleen, or heart. These findings give some idea of the uniqueness of the occurrence of GABA in the mammalian central nervous system. Like the monoamines, GABA also appears to have a discrete distribution within the central nervous system. However, unlike the monoamines, the concentration of GABA found in the central nervous system is in the order of μmoles/gm rather than nmoles/gm. It is interesting that brain also contains large amounts of glutamic acid (8–13 μmoles/gm), which is the main precursor of GABA and is itself a neurotransmitter candidate. In the rat, the corpora quadrigemina and the diencephalic regions contain the highest levels of GABA, while much lower concentrations are found in whole cerebral hemispheres, the pons, and medulla; white matter contains relatively low concentrations of GABA. The discrete localization of GABA in the brain of anesthetized monkeys and postmortem samples of human brain is summarized in Table 7-1. It should be noted that endogenous levels of GABA increase rapidly postmortem: a 30 to 45 percent increase in GABA occurs within two minutes after

TABLE 7-1. Regional distribution of GABA in monkey and human brain

	Region of brain	μmoles/g frozen tissue ± S.E.M. Rhesus monkey	Human
Highest	Substantia nigra	9.70 ± 0.63	5.31
	Globus pallidus	9.54 ± 0.91	5.69
	Hypothalamus	6.19 ± 0.13	3.72
High	Inferior colliculus	4.70 ± 0.29	
	Dentate nucleus	4.30 ± 0.47	
	Superior colliculus	4.19 ± 1.09	
	Periaqueductal gray	4.02 ± 0.56	
Medium	Putamen	3.62 ± 0.21	
	Pontine tegmentum	3.34 ± 0.43	
	Caudate nucleus	3.20 ± 0.18	3.03
	Medial thalamus	3.00 ± 0.14	2.52
Low	Lateral thalamus	2.68 ± 0.09	
	Occipital cortex	2.68 ± 0.23	2.32
	Anterior thalamus	2.50 ± 0.23	
	Medullary tegmentum	2.27 ± 0.22	
	Inferior olivary nucleus	2.25 ± 0.15	
	Frontal cortex	2.10 ± 0.01	2.09
	Motor cortex	2.09 ± 0.10	
	Cerebellar cortex (not pure gray)	2.03 ± 0.23	2.33
Lowest	Centrum semiovale (pure white)	0.31 ± 1.0	

SOURCES: S. Fahn and L. J. Côté (1968); and Perry *et al.* (1971)

death in the rat if the tissue is not instantly frozen *in situ*. The origin of this sudden increase is uncertain although it is believed to result in part from a transient activation of glutamic acid decarboxylase (GAD).

Progressive increases in GABA levels and in glutamic acid decarboxylase activity appear to occur in various regions of the brain during development. The high levels of GABA found in the various regions of the brain of the Rhesus monkey appear to correlate well with the activity of glutamate decarboxylase, the enzyme responsible for the conversion of L-glutamate to GABA. This is not the case for the degradative enzyme GABA-transaminase, since the globus pallidus and the substantia nigra, which have the highest concentration of GABA, have a relatively low transaminase activity. Yet there does not appear to be a consistent inverse relationship between GABA concentration and transaminase activity. Thus, some areas of brain, such as the dentate nucleus and the inferior colliculus, which have relatively high concentrations of GABA also have large amounts of transaminase activity.

Since GABA does not easily penetrate the blood—brain barrier, it is difficult if not impossible to increase the brain concentrations of GABA by peripheral administration, unless one alters the blood–brain barrier. Some investigators have tried to circumvent this problem by the administration of GABA-lactam (2-pyrrolidinone) to animals in the hope that this less polar and more lipid soluble compound would penetrate more easily into the brain and be hydrolyzed to yield GABA. Although the idea seems plausible, it does not succeed: the GABA-lactam that reaches the central nervous system is not hydrolyzed to any extent. A more recent successful attempt has been the development of progabide (Figure 7-5). This agent does penetrate the brain and is subsequently metabolized into GABA.

Metabolism

There are three primary enzymes involved in the metabolism of GABA prior to its entry into the Krebs cycle. The relative activity of enzymes involved in the degradation of GABA suggests that, similar to monoamines, they play only a minor role in the termination of the action of any neurally released GABA.

Figure 7-1 outlines the metabolism of GABA and its relationship to the Krebs cycle and carbohydrate metabolism. As mentioned

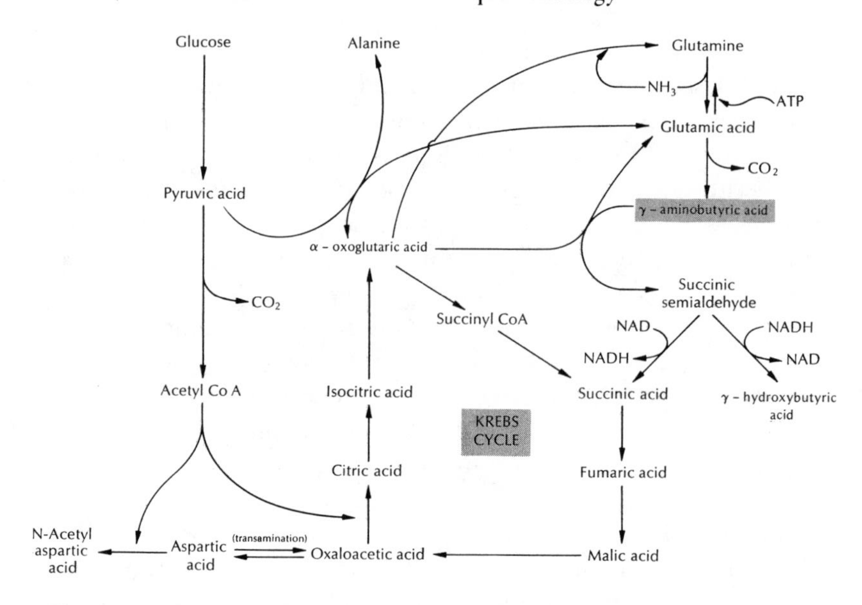

FIGURE 7-1. Interrelationship between γ-aminobutyric acid and carbohydrate metabolism.

previously, GABA is formed by the α-decarboxylation of L-glutamic acid, a reaction catalyzed by glutamic acid decarboxylase, an enzyme that occurs uniquely in the mammalian central nervous system and retinal tissue. The precursor of GABA, L-glutamic acid can be formed from α-oxoglutarate by transamination or reaction with ammonia. GABA is intimately related to the oxidative metabolism of carbohydrates in the central nervous system by means of a "shunt" involving its production from glutamate, its transamination with α-oxoglutarate by GABA-α-oxoglutarate transaminase (GABA-T) yielding succinic semialdehyde and regenerating glutamate, and its entry into the Krebs cycle as succinic acid via the oxidation of succinic semialdehyde by succinic semialdehyde dehydrogenase. In essence, then, the "shunt" bypasses the normal oxidative metabolism involving the enzymes α-oxoglutarate dehydrogenase and succinyl thiokinase.

From a metabolic standpoint, the significance of the "shunt" is unknown; energetically at least it is less efficient than direct oxidation through the Krebs cycle (3 ATP equivalents versus 3 ATP+1 GTP for the Krebs cycle). Experimentally, it has been quite difficult to make an adequate assessment of the quantitative significance of the "shunt" in the oxidation of α-oxoglutarate *in vivo*. Studies with [^{14}C]-labeled glucose both *in vivo* and *in vitro* have indicated that the carbon chain of GABA can be derived from glucose. Therefore, in some cases the incorporation of radioactivity into the "shunt" metabolites following the injection of uniformly labeled [^{14}C]-glucose has been used to assess the functional aspects of this "shunt." However, these experiments are somewhat inconclusive since they do not necessarily indicate the rate of flux through this pathway. Some relatively indirect experiments have given results suggesting that about 10 to 40 percent of the total brain metabolism may funnel through this "shunt."

Glutamic Acid Decarboxylase (GAD)

GAD is the enzyme responsible for the conversion of L-glutamic acid to GABA. At the present time, no convincing evidence has been gathered to indicate that significant amounts of GABA are formed by reactions other than the decarboxylation of L-glutamic acid. No reversal of this reaction has been demonstrable either *in vivo* or *in vitro*. In mammalian organisms this relatively specific decarboxylase is found primarily in the central nervous system, where it occurs in higher concentrations in the gray matter. In general, the localization of this enzyme in mammalian brain correlates quite well with the GABA content. The brain enzyme has been purified to homogeneity and its properties studied in detail. It has a pH optimum of about 6.5 and requires pyridoxal phosphate as a cofactor. The purified enzyme is inhibited by structural analogs of glutamate, carbonyl (pyridoxal phosphate) trapping agents, sulfhydryl reagents, thiol compounds, and anions such as chloride. This inhibition by anions may be interesting from a physiological standpoint. The possibility has been raised that variations in the chloride concentration at a given nerve ending could control GABA formation. Although there is no evidence of wide variations in chloride content under normal physiological conditions, it is of interest that in snail brain the D cells (depolarized by acetylcholine) and the H cells (hyperpolarized by acetylcholine) have different chloride contents.

Although the properties of GAD have been actively studied for twenty-five years, little was known until recently about the molecular mechanisms responsible for regulation of its *in vivo* activity in the CNS. Studies directed at the mechanism involved in the postmortem increase in GABA levels have been in part responsible for disclosing some interesting properties of mammalian GAD, which may provide some clues to its *in vivo* regulation. Since GAD appears to be only partially saturated by its cofactor, pyridoxal-5′-phosphate in intact brain, it is apparent that conditions that alter the degree of saturation of this enzyme *in vivo* will influence its activity. Although most recent data suggest that pyridoxal phosphate is tightly bound to GAD, glutamate and adenine nucleotides can promote dissociation and block association of this important cofactor, respectively. Thus, it is conceivable that alterations in the availability of one or both of these endogenous substances could influence GAD activity and GABA formation *in vivo*. The possibility that the association and dissociation of pyridoxal phosphate from GAD may be important for the *in vivo* regulation of GABA synthesis is supported by the observation that GAD is no more than about 35 percent saturated with pyridoxal phosphate *in vivo* despite the fact that the levels of this cofactor in brain appear to be sufficiently high to fully activate this enzyme. The degree of saturation of GAD with pyridoxal phosphate *in vivo* may be determined in part by a balance between the rate of dissociation of pyridoxal phosphate from the enzyme brought about by glutamate and the rate of association of the cofactor with the apoenzyme which is inhibited by nucleotides. This possibility is consistent with the observation that a postmortem activation of GAD occurs in brain only when postmortem changes in ATP are allowed to take place.

Recently another mammalian GAD (Type 2) has been described.

This enzyme differs from the brain GAD (Type 1) in that it is insensitive to inhibition by chloride ions and is stimulated rather than inhibited by certain carbonyl trapping agents. In contrast to the brain GAD, this enzyme in addition to being found in CNS white matter is also found in nonneuronal tissue such as kidney and human glial cells grown in cluture. The existence of GAD Type 2 has been questioned by some investigators since the product of the reaction was not convincingly demonstrated in many experiments. However, a nonneuronal GAD, present in glial cells as well as in blood vessels and several peripheral tissues, has now been highly purified from beef heart muscle and found to be immunologically distinct from the brain enzyme, thus documenting the existence of two forms of this enzyme.

GABA-Transaminase (GABA-T)

GABA-T, unlike the decarboxylase, has a wide tissue distribution. Therefore, although GABA cannot be formed to any extent outside the central nervous system, exogenous GABA can be rapidly metabolized by both central and peripheral tissue. However, since endogenous GABA is present only in nanomolar amounts in cerebrospinal fluid, it is unlikely that a significant amount of endogenous GABA leaves the brain intact. The brain transaminase has a pH optimum of 8.2 and also requires pyridoxal phosphate. It appears that the coenzyme is more tightly bound, to this enzyme than it is to GAD. The brain ratio of GABA-T/GAD activity is almost always greater than one. Sulfhydryl reagents tend to decrease GABA-T activity, suggesting that this enzyme requires the integrity of one or more sulfhydryl groups for optimal activity. The transamination of GABA catalyzed by GABA-T is a reversible reaction, so if a metabolic source of succinic semialdehyde were made available it would be theoretically possible to form GABA by the reversal of this reaction. However, as indicated below, this does not appear to be the case *in vivo* under normal or experimental conditions so far investigated.

Recent studies with more sophisticated cell fractionation techniques and electron microscopic monitoring of the fractions obtained have borne out the original claims that both GAD and GABA-T are particulate to some extent. GAD was found associated with the synaptosome fraction, whereas the GABA-T was largely associated with mitochondria. Further studies on the mitochondrial distribution of GABA-T have suggested that the mitochondria released from synaptosomes have less activity than the crude unpurified mitochondrial fraction, and it has been inferred that the mitochondria within nerve endings have little GABA-T activity. This finding has led to the speculation that GABA is metabolized largely at extraneuronal intercellular sites or in the postsynaptic neurons. Gabaculine is the most potent GABA-T inhibitor currently available. Similar to γ-acetylenic and γ-vinyl GABA, this agent is a catalytic inhibitor of GABA-T and, unfortunately, will also inhibit GAD. However, gabaculine has a fair degree of specificity since it is about 1000-fold less effective as a GAD inhibitor than as a GABA-T inhibitor.

Succinic Semialdehyde Dehydrogenase (SSADH)

Brain succinic semialdehyde dehydrogenase (SSADH) has a high substrate specificity and can be distinguished from the nonspecific aldehyde dehydrogenase found in brain. The enzyme purified from human brain has a pH optimum of about 9.2 and quite a low Michaelis constant (K_m for succinic semialdehyde of 5.3×10^{-6}). SSADH from a rat brain has a similarly low K_m for succinic semiehyde of 7.8×10^{-5} and for NAD of 5×10^{-5}. The high activity of this enzyme and the low Michaelis constant, which allow the enzyme to function effectively at low substrate concentrations, probably account for the fact that succinic semialdehyde (SSA) has not even been detected as an endogenous metabolite in neural tissue despite the active metabolism of GABA *in vivo*. In contrast to SSADH isolated from bacterial sources where NADP is several times more active as a cofactor than NAD, the enzyme from monkey and human brain demonstrates a specificity for NAD as a cofactor. The regional distribution of this enzyme has been studied in human brain and found to parallel the distribution of GABA-T activity, although the dehydrogenase is about 1.5 times as active. The greatest activity was found in the hypothalamus, basal ganglia, cortical gray matter, and mesencephalic tegmentum. SSADH also has a marked heat activation at 38°C, and Pitts suggested that its activity *in vivo* might be regulated by temperature such that fever might result in an increased flux through the GABA shunt. This appears unlikely, however, since this step is usually not considered to be rate limiting in the conversion of GABA to succinic acid, and thus small changes in its activity would not be reflected in an overall change in GABA metabolism. A sensitive and specific assay for SSADH is based on the fluorescence of NADH formed in the conversion of succinic semialdehyde to succinic acid. This method is sensitive enough to assay samples as small as 0.05 μg of freeze-dried brain tissue.

Since GABA's rise to popularity, the literature has been inundated with reports purporting to demonstrate that many pharmacological and physiological effects can be ascribed to and correlated well with changes in the brain levels of this substance. Since both GAD and GABA-T are dependent on the coenzyme pyridoxal phosphate, it is not surprising that pharmacological agents or pathological conditions affecting this coenzyme can cause alterations in the GABA content of the brain. Epileptiform seizure can be produced by a lack of this coenzyme or by its inactivation. Conditions of this sort also lead to a reduction in GABA levels, since GAD appears to be preferentially inhibited over the transaminase, presumably due to the fact that GAD has a lower affinity for the coenzyme than does GABA-T. A diet deficient in vitamin B_6 in infants can lead to seizures that respond successfully to treatment consisting of addition of pyridoxine to the diet. However, it must be remembered that many other enzymes, including some of those involved in the biosynthesis of other bioactive substances, are also pyridoxal-dependent enzymes. A number of observations, in fact, indicate that there is no simple correlation between GABA content and convulsive activity. Administration of a variety of hydrazides to animals has uniformly resulted in the production of repetitive seizures following a rather prolonged latent period. The finding that the hydrazide-induced seizures could be prevented by parenteral administration of various forms of vitamin B_6 led to the suggestion that some enzyme system requiring pyridoxal phosphate as a coenzyme was being inhibited and that the decrease in the activity of this enzyme was somehow related to the production of the seizures observed. At this time attention focused on GABA and GAD because of their unique occurrence in the central nervous system and because GAD had been shown to be inhibited by carbonyl trapping agents *in vitro*. The hydrazide-induced seizures were accompanied by substantial decreases in the levels of GABA and reductions in GAD activity in various areas of the brain studied. (This decrease in GABA produced by thiosemicarbazide is now believed to be due at least in part to a decrease in the rapid postmortem increase in GABA mentioned previously.) The direct demonstration of the reversal of the action of the convulsant hydrazides by GABA itself proved to be difficult, because of the lack of ability of GABA to pass the blood–brain barrier in adult mammalian organisms. However, a preferential inhibition of GABA-T could be achieved *in vivo* with carbonyl reagents such as hydroxylamine (NH_2OH) or amino-oxyacetic acid which resulted in an increase in GABA levels (up to 500 percent of control) in the central nervous system. Although these agents caused a decrease in the susceptibility to the seizures induced by some agents, such as metrazole, they did not exert any protective effect against the hydrazide-induced convulsions, even though they prevented the depletion of GABA and in some cases even increased the GABA levels above the controls. In fact, administration of very high doses of only amino-oxyacetic acid instead of producing the normally observed sedation caused some seizure activity in spite of the extremely high brain levels of GABA.

An interesting finding with amino-oxyacetic acid is that administration of this compound to a strain of genetically spastic mice in a single dose of 5–15 mg/kg results in a marked improvement of their symptomatology for 12 to 24 hr. This improvement is associated with an increase in GABA levels, but the GABA level increases with a similar time course and to the same extent as in normal control mice. All studies to date indicate that the principal genetic defect in these mice is not in the operation of the GABA system. However, the drug-induced increase in GABA may serve to quell an excess of or imbalance in excitatory input in some unknown area of the central nervous system.

Hydrazinopropionic acid has been described as a potent inhibitor of GABA-T. It has been suggested that this compound inhibits GABA-T because of its close similarity to GABA with respect to structural configuration, molecular size, and molecular charge distribution. Its inhibitory action cannot be reversed by the addition of pyridoxal phosphate. Hydrazinopropionic acid is about 1000 times more potent than aminooxyacetic acid in inhibiting mouse brain transaminase.

Alternate Metabolic Pathways

In addition to undergoing transamination and subsequently entering the Krebs cycle, GABA can apparently undergo various other transformations in the central nervous system, forming a number of compounds whose importance, and in some cases natural occurrence, has not been conclusively established. Figure 7-2 depicts a variety of derivatives for which GABA may serve as a precursor. Perhaps the simplest of these metabolic conversions is the reduction of succinic semialdehyde (a product of GABA transamination) to γ-hydroxybutyrate (GHB). The transformation of GABA to GHB has been demonstrated in rat brain both *in vivo* and *in vitro*. Recent studies have demonstrated that GHB administered in physiologically relevant concentrations can also be converted to GABA by transamination.

GABA Receptors

The terminology "GABA receptor" usually refers to a GABA recognition site on postsynaptic membranes which, when coupled with GABA or an appropriate agonist, causes a shift in membrane permeability to inorganic ions, primarily chloride. This change in

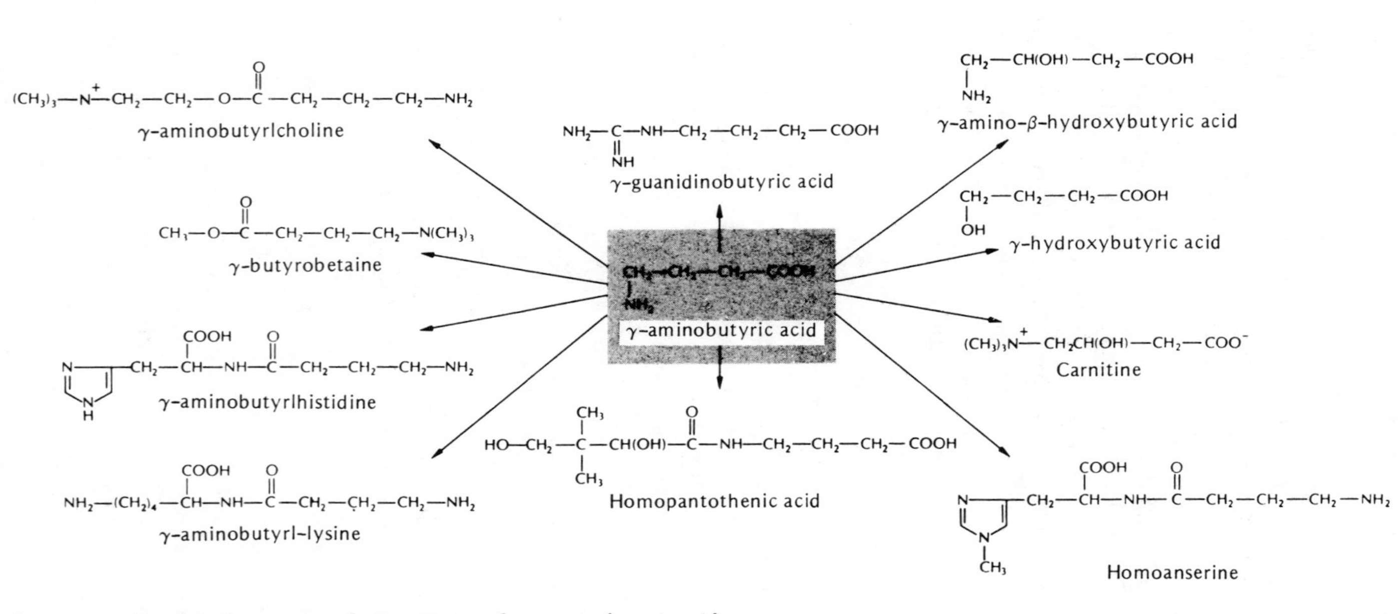

FIGURE 7-2. Possible alternate metabolic pathways for γ-aminobutyric acid.

chloride permeability results in hyperpolarization of the receptive neuron in the case of postsynaptic inhibition or depolarization in the case of presynaptic inhibition. However, it is clear that GABA can attach to a vast number of sites in central nervous system tissue, some of which may be physiologically relevant but not appropriately referred to as GABA receptors. These sites include the high-affinity GABA transport site and enzymes involved in GABA metabolism such as GAD and GABA-T. It has been long suspected that pharmacologically and functionally distinct types of GABA receptors exist in the CNS and although electrophysiological studies have demonstrated the existence of bicuculline insensitive GABA receptors, these receptors have been difficult to characterize further because of the lack of appropriate pharmacological agents. For this reason many investigators have defined GABA receptors strictly on the basis of their sensitivity to the GABA antagonists, bicuculline and picrotoxin. Even though GABA receptors have been extensively investigated in recent years, little precise information is available about the functional consequences of GABA receptor activation. However, in view of the large numbers of GABA receptors in the central nervous system, the indications are that these receptors regulate a variety of physiological, behavioral, and biochemical effects.

Recent studies have suggested that antianxiety drugs can alter GABA receptor function and the theory has been proposed that the anxiolytic action of the benzodiazepines is mediated by the facilitation of GABAergic transmission. Since 1,4-benzodiazepines do not mimic the action of GABA when applied directly to receptors, the facilitation of GABAergic transmission is believed to occur as a result of the ability of benzodiazepines to modify the affinity of GABA for its own receptor or the coupling of the GABA receptor to the chloride ion channel thus enhancing GABAergic transmission. Several laboratories are actively involved in working out the molecular mechanism by which 1,4-benzodiazepines facilitate GABAergic transmission via an alteration of GABA receptor function.

There is strong evidence for a multiplicity of GABA receptors. Receptors for GABA are found both presynaptically (axoaxonic) and postsynaptically (axodendritic and axosomatic). Activation of these receptors can be studied *in vivo* using *in vivo* electrophysiological or behavioral methods or *in vitro* using isolated tissue preparations. These studies of receptor interactions are complimented by neurochemical investigations of GABA binding to a variety of CNS membrane preparations. Although the latter studies provide direct information on the kinetics of binding interactions, only relatively indirect information can be gleaned concerning the relevance of these interactions to physiological processes. Binding to the GABA recognition sites is merely the initial step in a chain of events that alters the permeability of synaptic membranes to ions and ultimately influences membrane potential and neuronal excitability.

Several classes of centrally acting drugs appear to interact with GABA receptors. Benzodiazepines and some of the barbiturates appear to modulate the activation of some GABA receptors by potentiating the coupling between these receptors and ionophores. Many convulsants are GABA antagonists acting on GABA recognition sites (bicuculline) and/or on associated ionophores (picrotoxin). Other convulsants act presynaptically to alter the availability of synaptic GABA.

These studies suggest that it is convenient to consider GABA receptors as part of complex domains consisting of receptors, modulator sites, and ionophores, with all three domains influencing the final outcome of the initial activation of the GABA receptors. In general terms, activation of GABA receptors results in an increased membrane permeability to chloride ions, which is associated with a hyperpolarizing response on nerve cell bodies and a depolarizing response on presynaptic terminals and nerve fibers.

Binding studies have demonstrated two distinct subtypes of GABA receptors that vary in their affinity for the agonist ligand but that possess a similar pharmacology. These GABA sites are referred to as high- and low-affinity binding sites based on their different dissociation constants. A subgroup of GABA receptors, probably the low-affinity receptors, appear to be the receptors intimately associated with the benzodiazepine binding site. Based on their properties, two major subpopulations of GABAergic receptors have been defined and designated GABA-A and GABA-B receptors. The GABA-A receptor has been shown to be bicuculline sensitive, linked to the benzodiazepine receptor and coupled to chloride channels. The GABA-B receptor is bicuculline insensitive, influenced by guanyl nucleotides and linked to calcium channels.

Pharmacology of GABAergic Neurons

Drugs can influence GABAergic function by interaction at many different sites both pre- and postsynaptic (Fig. 7-3). Drugs can influence presynaptic events and modify the amount of GABA that ultimately reaches and interacts with postsynaptic GABA receptors. In most cases, presynaptic drug effects do not involve an interaction with "GABA receptors." The most extensively studied presynaptic drug actions involved inhibitory effects exerted on enzymes involved in GABA synthesis (GAD) and degradation (GABA-T) and the neuronal re-uptake of GABA. The major exception is the interaction of drugs with GABAergic autoreceptors to modulate both the physiological activity of GABA neurons and the release and synthesis of GABA in a manner analagous to the role played by dopamine autoreceptors in the regulation of dopaminergic function.

A great deal of emphasis has been directed recently to the study of drug interactions with GABA receptors. Drugs interacting at the level of GABA receptors can be assigned into two general categories, GABA antagonists and GABA agonists.

Figure 7–3 depicts possible sites of drug interaction in a hypothetical GABAergic synapse. The structures of compounds that act at GABAergic synapses are depicted in Figures 7–4 and 7–5. Picrotoxinin is the active component of picrotoxin.

GABA Antagonists

The action of GABA at the receptor-ionophore complex may be antagonized by GABA antagonists either directly by competition with GABA for its receptor or indirectly by modification of the re-

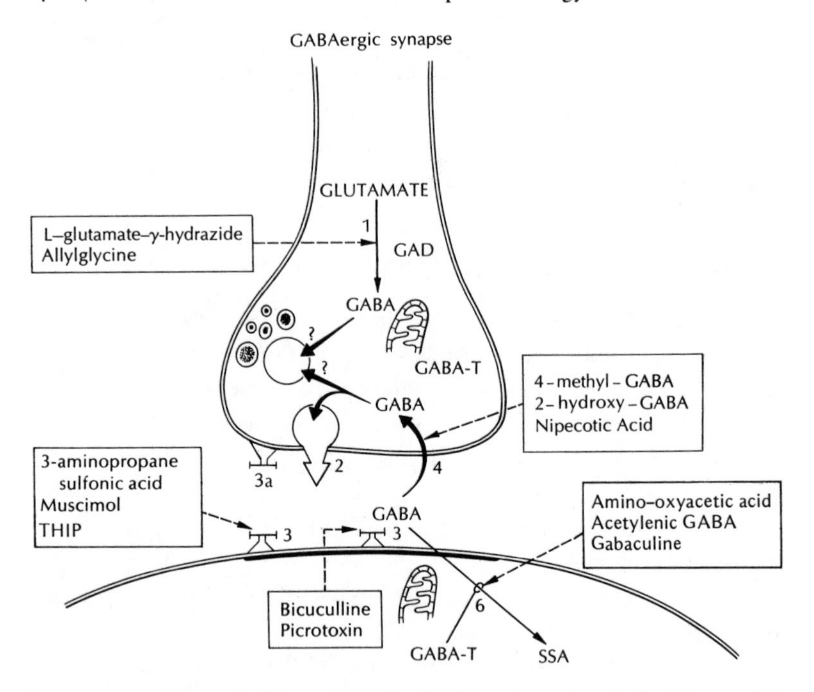

FIGURE 7-3. Schematic illustration of a GABAergic neuron indicating possible sites of drug action.

Site 1: *Enzymatic synthesis:* Glutamic acid decarboxylase (GAD-I) is inhibited by a number of various hydrazines. These agents appear to act primarily as pyridoxal antagonists and are therefore very nonspecific inhibitors. L-glutamate-γ-hydrazide and allylglycine are more selective inhibitors of GAD-I, but these agents are also not entirely specific in their effects.

Side 2: *Release:* GABA release appears to be calcium dependent. At present no selective inhibitors of GABA release have been found.

Site 3: *Intereaction with postsynaptic receptor:* Biculline and picrotoxin block the action of GABA at postsynaptic receptors. 3-Aminopropane sulfonic acid and the hallucinogenic isoxazole derivative, muscimol, appear to be effective GABA agonists at postsynaptic receptors and autoreceptors.

Site 3a. *Presynaptic autoreceptors:* Possible involvement in the control of GABA release.

Site 4: *Re-uptake:* In brain GABA appears to be actively taken up into presynaptic endings by a sodium-dependent mechanism. A number of compounds will inhibit this uptake mechanism such as 4-methyl-GABA and 2-hydroxy-GABA, but these agents are not completely specific in their inhibitory effects.

ceptor or by inhibition of the GABA activated ionophore. The two classical GABA antagonists (Fig. 7-4), bicuculline and picrotoxin, appear to act by different means. Bicuculline acts as a direct competitive antagonist of GABA at the receptor level while picrotoxin acts as a noncompetitive antagonist, presumably due to its ability to block GABA-activated ionophores. Although early studies raised some doubts concerning the usefulness of bicuculline as a selective GABA antagonist, these expressions of skepticism have been largely resolved and appear primarily related to the instability of bicuculline at 37°C and physiological pH. Under normal physiological conditions bicuculline is hydrolyzed to bicucine, a relatively inactive GABA antagonist with a short half-life of several minutes. The quaternary salts now used for most electrophysiological experiments (bicuculline methiodide and bicuculline methochloride) are much more water soluble and stable over a broad pH range of 2–8. It should be noted, however, that these quaternary salts are not suitable for systemic administration because of their poor penetration into the central nervous system.

GABA Agonists

Elctrophysiological studies have demonstrated that there are a wide variety of compounds that are capable of directly activating bicuculline sensitive GABA receptors. These agonists can be readily subdivided into two groups based upon their ability to penetrate the blood–brain barrier, dictating whether they will be active or inactive following systemic administration. Agents such as 3-aminopropanesulfonic acid, β-guanidinoproprionic acid, 4-aminotetrolic acid, *trans*-4-aminocrotonic acid, and *trans*-3-aminocyclopentane-1-carboxylic acid are all effective direct acting GABA

Site 5: *Metabolism:* GABA is metabolized primarily by transmination by GABA-transaminase (GABA-T), which appears to be localizd primarily in mitochondria. Amino-oxyacetic acid, gabaculline, and acetylenic GABA are effective inhibitors of GABA-T.

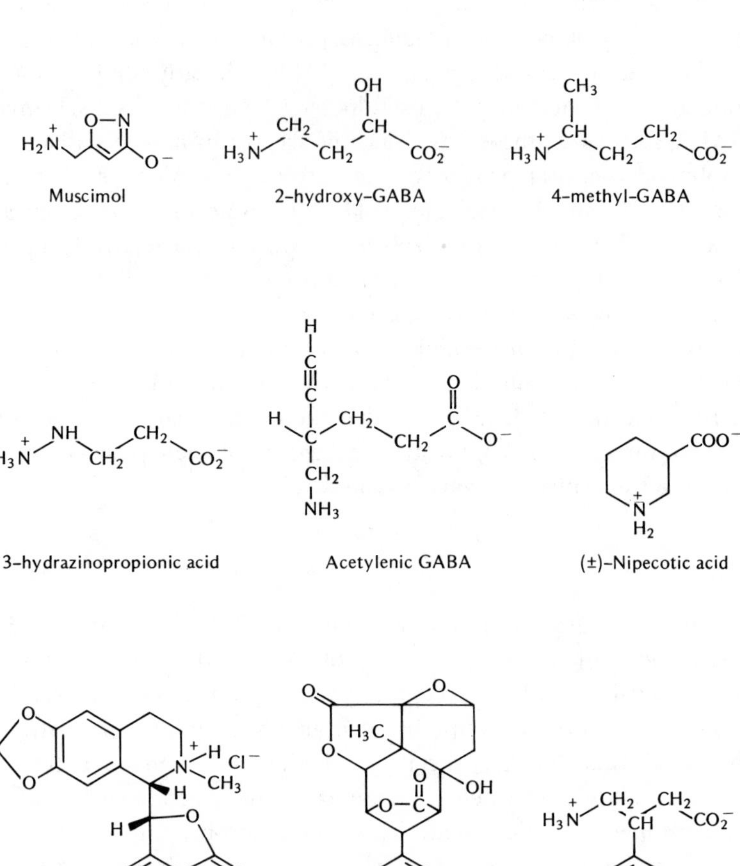

FIGURE 7-4. Structures of compounds that act at GABAergic synapses.

agonists. However, entry of these agents into the brain following systemic administration is minimal. In addition, compounds such as *trans*-4-aminotetrolic acid and 4-aminocrotonic acid also inhibit GABA-T and GABA uptake and, therefore, their action is not totally attributable to their direct agonist properties.

In contrast to this class of direct acting GABA agonists, the second group listed below readily pass the blood–brain barrier and are active following systemic administration. Muscimol (3-hydroxy-5-aminomethylisoxazole) is the agent in this group which has been most extensively studied. Some other agents in this group include (5)-(-)-5-(1-aminoethyl)-3-isoxazole, THIP (a bicyclic muscimol analogue), SL-76002 [a(chloro-4-phenyl)fluro-5-hydroxy-2 benzilideneamino-4H butyramide], and kojic amine (2-aminomethyl-3-hydroxy-4H-pyran-4-one).

In addition to classification based on their ability to penetrate the blood–brain barrier, GABAergic substances may be further divided into those compounds which stimulate GABA receptors directly or those that indirectly cause an activation of GABA receptors by several different mechanisms. For example, agents such as muscimol, isoguvacine, and THIP are true GABA mimetic agents that interact directly with GABA receptors. Indirectly acting GABA mimetics act to facilitate GABAergic transmission by increasing the amount of endogenous GABA which reaches the receptor or by altering in some manner the coupling of the GABA receptor-mediated change in chloride permeability. Thus many drugs often classified as indirect GABA agonists act presynaptically to modify GABA release and metabolism rather than interacting directly with GABA receptors. For this reason, drugs like gabaculine (a GABA-T inhibitor), nipecotic acid (a GABA uptake inhibitor), and baclofen (an agent which, in addition to many other actions, causes release of GABA from intracellular stores) are often classified incorrectly as GABA agonists. The benzodiazepines mentioned earlier also appear to potentiate the action of tonically released GABA at the receptor by displacement of an endogenous inhibitor of GABA receptor binding, allowing more endogenous GABA to reach and bind receptors. Thus, benzodiazepines are sometimes also classified

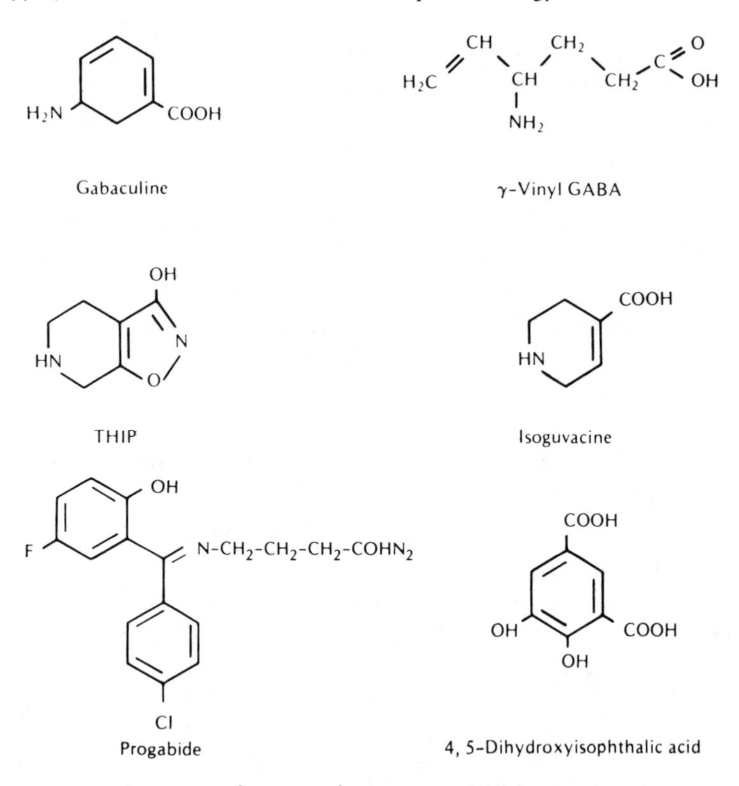

FIGURE 7-5. Structures of compounds that act at GABAergic synapses.

as GABA agonists. A GABA-like action can also be elicited by agents which bypass GABA receptors and influence GABA ionophores. Pentobarbital has been suggested to act at the level of the GABA ionophore, but it is unclear whether its CNS depressant effects are explainable by this action.

The structures of some of the more potent and widely used direct-acting GABA agonists are illustrated in Figures 7-4 and 7-5. Included are muscimol, isoguvacine, THIP, and (+)-*trans*-3-aminocyclopentane carboxylic acid. Useful therapeutic effects have not yet been obtained by use of agents of this sort that have direct GABA mimetic effects (such as muscimol), that inhibit the active re-uptake of GABA (nipecotic acid) or that alter the rate of synthesis or degradation of GABA (such as amino-oxyacetic acid and gabaculine).

However, useful therapeutic effects are achieved with the anxiolytic benzodiazepines (such as Valium (diazepam) and Librium (chlordiazepoxide), which may exert their actions by facilitating GABAergic transmission.

Endogenous Inhibitors

The concept of endogenous inhibitors or modulators of GABA receptors arose from studies on the binding of radioactive GABA to rat brain membranes. These studies revealed that the apparent affinity of GABA binding and the number of apparent binding sites were greatly influenced by the methods used to prepare the membranes. The availability of these latent GABA binding sites appeared to require the removal of endogenous substances that inhibit GABA binding and are normally incorporated into synaptic membranes. A variety of such endogenous inhibitors appear to exist and they have been collectively called GABARINS (GABA Receptor Inhibitors). It has been speculated that GABARINS may function to control the affinity and availability of certain GABA receptors, ultimately altering the properties of the GABA–receptor–ionophore complexes and possibly be involved in desensitization and supersensitivity phenomena. The interaction between GABA receptors and GABARINS could also provide another important site at which drugs could act to alter the functions of GABAergic synapses.

Neurotransmitter Role in Invertebrates

A strong case can be made for a transmitter role for GABA at inhibitory nerve endings to the muscle or to stretch-receptor neurons in crustacea. The evidence can be summarized as follows:

1. Externally applied GABA duplicates the effects of inhibitory nerve stimulation at the crustacean neuromuscular junction as well as at other crustacean synapses.
2. Both exogenous GABA and the endogenous inhibitory transmitter substance, when in contact with the postsyn-

aptic membrane or receptor site, cause this membrane to become more permeable to chloride ion. This increase in chloride permeability in both cases can be blocked by picrotoxin (an antagonist of the GABA-receptor Cl^- channel) and bicuculline (an antagonist of the GABA recognition site).

3. GABA appears to be the most potent inhibitory compound extractable from the lobster nervous system.
4. Inhibitory axons and cell bodies in the lobster nervous system have a much higher content of GABA than excitatory axons or cell bodies, the GABA ratio found being on the order of 100:1. The endogenous GABA content of the inhibitory axon is approximately 0.1 M.
5. The enzymes necessary for the formation and destruction of GABA are present in the inhibitory axon and cell body.
6. A specific GABA uptake mechanism, which could be involved in transmitter inactivation, analogous to the monoamine systems, has been described in the crayfish stretch receptor and the lobster nerve-muscle preparation. However, it is not known if the uptake is primarily into presynaptic nerve terminals.
7. It has also been demonstrated that GABA is released by the lobster nerve-muscle preparation during stimulation of the inhibitory axons. This release was shown to be frequency dependent. Moreover, stimulation of the excitatory nerve to the same muscle does not liberate GABA. Also, GABA release induced by stimulation of the inhibitory fibers can be blocked by lowering the Ca^{2+} content of the medium, which also blocks neuromuscular transmission.

These observations thus fully satisfy the criteria for identification of GABA as an inhibitory transmitter in the crustacea. The physiological and pharmacological observations made in invertebrates have also provided useful models for developing appropriate tests for evaluating the role of GABA as a transmitter in the mammalian CNS.

Neurotransmitter Role in the Mammalian Central Nervous System

The evidence from the mammalian central nervous system to support a role of GABA as an inhibitory transmitter is not as complete as that generated from the crustacean. With substances such as the amino acids, which may play a dual role (both in metabolism and as neurotransmitters), it is only to be expected that it will be more difficult to obtain conclusive evidence as to their role as neurotransmitters.

Neurophysiological and biochemical studies have indicated that the majority of GABA found in brain may serve a neurotransmitter function. In recent years enough evidence has accumulated to allow the conclusion that GABA functions as an inhibitory transmitter in the mammalian central nervous system. GABA is considered an inhibitory transmitter even though this substance can induce both a hyperpolarizing and depolarizing response. Both actions are thought to be a result of GABA receptor mediated change in chloride conductance. The hyperpolarization of neurons produced by an increase in membrane permeability to chloride ions appears identical to the changes induced by glycine-like amino acids. The depolarization induced by GABA of primary afferent terminals is also thought to be associated with an increased permeability to chloride ions, but this action is not produced by glycine-like amino acids. The depolarizing response to GABA predominates in spinal cord cells while the GABA-induced hyperpolarization is typical of cerebral cortical cells. These effects establish possible relationships between GABA and postsynaptic (hyperpolarizing) inhibition and between GABA and presynaptic inhibition.

The elementary criteria for identification of a compound as a putative transmitter have already been mentioned. In short, the agent in question must be produced, stored, released, exert its appropriate action, and then be removed from its site of action. There is sufficient evidence for the production, storage, and pharmacological activity of GABA consistent with its suggestive role as an inhibitory transmitter, but it has not yet been possible to demonstrate an association of GABA with specific inhibitory pathways in the cortex. Until recently there was also no evidence that GABA could be released from mammalian brain under physiological conditions. Several reports now claim to have demonstrated the spontaneous release of GABA from the surface of the brain. However, only one group of investigators has claimed that the amount of GABA released from the cortex is dependent upon the activity of the brain. This group has found that in cats showing an aroused EEG, either following cervical cord section or in awake animals receiving local anesthesia, small amounts of GABA will leak out of the cerebral cortex and can be recovered by a superfusion technique (cortical cup) provided the pia-arachnoid membrane has been punctured. Jasper and co-workers demonstrated that this release of GABA into cortical cups occurs about three times more rapidly from the brains of cats showing an EEG sleeplike pattern with marked spindle activity following midbrain section than in cats showing an EEG awake pattern. When a continuous waking state was maintained by periodic stimulation of the brain stem reticular formation, no measurable amount of GABA could be found in the perfusates. Iversen and co-workers have also demonstrated that GABA is released from the surface of the posterior lateral gyri of cats during periods of cortical inhibition produced by either stimulation of the ipsilateral geniculate or by direct stimulation of the cortex.

The vertebrate neurons most clearly identified by physiological, chemical, and immunochemical criteria as GABAergic are the cerebellar Purkinje cells. When GABA is applied iontophoretically to Deiters' neurons, it induces IPSP-like changes and therefore mimics the action of the inhibitory transmitter released from axon terminals of the Purkinje cells. GABA is also known to be concentrated in Purkinje cells, which exert a monosynaptic inhibitory effect on neurons of Deiters' nucleus. Obata and Takeda observed that stimulation of the cerebellum, which presumably activates the Purkinje cell axons, which have their terminals in cerebellar subcortical nuclei adjacent to the fourth ventricle, induces about a threefold increase in the amount of GABA released into the ventricular perfusate. However, the extremely high voltage employed, the relatively crude perfusion system, and the lack of critical identification of GABA make the significance of these otherwise interesting observations on release somewhat difficult to interpret, although they are consistent with the hypothesis that GABA is an inhibitory transmitter in this particular pathway.

In addition, other experiments have indicated that labeled GABA can be released from brain slices by electrical stimulation or by the addition of high potassium to the incubation medium. However, one should bear in mind that all the above-mentioned release experiments are somewhat gross in nature and do not provide evidence of the neuronal system involved in the release of GABA. Indeed, these experiments do not even indicate that this release is in any way associated with an inhibitory synaptic event. In fact, at the present time no suitable test system in the CNS has been developed that will be as appropriate as the lobster nerve–muscle preparation. The student is advised against holding his or her intellectual breath until it has.

Most of the investigations analyzing the mechanism of GABA's depressive action on the central nervous system have been conducted on spinal cord motoneurons. Until quite recently, GABA had not shown much evidence of closely mimicking the inhibitory transmitter. In fact, some years ago Curtis and his associates discounted the hypothesis that GABA might be the main inhibitory transmitter in the mammalian central nervous system. They noted that the action of GABA ionophoresed onto spinal neurons differed significantly from that of the inhibitory transmitter; whereas natural postsynaptic inhibition was associated with membrane hyperpolarization, experiments with coaxial pipettes failed to detect any hyperpolarization of motoneurons during the administration of GABA. Since at that time there was no evidence that GABA had a hyperpolarizing action in other parts of the central nervous system, Curtis concluded that GABA could not be the main central inhibitory transmitter. Recently, other investigators have demonstrated that in the cortical neurons of the cat GABA imitates the action of the cortical inhibitory transmitter, at least qualitatively; it usually raises the membrane potential and increases conductance of cortical neurons, just like the normal inhibitory synaptic mechanism. When the latter inhibitory effect is artifically reversed by the administration of chloride into the neuron, the action of GABA is also reversed in a similar way. Similar results have been obtained on neurons in Deiters' nucleus. Therefore, at least as far as the cortex and Deiters' nucleus are concerned, there appears to be much support for the hypothesis that GABA is an inhibitory transmitter.

There appears to be no pharmacologically sensitive mechanism for the rapid destruction of GABA similar to the cholinesterase mechanism for destruction of acetylcholine. Thus aminooxyacetic acid or hydroxylamine given intravenously or iontrophoresed does not appear to prolong significantly the duration of action of iontophoretically applied GABA. The exception is Deiters' nucleus, where the hyperpolarization produced by GABA on the neurons in Deiters' nucleus and the IPSP are prolonged in both cases by hydroxylamine, an agent of dubious pharmacological specificity. No functional re-uptake mechanism for GABA within a mammalian system has been conclusively demonstrated either. However, GABA is actively and efficiently taken up by brain slices and synaptosomes, and this implies that some uptake phenomena are functional in the intact brain for recycling GABA and terminating its action. This mechanism could also explain why it has been difficult to collect GABA released from neural tissue. Unfortunately, no drug has been discovered that will effectively block the uptake of GABA into nervous tissue without exerting a number of other unwanted pharmacological actions. Therefore, it is difficult to evaluate the functional significance of this uptake mechanism. The development of a selective inhibitor of GABA uptake might prove to be quite useful in the study of GABA systems in the CNS. Some progress in the development of more specific agents has been made recently in this area.

Several other potentially useful approaches can be identified: (1) It would be helpful to have a specific inhibitor of GAD, which would readily gain access to the CNS so that GABA formation could be selectively blocked. At present, no really specific inhibitors for GAD have been developed that will inhibit this enzyme *in vivo* without also inhibiting GABA-T to some extent as well as a number of other B_6-dependent enzymes. (2) Another real advance would be the development of a histochemical method for the visualization of GABA or GABA-related enzymes at both the light and electron microscope level. The recently developed immunohistochemical method for GAD and for GABA is an important step in this direction. This methodology is already providing useful cytological information in mapping out GABA-containing and GAD-containing systems in the CNS.

Until recently it had been difficult to determine if any physiological or even pharmacological alteration of animal behavior produced changes in the function of the GABA system in the intact animal. This was primarily because a simple and reliable technique had not been developed for the measurement of GABA turnover *in vivo*. Thus we were limited to looking at changes in the brain levels of GABA in order to glean some evidence concerning the activity of this system. Since the brain undoubtedly has many homeostatic mechanisms for the maintenance of GABA levels under widely varying conditions of activity, it would seem that changes in the activity of the GABA system, at least within the physiological range, would not necessarily lead to substantial changes in GABA concentration within the brain. Thus, development of any technique that could reliably evaluate the turnover of GABA *in vivo* would undoubtedly add to a better understanding of the functional importance of GABA in the central nervous system.

Bertilsson and Costa developed a sensitive, although quite laborious, method for measuring GABA turnover in the CNS. They estimated GABA turnover by applying principles of steady state kinetics to the changes with time in the ^{13}C-enrichment of glutamic acid and GABA after systemic administration of ^{13}C-labeled glucose. Depending on the brain region analyzed, the turnover time varied from about 2 to 15 minutes. Table 7–2 compares the steady-state levels and turnover rates of GABA to two other transmitters found in the striatum, ACh and dopamine. The turnover rate for GABA is about 15 times higher than ACh and about 1000 times higher than dopamine turnover.

TABLE 7-2. Steady-state levels and turnover rate of GABA, acetylcholine, and dopamine in the rat caudate nucleus

Transmitter	*Levels (nmoles/mg protein)*	*Turnover rate (nmoles/mg protein/hr)*
GABA	20 ± 0.8	189 ± 10
Acetylcholine	0.56 ± 0.031	11 ± 0.05
Dopamine	0.60 ± 0.05	0.23 ± 0.065

SOURCE: E. Costa, *Neuroregulators and Psychiatric Disorders* (1977).

By means of this technique, Costa and co-workers have demonstrated that drugs such as morphine administered systemically and β-endorphin administered intraventricularly cause marked changes in the turnover of brain GABA (Table 7–3). Treatment of rats with morphine or β-endorphin causes a decrease in the turnover of GABA in the caudate nucleus and a large increase in the turnover of GABA in the globus pallidus and substantia nigra. These alterations were shown to be dose related and inhibited with the opiate antagonist, naltrexone. These observations are consistent with the hypothesis that opiate receptors regulate GABAergic interneurons located in the caudate nucleus causing an inhibition in the activity of these GABAergic neurons. These caudate interneurons are thought to be in contact with caudatal-pallidal but not with caudatal-nigral GABAergic neurons. When these interneurons are inhibited by morphine, there is an increase in GABA metabolism in the globus pallidus, and this change is correlated with the development of extrapyramidal rigidity. Antipsychotic drugs can induce similar rigidity and their administration to rats is also associated with an increase in GABA turnover in the globus pallidus. These pharmacological observations are consistent with the possibility that hyper-GABAergic activation in the globus pallidus might be in part responsible for extrapyramidal rigidity induced by drug administration. The molecular mechanism by which these drugs cause selective changes in brain GABA turnover is still unknown.

TABLE 7-3. Selective alterations in GABA turnover rate following administration of morphine and β-endorphin

	GABA turnover rate (% control)		
Drug	*Caudate nucleus*	*Globus pallidus*	*Substantia nigra*
Morphine (70 μmol/kg s.c.)	34	130	220
β-Endorphin (2.9 nmol/intraventricularly)	55	260	300

SOURCE: Data taken in part from Moroni *et al.* (1979).

A number of different techniques have been used to glean knowledge concerning the neuronal pathways in brain using GABA as a transmitter. These include the binding of specific GABA receptor agonists in different brain regions and the autoradiographic localization of labeled GABA taken up through the high-affinity uptake process, the calcium-dependent, depolarization-induced release of GABA from specific brain regions, the measurement of GABA concentration and GAD activity in different brain regions, and the changes in these various parameters after surgical or chemical lesions of afferent pathways. However, the most powerful technique has been the immunohistochemical localization of GAD-like and GABA-like activity in brain by means of specific antibodies directed against GAD and GABA. In the most recent approach, antibodies have been raised in rabbits against GABA conjugated to bovine serum albumin and this antisera has been used to map GABA-like neurons in the brain. In general, the distribution of GABA-like and GAD-like activity parallels each other and detailed maps of GABA- and GAD-containing neurons in brain are being produced. The GABA antisera appears to provide better resolution of GABA-containing cell bodies while the GAD antisera provides better resolution of GAD-containing terminal fields. GAD localization studies have already given additional support to a role for GABA as an inhibitory transmitter in certain cerebellar, cortical, and striatonigral neurons by localizing GABA synthesis to specific nerve terminals.

Studies employing the above techniques have demonstrated that in the central nervous system of mammals, GABA is found primarily in inhibitory interneurons with short axons. For example, the majority of GAD-like and GABA-like activity found in cerebral cortex, hippocampus, and spinal cord is present in inhibitory interneurons. At the present time, only two long axonal projections from one brain region to another are known that utilize GABA as a transmitter. These are the two systems alluded to above: the Purkinje cells of the cerebellum and their projections to vestibular and cerebellar nuclei and a system of neurons in the striatum with axons projecting to the substantia nigra. As noted in Table 7–1, this latter region of brain (substantia nigra) contains the highest known concentration of GABA. It seems likely that these two immunocytochemical approaches will continue to be employed successfully and will ultimately permit a detailed anatomical description of the GABA pathways throughout the CNS. Once the anatomy of the GABA-containing neuronal systems in the mammalian CNS is more clearly defined, it should be possible to learn a great deal more about the functional aspects of these systems.

In summary, numerous biochemical observations have been made on brain GABA systems over the years which are consistent with a neurotransmitter role for this substance in mammalian brain. Similar to other neurotransmitters or neurotransmitter candidates. GABA as well as its biosynthetic enzyme (GAD) have a discrete nonuniform distribution in brain. Brain contains a high-affinity, sodium-dependent transport system, and storage of GABA can be demonstrated in selected synaptosomal populations. The release of endogenous or radioactively labeled exogenously accumulated GABA can be evoked by the appropriate experimental conditions. Most recently, the presence of GABA-containing neurons has been verified and the anatomical distribution of GABAergic neurons mapped out by use of autoradiographic and immunocytochemical techniques.

However, the most compelling evidence that GABA plays a neurotransmitter role in mammalian brain has been provided by intracellular recording studies demonstrating that GABA causes a hyperpolarization of neurons similar to that evoked by the naturally occurring transmitter substance.

Glycine

As an Inhibitory Transmitter

Glycine structurally is the simplest amino acid and is found in all mammalian body fluids and tissue proteins in substantial amounts. Although glycine is not an essential amino acid, it is an essential intermediate in the metabolism of protein, peptides, one-carbon fragments, nucleic acids, porphyrins, and bile salts. It is also believed to play a role as a neurotransmitter in the CNS. Over the past two decades, numerous neurochemical studies have attempted to separate and distinguish between the general "metabolic" and "transmitter" functions of glycine within the CNS. These studies have confirmed glycine's role as a neurotransmitter in the spinal cord and have suggested that this amino acid may play a similar role in more rostral portions of the CNS and also in the retina. Thus, glycine appears to have a more circumscribed function in the central nervous system than GABA since the inhibitory role for this substance is restricted to the spinal cord, lower brainstem, and perhaps the retina. Glycine also appears to be an exclusively vertebrate transmitter, making it unique among the transmitter substances. However, within the last several years very little progress has been achieved in developing pharmacological tools that act selectively on glycine systems or in generating more information concerning glycine metabolism in neuronal tissue.

Our knowledge of the metabolism of glycine in nervous tissue is still somewhat rudimentary despite the fact that the process has been studied extensively in other tissues. For example, we still do not

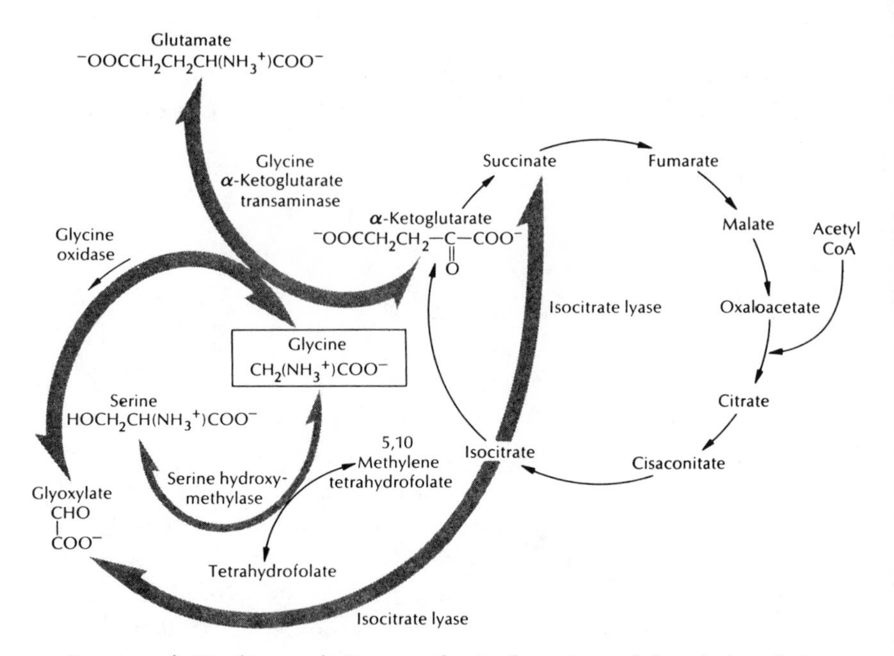

FIGURE 7-6. Possible metabolic routes for the formation and degradation of glycine by nervous tissue. (Modified after Roberts and Hammerschlag, in: *Basic Neurochemistry,* Little, Brown, Boston, 1972, p. 151)

know whether biosynthesis is important for the maintenance of glycine levels in the spinal cord or whether the neurons depend upon the uptake and accumulation of preformed glycine. As indicated in Figure 7–6, glycine can be formed from serine by a reversible folate-dependent reaction catalyzed by the enzyme, serine *trans*-hydroxy-methylase. Serine itself can also be formed in nerve tissue from glucose via the intermediates 3-phosphoglycerate and 3-phosphoserine. It is also conceivable that glycine might be formed from glyoxylate via a transaminase reaction with glutamate. Although not established definitively, it seems likely that serine serves as the major precursor of glycine in the CNS and that serine hydroxy-methyl transferase and D-glycerate dehydrogenase are the best candidates for the rate-limiting enzymes involved in the biosynthesis of glycine. Not only is our knowledge of the metabolism of glycine in nervous tissue minimal but at the present time only scanty information is available on the factors regulating the concentration of glycine in the CNS. Glycine in the spinal cord is labeled only slowly from radioactive glucose via one of the glycolytic intermediates indicated in Figure 7–6. It is also possible to label glycine by administration of labeled glyoxylate, which is readily transaminated to glycine by nervous tissue.

Substantial evidence now exists to indicate that glycine may play a role as an inhibitory transmitter in the mammalian spinal cord. From the evidence available we will try to assess how well this compound fulfills the criteria necessary to categorize this substance as a central nervous system inhibitory transmitter. First, we can say that this amino acid is found in relatively high concentrations in the spinal cord compared with other amino acids. Table 7–4 illustrates the occurrence of glycine and other free amino acids in the spinal cord, and from these date it is apparent that glycine is more concentrated in the spinal gray matter than in the spinal white matter. The concentration found in the spinal gray matter is much higher than the level in whole brain or spinal roots. This high level of glycine in the ventral horn, together with the comparatively low content in ventral root fibers, initially suggested that glycine may be associated with the inhibitory interneurons, and subsequently this was shown to be the case. From iontophoretic studies the compounds in Table 7–4 can be divided into four main groups.

1. Excitatory — glutamate, aspartate
2. Inhibitory — GABA, glycine, alanine, cystathionine, and serine
3. Inactive — glutamine, leucine, threonine, and lysine
4. Untested — arginine and methyl histidine

Of the amino acids in group two, cystathionine and serine appear to have a weaker and relatively sluggish action on spinal neurons compared with that of GABA and gycine. Therefore, in view of the distribution and the relative inhibitory potency of the com-

TABLE 7-4. Amino acid concentration of cat spinal cord and roots (μmoles/g)

Amino acids	*Spinal gray*	*Dorsal root*	*Ventral root*
Alanine	0.62	0.36	0.30
Arginine	0.14	0.07	0.06
Aspartate	2.14	0.79	1.13
Cystathionine	3.02	0.02	0.01
GABA	0.84	Trace	Trace
Glutamate	4.48	3.33	2.09
Glutamine	5.48	1.46	1.10
Glycine	4.47	0.28	0.32
Leucine	0.16	0.06	0.05
Lysine	0.16	0.07	0.05
Serine	0.41	0.32	0.23
Threonine	0.21	0.18	0.15

SOURCE: M. H. Aprison and R. Werman (1968), *Neurosciences Research 1*, 157, Academic Press, New York.

pounds in this group, it was suggested by Aprison and Werman that GABA and glycine were the most likely candidates for spinal inhibitory transmitters. Glycine administered by iontophoretic techniques was consistently found to diminish the firing and excitability of both spinal motoneurons and interneurons. This is interesting in itself since glycine is quite ineffective as an inhibitor of cortical neurons. In addition, the hyperpolarization and the changes in membrane permeability produced by glycine seem to be quite similar to those produced by the spinal inhibitory transmitter. Alteration in K^+ or Cl^- ion concentrations affect the inhibitory postsynapic potentials and the glycine-induced potentials in the same fashion. Strychnine, a compound that has been shown to reduce spinal postsynapic inhibition, also blocks the effects of glycine on spinal motoneurons. It was originally reported that strychnine reversibly blocks the action of the natural inhibitory synaptic transmitter(s), as well as glycine and β-alanine, but does not have any effect on the hyperpolarizing action of GABA in cats anesthetized with pentobarbital. However, similar experiments on decerebrated, unanesthetized cats have indicated that the qualitative dichotomy between glycine and β-alanine and GABA originally reported does not usually hold. Instead, only a quantitative difference in the interaction of strychnine and the amino acids above is reported. It is of interest that except for experiments with strychnine, there has been little pharmacological manipulation of neurons that release glycine. In all these experiments, strychnine antagonized glycine more potently than GABA. However, GABA also inhibits these spinal neurons and is present endogenously, and in the course of distinguishing between the relative merits of GABA and glycine as inhibitory transmitter candidates in the spinal cord, it was of interest to see if a loss of function of inhibitory interneurons would be accompanied by a change in one or both of these compounds. Anoxia of the lumbosacral cord produced by clamping the thoracic aorta seems to destroy the interneurons preferentially, while leaving 80 percent or more of the motoneurons intact. Thus, the presynaptic cells responsible for inhibition (and other functions) are lost and therefore the concentration of transmitter associated with those cells would be expected to be decreased. When the effect of anoxia on GABA and glycine content of the cat spinal cord was analyzed, glycine was the only potential inhibitory amino acid markedly decreased. Thus the distribution of glycine in the spinal cord of the cat appears to be related to the inhibitory interneurons. Some preliminary results with perfused toad spinal cord have indicated that glycine release appears to occur during stimulation of the dorsal roots. However, the release of glycine from interneurons in the cord has not yet been demonstrated. In summary, glycine thus satisfies many of the criteria sufficiently to warrant its consideration as a possible inhibitory transmitter in the cat spinal cord.

1. It occurs in the cat spinal cord associated with interneurons.

2. When administered iontophoretically, it hyperpolarizes motoneurons to the same equilibrium potential as postsynaptic inhibition.
3. The permeability changes of the postsynaptic membrane induced by glycine appear to be similar to those associated with postsynaptic inhibition.
4. Strychnine, a drug that blocks the action of glycine, also blocks postsynaptic inhibition.
5. Stimulation of dorsal roots causes release of glycine from perfused spinal cord.

The mechanisms by which the action of iontophoretically applied glycine or that of the putative spinal inhibitory transmitters are terminated are not known at this time. Recently, it has been demonstrated that there is an active and efficient sodium-dependent uptake mechanism for glycine in rat spinal cord and that labeled glycine taken up into slices can be released by electrical or potassium stimulation. However, whether or not this uptake mechanism is the physiological mechanism by which the action of administered or release glycine is terminated remains to be determined. Also, to date very little appears to be known with regard to the factors controlling the release of glycine from the spinal cord. Again, as with GABA, the efficient uptake process may explain why it is difficult to detect glycine release from the central nervous system. The main problem (as with GABA, glutamate, etc.) is that there is no distinct neuronal pathway that may be isolated and stimulated; thus all the induced activity is very generalized, making the significance of any demonstrable release (metabolite or excess transmitter) very difficult to interpret.

To summarize, then, probably the most critical missing piece of evidence to establish an even more complete identification of the inhibitory role of glycine in the spinal cord is the demonstration that glycine is the main substance contained in the terminals of the interneurons that synapse on the motoneurons and that it is glycine that is released from these terminals when direct inhibition is produced. Some partial support for the localization of glycine has been provided by autoradiographic localization of glycine uptake sites as visualized by electron microscopy. However, demonstration of discretely evoked release of glycine from spinal cord interneurons has been more difficult to obtain.

Glutamic Acid

It has been recognized for many years that certain amino acids such as glutamate and aspartate occur in uniquely high concentrations in the brain and that they can exert very powerful stimulatory effects on neuronal activity. Thus, if any amino acid is involved in regulation of nerve-cell activity, as an excitatory transmitter or otherwise, it seems unnecessary to look beyond these two candidates.

The excitatory potency of glutamate was first demonstrated on crustacean muscle and later by direct topical application to mammalian brain. However, except for the invertebrate model where substantial evidence has accumulated to support a role for glutamate as an excitatory neuromuscular transmitter, its status as a neurotransmitter in mammalian brain was uncertain for many years. This is probably in part explainable by the fact that glutamate (and also aspartate) is a compound that is involved in intermediary metabolism in neural tissue. For example, glutamate plays an important function in the detoxification of ammonia in brain, is an important building block in the synthesis of proteins and peptides including glutathione and also plays a role as a precursor for the inhibitory neurotransmitter, GABA. Thus it has been extremely difficult to dissociate the role this amino acid plays in neuronal metabolism, and as a precursor for GABA, from its possible role as a transmitter substance. The transport of circulating glutamate to the brain normally plays only a very minor role in regulating the levels of brain glutamate. In fact, the influx of glutamate from the blood across the blood–brain barrier is much lower than the efflux of glutamate from the brain.

In view of the many different roles assigned to brain glutamate in addition to its role as a putative transmitter, it is not surprising that its synthesis and metabolism are compartmentalized in a very complex fashion. Compartmentation studies have indicated that a number of immediate precursors for glutamate synthesis exist: (1) from 2-oxoglutarate and aspartate by aspartate aminotransferase (2) from glutamine by phosphate-activated glutaminase and (3) from 2-oxoglutarate by ornithine-aminotransferase. The relative contribution of glutamine or glucose to transmitter glutamate synthesis in brain is still an open question. It may be that the transmitter pool of glutamate will utilize any glutamate available, independent of its source, and that the synthesis of glutamate itself is not under normal conditions a rate-limiting factor. In view of the numerous "metabolic" and precursor roles served by excitatory amino acids including glutamate, the presence of these amino acids in brain is not restricted to a particular type of neuron as is acetylcholine and the catecholamines, norepinephrine, epinephrine, and dopamine. Thus, even though neuronal systems beleived to utilize glutamate or aspartate as transmitter substances are being described in the CNS, it seems quite unlikely that it will be possible to accurately map out these systems by simply following the presence of glutamate or aspartate or their synthesizing enzymes. Mapping will require the use of these techniques in conjunction with other less direct approaches.

Despite its ubiquitous occurrence in high levels throughout the central nervous system, the unequal regional distribution of glutamate in the spinal cord (higher in dorsal than corresponding ventral regions) is one of the main observations in support of the proposal that this common amino acid functions as an excitatory transmitter released from primary afferent nerve endings.

A unique approach towards investigating the possible role of glutamate as a transmitter in mammalian CNS has been the employment of a virus that selectively destroys granule cells in the cerebellum. After viral infection had destroyed more than 95 percent of the granule cells, a marked decrease in cerebellar glutamate was observed, but no significant decrease in other amino acids, including aspartate, was detected. The high-affinity, sodium-dependent uptake of glutamate was also significantly reduced following destruction of the granule cells. The selective decline of glutamic acid levels and uptake does not prove that glutamic acid is a neurotransmitter in the granule cells, although it is consistent with this hypothesis, especially since the granule cell transmitter is known to be excitatory.

A recent technical advance that has aided in the elucidation of the excitatory effects of glutamate and related amino acids is the discovery of several conformationally restricted analogs of glutamic acid that exhibit marked potency and specificity in depolarizing central neurons. These agents include quisqualic acid (isolated from seeds), ibotenic acid (isolated from mushrooms), and kainic acid (isolated from seaweed). (See Fig. 7–7.) Of these agents, kainic acid has been the most extensively studied. Kainic acid is about fifty fold more potent in depolarizing neurons in the mammalian CNS than glutamate and thus is one of the most potent of the aminoacid neuroexcitants. Extensive studies with this substance have revealed that direct injection of kainic acid in discrete brain regions results in selective destruction of the neurons that have their cell bodies at or near the site of injection. Axons and nerve terminals appear to be more resistant to the destructive effects of kainic acid than the cell soma. Although it was originally believed that the selective neurotoxic effects of kainic acid were due simply to excessive excitation of all neuronal soma exposed to this agent, it is now clear that the mechanism of its neurotoxic action is more complex. The neurotoxic action appears to involve an interaction between specific receptors for kainic acid and certain types of presynaptic input. For example, studies in striatum and hippocampus indicate that kainic acid toxicity depends on an intact glutaminergic innervation of the injected brain region.

Kainic acid is now being used extensively as a research tool in neurobiology to produce selective lesions in the brain of experimental animals and to explore the physiological pharmacology of excitatory transmission.

The status of glutamate and aspartate as neurotransmitters has suffered through many cycles of acceptability and nonacceptability in the past decade. Although the case for glutamate as a transmitter in the mammalian central nervous system is considerably weaker

ANTAGONISTS

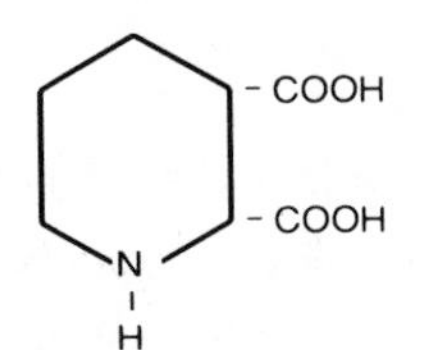

AGONISTS

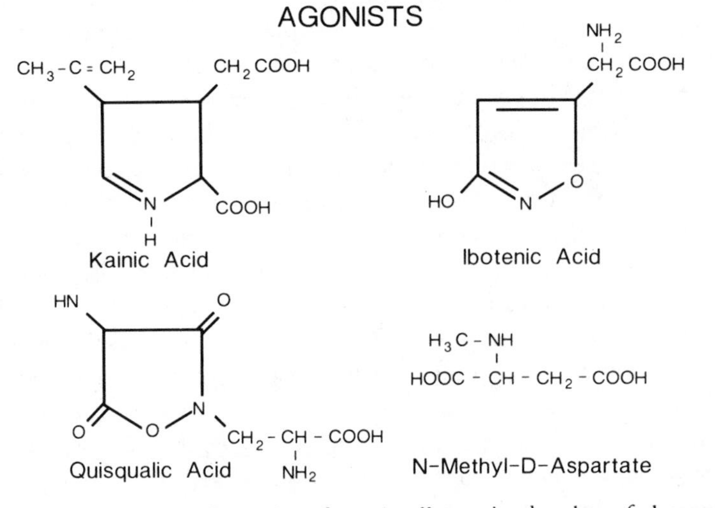

FIGURE 7-7. Structures of several conformationally restricted analogs of glutamate and several antagonists.

than that for acetylcholine, catecholamines, 5-HT, GABA, and glycine, its case has been fortified in the last several years. In the vertebrate central nervous system, iontophoretic application of both glutamate and aspartate to nerve cells does produce depolarization and an increase in firing rate. In this connection, as was mentioned above, both GABA and L-glutamate are released from the cerebral cortex of the cat at a rate that appears to be dependent upon its state of CNS activation—more GABA and less L-glutamic acid being released under conditions associated with "sleeplike" EEG patterns, with the reverse situation occurring under arousal conditions. This differential release does not appear to result from alterations in cortical blood flow, since the release of other amino acids such as L-glutamine and L-aspartic acid does not vary similarly with different conditions. However, at this time it was not demonstrated that the origin of these amino acids was in fact from neurons or, for that matter, even from brain cells. The possibility could not be ruled out that these amino acids were derived from blood or cerebrospinal fluid. Although these experiments can be taken as supportive evidence for a role of these agents as transmitters, they by no means establish this role.

In spite of the above evidence suggesting that glutamate may be an excitatory transmitter in the CNS, little was known until recently about the biosynthesis and release of the pool of releasable "transmitter glutamate." Utilizing the molecular layer of the dentate gyrus of the hippocampal formation to provide a definitive system in which the major input appears to be glutaminergic, Cotman and co-workers have addressed these questions. Glutamate was shown to be released by depolarization from slices of the dentate gyrus in a Ca^{++}-dependent manner and lesions of the major input to the dentate gyrus originating from the entorhinal cortex diminished this release and the high-affinity uptake of glutamate. Glutamate biosynthesis in the releasable pools was rapidly regulated by the activity of glutaminase and by the uptake of glutamine. These properties are all consistent with properties expected of a neurotransmitter, and the observations strengthened the premise that glutamate may be an important neurotransmitter in the molecular layer of the dentate gyrus. Futhermore, these studies demonstrate that the regulation of glutamate synthesis and release share many properties in common with those discussed elsewhere for other transmitters. For example, similar to acetylcholine synthesis, the synthesis of glutamate is regulated in part via the accumulation of its major precursor glutamine, and newly synthesized glutamate, like acetylcholine, is preferentially released. In addition, the synthesis of glutamate is regulated by end-product inhibition. This is similar to the mechanism by which the rate-limiting enzyme in catecholamine synthesis, tyrosine hydroxylase, is regulated in catecholaminergic neurons by dopamine and norepinephrine. It is interesting that these similarities are demonstrable despite the involvement of glutamate in general brain metabolism.

These studies on the dentate gyrus as well as many others have demonstrated that lesions studies coupled with biochemical measures of alterations in high-affinity, sodium-dependent uptake of glutamate in brain synaptosomes provide a useful technique for mapping out glutaminergic neuronal systems in brain although this technique cannot be used to differentiate between glutamate and aspartate containing neurons. Retrograde transport of D-aspartate after microinjection in terminal regions is another method that may be of value in tracing glutamate pathways. The success and specificity of this method is also dependent on the specificity of the uptake process and therefore not capable of discriminating between glutamate and aspartate containing neurons.

It is somewhat disturbing that glutamate, aspartate, and synthetic derivatives of these dicarboxylic acids result in almost universal activation of unit discharge, and the endogenous excitants appear to be almost ubiquitous in the nervous system without the expected asymmetric distribution. It has also been reported that glutamate does not bring the cell-membrane potential to the same level as the natural excitatory transmitter. In addition, both the D and the naturally occurring L isomers of excitatory amino acids are active, although in the case of glutamate the D isomer is often reported to be somewhat less active. These findings have led some investigators to suggest that the response to amino acids represents a nonspecific receptivity of the neuron to these agents and is therefore not necessarily indicative of a transmitter function. However, based upon the current available evidence, glutamate appears to have satisfied four of the main criteria for classification of this substance as an excitatory neurotransmitter in the mammalian CNS: (1) it is localized presynaptically in specific neurons; (2) it is released by a calcium-dependent mechanism by physiologically relevant stimuli in amounts sufficient to elicit postsynaptic responses; (3) a mechanism exists (re-uptake) that will rapidly terminate its transmitter action; and (4) it demonstrates pharmacological identity with the naturally occurring transmitter.

Recent studies have provided new insight into the synaptic mechanisms controlled by excitatory amino acids and their involvement in brain function. Based upon the rank order of potency of agonists and the selective antagonism of these agonists in electrophysiological studies, multiple excitatory amino acid receptor types have been proposed. The most common classification scheme describes these receptor subtypes named for the agonist that specifically elicits a characteristic electrophysiological response. Three excitatory amino acid receptor subtypes have now been defined and are termed *N*-methyl-D-aspartate (NMDA), and the non-NMDA receptors, the latter subtype being further subdivided into quisqualate and kainate preferring receptors (Table 7–5). New data are now emerging on these receptor subtypes including their anatomical and subsynaptic localization and their involvement in defined synaptic responses. The availability of specific antagonists has rendered the NMDA excitatory receptor subtype particularly amenable to study, and significant progress has been made regarding the involvement of NMDA receptors in neural function and the potential therapeutic value of NMDA receptor antagonists. Thus, the NMDA-preferring receptor is the most extensively characterized excitatory amino-acid receptor subtype. The NMDA receptor is known to participate in long-term potentiation, epileptic seizure activity, and neuronal degeneration, and its function in the membrane appears to be regulated by a Mg^{++} and voltage dependent switch. Table 7–5 summarizes the pharmacology, distribution, and possible functions of these three excitatory receptor subtypes.

The hypothesis that excitatory amino-acid receptor mechanisms may underlie the pathogenesis of neurodegenerative disorders has

TABLE 7-5. Excitatory amino-acid receptor subtypes

		Non-N-*methyl*-d-*aspartate*	
Receptor subtypes	N-*Methyl*-D-*aspartate*	*Quisqualate*	*Kainate*
Selective agonists	*N*-Methyl-D-Aspartate 1-Amino-1,3-dicarboxy-cyclopentane Ibotenate	Quisqualate α-Amino-3-hydroxy-5-methyl-isoxasole-4-propionic acid	Kainate Domoate
Selective antagonists	D-AP5 D-AP7 Asp-AMP	γ-D-Glutamyltaurine GAMS	γ-D-Glutamyltaurine GAMS
Possible synaptic pathways	Excitatory interneurons in spinal cord Schaffer collateral fibres in hippocampus (CA3-CA1)	Cerebellar parallel fibers to Purkinje cells Sensory nerves activated by hair movements to neurones in cuneate nucleus and trigeminal nucleus Perforant path to hippocampal dentate granule cells Cerebral cortex to cuneate neurones Cerebral cortex to dopaminergic caudate neurones Optic tract fibers to LGN cells	

Receptor function		—*Mediate transmission at excitatory synapses*—	
	Induction of long-term potentiation Excitotoxicity		Excitotoxicity
Therapeutic Indications for antagonists	Convulsive and neurodegenerative disorders	Not known at present	
Radioligands	L-Glutamate D-AP5	L-Glutamate AMPA	L-Glutamate Kainate

Abbreviations:

AMPA—α-amino-3-hydroxy-5-methylisoxasole-propionic acid
Asp-AMP—β-D-aspartylaminomethylphosphonate
D-AP5—D-2-amino-5-phosphonovalerate
D-AP7—D-2-amino-7-phosphonoheptanoate
GAMS—γ-D-glutamylaminomethyl-sulphonate

SOURCE: Data taken in part from Fagg (1985) and Watkins (1984).

recently had its focus directed to the NMDA receptor subtype with the finding that the neurotoxic action of an endogenous substance, quinolinic acid, can be blocked by the selective antagonist of this receptor, D-AP7. Also, work from Meldrum's group demonstrating that neuronal damage in the rat hippocampus resulting from occlusion of the carotid arteries is abolished by the NMDA receptor antagonists together with similar observations made on primary hippocampal cells in culture in which cell death was induced by exposure to an anoxic environment suggests that ischaemic neuronal damage appears to result not simply from anoxia *per se*, but perhaps via the release of an excitatory amino acid and activation of the NMDA receptor subtype. Additional studies are required before definitive conclusions can be made. Future studies characterizing the various subtypes of excitatory amino-acid receptors will depend increasingly on the availability of selective antagonists for the other receptor subtypes as well as radioligands specific for these receptors. It is likely that, in the future, research on glutamate will continue to be dominated by studies pertaining to some aspects of its neurotransmitter function with focus on further characterization of glutamate receptors and identification of neurons that utilize glutamate as a neurotransmitter.

Selected References

Cotman, C. W., A. C. Foster, and T. Lanthorn (1981). An overview of glutamate as a neurotransmitter. *Adv. Biochem. Psychopharmacol.* 271.

Daly, E. C., and M. H. Aprison (1983). Glycine. In *Handbook of Neurochemistry*, Vol. 3 (A. Lajtha, ed.), p. 467. Plenum, New York.

Enna, S. J. (1983). Biochemical and electrophysiological characteristics of mammalian GABA receptors. In *International Review of Neurobiology*, Vol. 24, p. 181. Academic Press, New York.

Fagg, G. E., and A. C. Foster (1983). Amino acid neurotransmitters and their pathways in the mammalian central nervous system. *Neurosci.* *9*, 701.

Fagg, G. E. (1985). L-Glutamate, excitatory amino acid receptors and brain function. *Trends in Neurosci.* 8, 207.

Fahn, S., and L. J. Cote (1968). Regional distribution of γ-aminobutyric acid (GABA) in brain of the Rhesus monkey. *J. Neurochem.* 15, *209*.

Fonnum, G. (1984). Glutamate: A neurotransmitter in mammalian brain. *J. Neurochem.* 42, 1.

Greennamyre, J. T., J. M. M. Olson, J. B. Penney, Jr., and A. B. Young (1985). Autoradiographic characterization of N-methyl-D-aspartate-quisqualate- and kainate-sensitive glutamate binding sites. *J. Pharmacol. and Exp. Thera.* 233, 254.

Meldrum B. (1985). Excitatory amino acids and anoxic/ischaemic brain damage. *Trends in Neurosci.* 8, 47.

Moroni, F., E. Peralta, and E. Costa (1979). Turnover rates of GABA in striatal structures: Regulation and pharmacological implications. In GABA-*Neurotransmitters, Pharmacochemical, Biochemical and Pharmacological Aspects.* (P. Krogsgaard-Larsen, J. Scheel-Kruger, and H. Kofod, eds.), p. 95. Academic Press, New York.

Ottersen, O. P., and J. Storm-Mathisen (1984). Neurons containing or accumulation transmitter amino acids. In *Handbook of Chemical Neuroanatomy*, Vol. 3, Part 2 (A. Björklund, T. Hökfelt, and M. J. Kuhar, eds.), p. 141. Elsevier, Amsterdam.

Perry, T. L., K. Berry, S. Hansen, S. Diamond, and C. Mok (1971). Regional distribution of amino acids in human brain obtained at autopsy. *J. Neurochem.* 18, 513.

Roberts, P. J., J. Storm-Mathisen, and G. A. R. Johnston (1981). *Glutamate Transmitter in the Central Nervous System.* Wiley, Chichester.

Shank, R. P., and G. LeM. Campbell (1983). Glutamate. In *Handbook of Neurochemistry*, Vol. 3 (A. Lajtha, ed.), p. 381. Plenum, New York.

Sharif, N. A. (1985). Multiple synaptic receptors for neuroactive amino acid transmitters—New vistas. In *International Review of Neurobiology*, Vol. 26, p. 85. Academic Press, New York.

Tallman, J. F., and D. W. Gallagher (1985). The GABA-ergic system: A locus of benzodiazephine action. *Ann. Rev. Neurosci.* 8, 21.

Tapia, R. (1983). Regulation of glutamate decarboxylase activity. In *Glutamine, Glutamate, and GABA in the Central Nervous System*, p. 113. Alan R. Liss, New York.

Tapia, R. (1983). γ-Aminobutyric acid: Metabolism and biochemistry of synaptic transmission. In *Handbook of Neurochemistry* (2nd ed.), Vol. 3 (A. Lajtha, ed.), p. 423. Plenum, New York.

Vayer, P., P. Mandel, and M. Maitre (1985). Conversion of γ-hydroxybutyrate to γ-aminobutyrate *in vivo*. *J. Neurochem.* *45*, 810.

Watkins, J. C., and R. H. Evans (1981). Excitatory amino acid transmitters. *Ann. Rev. Pharmacol. Toxicol.* 21, 165.

Watkins, J. C. (1984). Excitatory amino acids and central synaptic transmission. *Trends in Pharmacological Sciences* 61, 373.

Wirklund, L., D. Toggenburger, and M. Cuenod (1982). Aspartate: Possible neurotransmitter in cerebellar climbing fibers. *Science 216*, 78.

8 | Acetylcholine

The neurophysiological activity of acetylcholine (ACh) has been known since the turn of the century and its neurotransmitter role since the mid-1920s. With this history, it is not surprising that the graduate student or medical student assumes that everything is known about this subject. Unfortunately, the lack of sophisticated technology for determining the presence of ACh in cholinergic tracts and terminals, which only recently has been partially overcome, has left this field far behind the biogenic amines. The structural formula of ACh is presented below:

$$(CH_3)_3N^{+}—CH_2CH_2—O—\overset{\overset{\displaystyle O}{\|}}{C}—CH_3$$

Acetylcholine

Assay Procedures

ACh may be assayed by its effect on biological test systems or by physicochemical methods. Popular bioassay preparations include the frog rectus abdominis, the dorsal muscle of the leech, the guinea pig ileum, the blood pressure of the rat (or cat), and the heart of Venus mercenaria. In general, bioassays tend to be laborious, subject to interference by naturally occurring substances, and on occasion to behave in a mysterious fashion (e.g., the frog rectus abdominis is not as sensitive to ACh in the summer months as in the winter, and it is not unusual to encounter a guinea pig ileum that will not respond to ACh). Nevertheless, bioassays at the present time represent one of the most sensitive (0.01 pmol in the toad lung) and, under properly controlled conditions, the most specific procedures for determining ACh. It is probably true to state that the neurochemically oriented investigator's natural fear and distrust of a bioassay have hampered progress in elucidating biochemical and biophysical aspects of ACh. This statement is supported by a consideration of the recent explosion of information on norepinephrine. This neurotransmitter can also be bioassayed, but it was only after the development of sensitive fluorometric and radiometric procedures for determining components of the adrenergic nervous system that the information explosion occurred.

Until about 1965 physicochemical methods for determining ACh were so insensitive as to be virtually useless in measuring endogenous levels of ACh. However, since then there have been reports published on enzymatic, fluorometric, gas chromatographic, chemiluminescent, and radioimmunoassay techniques that approach the sensitivity and specificity of the bioassays. In particular, the gas chromatographic procedure of Jenden has recently been coupled to mass spectroscopy to yield a high degree of sensitivity and specificity. Unfortunately, the cost of the apparatus precludes the use of this procedure by indigent investigators.

McCaman and Stetzler have modified a radiometric procedure devised by Reid *et al.* to yield a method for measuring ACh that will probably supplant most existing procedures. The principle of the method is that ACh is first hydrolyzed to acetate and choline by cholinesterase and choline is phosphorylated by choline kinase with [γ^{32}P]ATP to produce choline-^{32}P-phosphate. The mixture is then put through a small column containing an anionic resin whereby [γ^{32}P]ATP remains on the column and the labeled choline phosphate passes through and is collected into scintillation vials for a subsequent determination of the radioactivity. With this procedure, one can determine 0.05 pmols (5×10^{-14} moles) of ACh.

A new procedure reported by Potter *et al.* is not quite as sensitive (2 pmoles ACh) but could prove useful. This method utilizes reverse-phase HPLC to separate ACh and choline followed by hydrolysis of ACh by cholinesterase, oxidation of choline by choline oxidase, and the subsequent electrochemical detection of hydrogen peroxide that is generated from the oxidase activity.

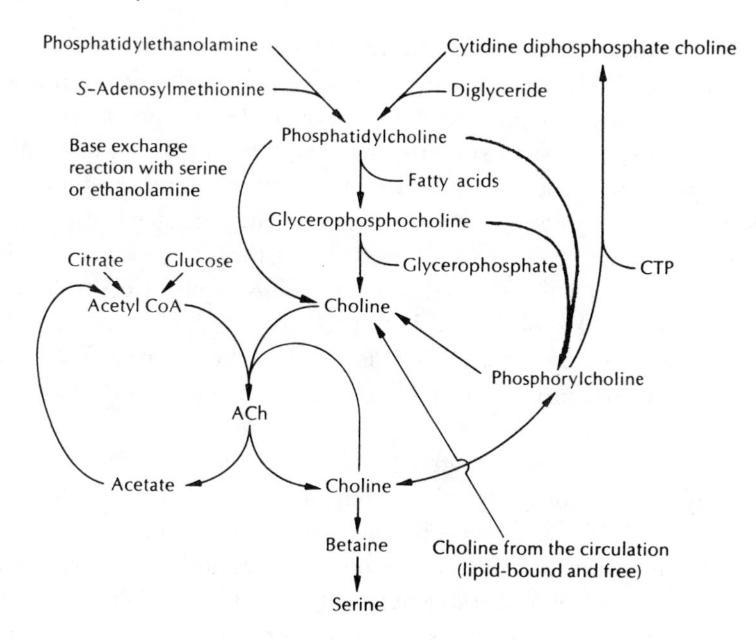

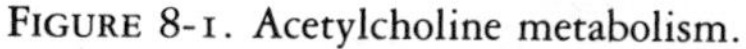
FIGURE 8-1. Acetylcholine metabolism.

SYNTHESIS

Acetylcholine is synthesized in a reaction catalyzed by choline acetyltransferase:

$$\text{Acetyl CoA} + \text{choline} \rightleftharpoons \text{ACh} + \text{CoA}$$

Before entering into a discussion of choline acetyltransferase, we should take note of Figure 8-1, which depicts the possible sources of acetyl CoA and choline. In theory, the acetyl CoA for ACh synthesis may arise from glucose, through glycolysis and the pyruvate oxidase system; from citrate, either by a reversal of the condensing enzyme (citrate synthetase) or by the citrate cleavage enzyme (citrate lyase); or from acetate through acetatethiokinase. In brain slices, homogenates, acetone powder extracts, and preparations of nerve-ending particles, glucose or citrate proved to be the best sources for

ACh synthesis, with acetate rarely showing any activity. In lobster axons, the electric organ of the *Torpedo*, corneal epithelium, and frog neuromuscular junction, acetate appears to be the preferred substrate. However, these systems are all *in vitro* and do not necessarily reflect the situation *in vivo*. Regardless of its source, acetyl CoA is primarily synthesized in mitochondria. Since, as detailed below, choline acetyltransferase appears to be in the synaptosomal cytoplasm, another still unsolved problem is how acetyl CoA is transported out of the mitochondria to participate in ACh synthesis. A probable carrier for acetyl CoA is citrate, which can diffuse into the cytosol and produce acetyl CoA via citrate lyase; a possible carrier is acetyl carnitine.

Evidence suggests that although phosphatidylethanolamine can be methylated to form phosphatidylcholine, free choline cannot be synthesized *de novo* in brain by successive methylations of ethanolamine. It is transported to the brain both free and in phospholipid form (possibly phosphatidylcholine) by the blood. Following the hydrolysis of ACh, about 35 to 50 percent of the liberated choline is transported back into the presynaptic terminal by a sodium-dependent, high-affinity active transport system, to be reutilized in ACh synthesis (see below). As outlined in Figure 8–1, the remaining choline may be catabolized or alternatively become incorporated into phospholipids, which can again serve as a source of choline. It should be noted that lesioning in the CNS or denervation in the superior cervical ganglion produces a decrease in choline content. A curious observation is that when brain cortical slices are incubated for two hours in a Krebs-Ringer medium, choline accumulates to a level about ten times its original concentration. Similarly, a rapid postmortem increase in choline has been observed. The precise source of this choline is unknown; a probable candidate is phosphatidylcholine.

Choline Transport

Recent studies of choline transport have produced a number of significant findings:

1. Choline crosses cell membranes by two processes referred to as high-affinity and low-affinity transport. High-affinity transport with a K_m for choline of 1–5 μM is saturable, presumably carrier mediated, and dependent on sodium; it is also dependent on the membrane potential of the cell or organelle so that any agent (e.g., K^+) that depolarizes the cell will concurrently inhibit high-affinity transport. Low-affinity choline transport with a K_m of 40–80 μM appears to operate by a passive diffusion process, linearly dependent on the concentration of choline, and virtually nonsaturable.
2. In contrast to the other neurotransmitters, ACh is taken up only via low-affinity transport; it is only choline that exhibits high-affinity kinetics.
3. Current evidence suggests that the high-affinity transport of choline is specific for cholinergic terminals and is not present in aminergic nerve terminals. Furthermore, it is kinetically coupled to ACh synthesis. It has been calculated that about 50 to 85 percent of the choline that is transported by the high-affinity process is utilized for ACh synthesis. Low-affinity transport, on the other hand, is found in cell bodies and in tissues such as the corneal epithelium, and it is thought to function in the synthesis of choline-containing phospholipids. On the other hand, tissues that do not synthesize ACh (e.g., fibroplasts, erythrocytes, photoreceptor cells) exhibit high-affinity choline transport which is coupled to phospholipid synthesis.
4. Hemicholinium-3 is an extremely potent inhibitor of high-affinity transport (K_m of 0.05–1 μM) but a relatively weak (K_m of 10–120 μM) inhibitor of low-affinity transport.
5. There are three obvious mechanisms for regulating the level of ACh in cells: feedback inhibition by ACh on choline acetyltransferase, mass action, and the availability of acetyl CoA and/or choline. Of these three possibilities, the major regulatory factor seems to be high-affinity choline transport. This view derives from early observations that choline is rate

limiting in the synthesis of ACh, coupled with recent findings in Kuhar's laboratory. Using the septal-hippocampal pathway, a known cholinergic tract, Kuhar and his associates have shown that changes in impulse flow induced via electrical stimulation or pentylenetetrazol administration (which increase impulse flow) or via lesioning or the administration of pentobarbital (which decrease neuronal traffic) will alter high-affinity transport of choline into hippocampal synaptosomes. In their studies, procedures that activated impulse flow increased the V_{max} of choline transport, while agents that stopped neuronal activity decreased V_{max}. In neither situation was the K_m changed, a result to be expected since the concentration of choline outside the neuron (5–10 μM) normally exceeds the K_m for transport (1–5 μM). Recent evidence, however, is accumulating to suggest that this relationship between impulse traffic and choline transport does not occur in all brain areas (e.g., in the striatum, where cholinergic interneurons abound). In addition, the endogenous concentration of ACh is also implicated in regulating the level of the transmitter in brain. Thus in several studies an increase in choline uptake following depolarization of a preparation has been attributed to the release of endogenous ACh on depolarization. Other studies, however, suggest that this increased choline uptake is not related to ACh release but rather to an increase in Na^+, K^+-ATPase activity.

Choline Acetyltransferase

With respect to the cellular localization of choline acetyltransferase, (CAT), the highest activity is found in the interpeduncular nucleus, the caudate nucleus, retina, corneal opithelium, and central spinal roots (3000–4000 μg ACh synthesized/g/hr). In contrast, dorsal spinal roots contain only trace amounts of the enzyme, as does the cerebellum.

Intracellularly, after differential centrifugation in a sucrose medium, choline acetyltransferase is found in mammalian brain predominantly in the crude mitochondrial fraction. This fraction contains mitochondria, nerve-ending particles (synaptosomes) with enclosed synaptic vesicles, and membrane fragments. When this fraction is subjected to sucrose density gradient centrifugation, the bulk of the choline acetyltransferase is found to be associated with nerve-ending particles. When these synaptosomes are ruptured by hypoosmotic shock, synaptosomal cytoplasm can be separated from synaptic vesicles. In a solution of low ionic strength, the choline acetyltransferase is absorbed to membranes and to vesicles, but in the presence of salts at physiological concentration the enzyme is solubilized and remains in the cytoplasm. *In vivo* the enzyme is most likely present in the cytoplasm of the nerve-ending particle. However, with the apparent kinetic coupling of transport and acetylation of choline, choline acetyltransferase may also either be present at terminal membranes or, as some recent evidence suggests, another form of the enzyme is present at this site. Support for this latter possibility derives from the observation that homocholine can be acetylated to form acetylhomocholine in a synaptosomal preparation but not with a purified preparation of choline acetyltransferase. Benishin and Carroll have shown that a particulate form of CAT which has been partially purified from lysed synaptosomes is able to acetylate homocholine; that this enzyme activity may be carnitine acetyltransferase (which can acetylate choline) has been ruled out. Thus it is possible that this particulate form of CAT is the physiologically relevant form of the enzyme.

A cell-free system of choline acetyltransferase was first described by Nachmansohn and Machado in 1943. Since that time, the enzyme from squid head ganglia, human placenta, *Drosophila melanogaster*, *Torpedo californica*, and brain has been purified and some of its characteristics have been defined.

As highly purified from rat brain, choline acetyltransferase has a molecular weight of 66 to 70 kiloDaltons: It has an apparent Michaelis constant (Km) for choline of 7.5×10^{-4} M and for acetyl CoA of 1.0×10^{-5} M. Currently a controversy exists on the extent of the reversibility of the reaction with recent estimates giving an equilibrium constant of 12. The enzyme is activated by chloride and is inhibited by sulfhydryl reagents. A variety of studies on the substrate specificity of the enzyme indicates that various acyl derivatives both of CoA and of ethanolamine can be utilized by the enzyme. The major gap in our knowledge of choline acetyltransferase is that as yet we do not know of any useful (i.e., potent and specific) direct inhibitor of the enzyme. Styrylpyridine derivatives inhibit the enzyme but suffer from the fact that they are light sensitive, somewhat insoluble, and possess varying degrees of anticholinesterase activity. Hemicholinium (HC-3) inhibits the synthesis of ACh, but indirectly, by preventing the transport of choline across cell membranes. Until recently, another problem with choline acetyltransferase is that it was difficult to raise specific antibodies to this enzyme. However, immunohistochemical procedures have now been sufficiently improved to permit confident interpretation of localization studies.

Acetylcholinesterase

Everybody agrees that ACh is hydrolyzed by cholinesterases, but nobody is sure just how many cholinesterases exist in the body. All cholinesterases will hydrolyze not only ACh but other esters. Conversely, hydrolytic enzymes such as arylesterases, trypsin, and chymotrypsin will not hydrolyze choline esters. The problem in deciding the number of cholinesterases that exist is that different species and organs sometimes exhibit maximal activity with different substrates. For our purposes we will divide the enzymes into two rigidly defined classes: acetylcholinesterase (also called "true" or specific cholinesterase) and butyrylcholinesterase (also called "pseudo" or nonspecific cholinesterase; the term "propionylcholinesterase" is sometimes used since in some tissues propionylcholine is hydrolyzed more rapidly than butyrylcholine). Although their molecular forms are similar, the two enzymes are distinct entities, probably encoded by specific genes. When distinguishing between the two types of cholinesterases, at least two critera should be used because of the aforementioned species or organ variation.

The first criterion is the optimum substrate. Acetylcholinesterase hydrolyzes ACh faster than butyrylcholine, propionylcholine, or tributyrin; the reverse is true with butyrylcholinesterase. In addition, acetyl-β-methyl choline (methachol) is only split by acetylcholinesterase. That this criterion for distinguishing between the two esterases is not inviolate and must be used along with other indices is illustrated by the fact that the chicken brain acetylcholinesterase will hydrolyze acetyl-β-methyl choline but will also hydrolyze propionylcholine faster than ACh. Also the beehead enzyme will not hydrolyze either ACh or butyrylcholine but will split acetyl-β-methyl choline.

A second criterion that is used in differentiating the cholinesterases is the substrate concentration versus activity relationship. Acetylcholinesterase is inhibited by high concentrations of ACh so that a bell-shaped substrate concentration curve results. This is observed also when butyrylcholine or propionylcholine is used. In contrast, butyrylcholinesterase is not inhibited by high substrate concentrations so that the usual Michaelis-Menten type of substrate concentration curve is obtained. The reason for this difference is that in acetylcholinesterase there is at least a two-point attachment of substrate to enzyme whereas with butyrylcholinesterase the substrate is attached at only one site.

The type of cholinesterase found in a tissue is often a reflection of the tissue. This fact is used as a discriminating index between cholinesterases. In general, neural tissue contains acetylcholinesterase while nonneural tissue usually contains butyrylcholinesterase. However, this is a generalization, and some neural tissue (e.g., autonomic ganglia) contains both esterases as do some extraneural organs (e.g., liver, lung). In the blood, erythrocytes contain only acetylcholinesterase while plasma contains butyrylcholinesterase, but plasma has primary substrates varying from species to species. Because of its ubiquity, cholinesterase activity cannot be used as the sole indicator of a cholinergic system in the absence of additional supporting evidence. To generalize on this point, until neuron-specific transmitter degradating enzymes are discovered, it is a neurochemical commandment that in order to delineate a neuronal tract, one must always assay an enzyme involved in the synthesis of a neurotransmitter and not one concerned with catabolism.

A final criterion that may be applied to differentiate between the esterases is their susceptibility to inhibitors. Thus the organophosphorus anticholinesterases such as diisopropyl phosphorofluoridate (DFP) are more potent inhibitors of butyrylcholinesterase whereas WIN8077 (Ambinonium) is about two thousand times better an inhibitor of acetylcholinesterase. The compound BW284C51 (1,5- bis-(4-allyldimethylammoniumphenyl) penta-3-one dibromide) is presumed to be a specific reversible inhibitor of acetylcholinesterase.

In discussing the various techniques that are used to classify the cholinesterases, we touched on some aspects of the molecular properties of the enzymes. Because very little work has been done on butyrylcholinesterase and because no physiological role for this enzyme (or enzymes) has been demonstrated, we will focus our attention on acetylcholinesterase. In sucrose homogenates of mammalian brain, subjected to differential centrifugation, acetylcholinesterase is found both in the mitochondrial and microsomal fractions. The latter, consisting of endoplasmic reticulum and plasma cell membranes, exhibits a higher specific activity. This localization of the enzyme is supported by electron microscopic and histochemical studies that fix the activity at membranes of all kinds both in the CNS and the peripheral nervous system.

Acetylcholinesterase, purified to homogeneity from *Electrophorus electricus*, appears to be composed of three tetrameric globular catalytic units attached to a filamentous, collagenlike tail (Fig. 8–2). Since the tail has a composition similar to basement membrane, it is thought to play a structural role in the localization of the enzyme. With its turnover time of 150 microseconds, equivalent to hydrolyzing 5000 molecules of ACh per molecule of enzyme per second, acetylcholinesterase ranks as one of the most efficient enzymes extant.

With respect to the topography of the enzyme, the twin-hatted diagram of the anionic and esteratic sites has been reproduced countless times and need not be presented again here. However, some discussion is in order since this was the first enzyme to be

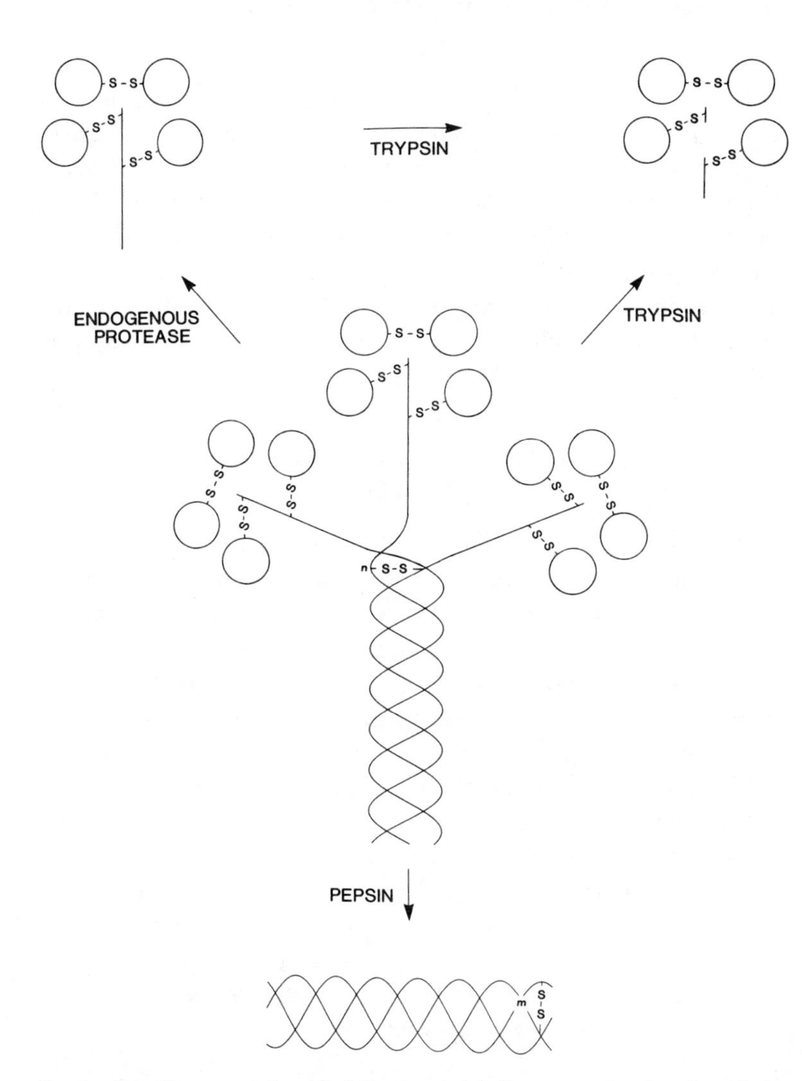

FIGURE 8-2. Representation of 18 S eel acetylcholinesterase showing degradation patterns after protease treatment. The molecule is depicted as possessing 12 catalytic subunits (circles), arranged as 3 tetrameric groups. Within each tetramer, 2 catalytic subunits are disulfide linked and the remaining 2 are covalently linked by a single disulfide bond to one tail subunit. Pepsin degrades the catalytic subunits and the noncollagenlike domains of the tail subunits to give triple-helical tail subunit fragments. An endogenous protease can cleave the tail subunits at a

dissected at a molecular level. For this initiation into molecular biology we owe a debt of gratitude to Nachmansohn and his colleagues, particularly Wilson. The active center of acetylcholinesterase has two main subsites. The first is an anionic site that attracts the positive charge in ACh and the second, about 5 Å distant, is an esteratic site that binds the carbonyl carbon atom of ACh. Current information suggests that the anionic site contains at least one carboxyl group, possibly from glutamate, and the esteratic site involves a histidine residue adjacent to serine. The overall reaction is written as follows:

$$E + ACh \longleftrightarrow E{\cdot}ACh \longrightarrow \underset{+\text{choline}}{E{\cdot}\text{Acetyl}} \xrightarrow{H_2O} E + \text{acetic acid}$$

Information on the architecture of the active center has been derived not only from kinetic studies using model compounds but from a group of inhibitors known as the anticholinesterases. (The pharmacology of these agents will be discussed in the last section of the chapter.) The anticholinesterases are classified as reversible and irreversible inhibitors of the enzyme. Like ACh, both types of inhibitor acylate the enzyme at the esteratic site. However, in contrast to ACh or to a reversible inhibitor such as physostigmine, the irreversible inhibitors, which are organophosphorus compounds, irreversibly phosphorylate the esteratic site. This phosphorylation has been shown to occur on the hydroxyl group of serine when DFP was incubated with purified acetylcholinesterase. Although the organophosphorus agents (referred to as nerve gases though they are actually oils) are classed as irreversible anticholinesterases, there is a slow detachment of the compounds from the enzyme. Wilson ob-

---

point that releases an 11 S enzyme tetramer which still retains an 8000 molecular weight residual tail subunit. Exposure of either this 11 S form or the 18 S form to trypsin generates an 11 S tetramer in which the residual tail subunit is cleaved between its disulfide linkages to the two catalytic subunits. (After Rosenberry *et al.*, *Neurochem.. Int.* 2, 135, 1980) served that hydroxylamine speeded up this dissociation and regenerated active enzyme. He then set about designing a nucleophilic agent with a spatial structure that would fit the active center of acetylcholinesterase, and he did in fact produce a compound that is very active in displacing the inhibitor. This is 2-pyridine aldoxime methiodide (PAM), which has been used with moderate success in treating poisoning from organophosphorus compounds used as insecticides. PAM, with its quaternary ammonium group, does not penetrate the blood–brain barrier well enough to overcome central actions of the anticholinesterase. For this reason, atropine is usually used as an antidote along with PAM. It will be recalled that atropine blocks the effect of ACh at neuroeffector sites and has nothing to do with acetylcholinesterase. Currently, the oxime of choice in dephosphorylating organophosphorous-inactivated cholinesterase is HI-6 (1 [2-(hydroxyimino) methyl pyridinium]-2-(4-carboxyamidopyridinium) dimethyl ether dichloride.

Although the hydrolysis of Substance P by purified serum cholinesterase has been described, evidence now suggests that this activity is due to contamination by dipeptidyl peptidase.

The Genesis of the Cholinergic Triad in Neurons

It has already been pointed out that choline acetyltransferase is found in the cytoplasm of nerve-ending particles, that acetylcholinesterase is associated with cell membranes of all kinds (synaptosomal, axonal, and glial), and that ACh is localized to some extent in synaptic vesicles. It is now pertinent to inquire how these three components of the cholinergic system arrived at their respective residences.

The phenomenon of axoplasmic transport has been described in Chapter 2. This is the process by which material synthesized in the cell body moves down axons by mechanisms as yet not understood. In the case of choline acetyltransferase a variety of investigators have by ligation and sectioning of peripheral nerves shown that this enzyme originates in the perikaryon and travels down to the nerve terminal. With acetylcholinesterase the situation is still not clear.

Again, by sectioning experiments and by the use of irreversible anticholinesterases, it has been demonstrated that the enzyme is formed in the cell body and then moves down an axon, possibly attached to neurotubules as carriers. On the other hand, local synthesis of the enzyme in axons has been reported. The latter results are slightly tainted by Schwann cell contamination of the neuronal preparation so that it cannot be stated with certainty whether the enzyme is synthesized by neuronal elements alone. Nevertheless, regardless of the possible contribution of satellite cells to axonal synthesis of acetylcholinesterase, it is clear that the enzyme originates primarily in the neuronal cell body.

Since choline acetyltransferase is found both in cell bodies and terminals, it is understandable that ACh is localized in these two regions and that some transmitter appears within cholinergic axons. At the nerve terminal, about half the ACh is found in synaptic vesicles and half in the synaptosomal cytoplasm. However, it should be noted that this "free" ACh pool in the cytosol, as determined by hypotonic lysis of synaptosomes, may be artifactual. It is entirely conceivable that this cytosolic pool may be ACh that is loosely bound to the terminal membrane or to labile synaptic vesicles. In the cholinergic vesicles of the bovine superior cervical ganglion, the concentration of ACh/vesicle is about 1600 molecules; in the *Torpedo* vesicular population, it is about 66,000 molecules/vesicle.

How synaptic vesicles are formed and why they cluster at this location are not understood (see Chapter 2).

Uptake, Synthesis, and Release of ACh

Superior Cervical Ganglion, Brain, and Skeletal Muscle

To date the only major and thorough study of ACh turnover in nervous tissue was done originally by MacIntosh, Birks, and colleagues, and more recently by Collier using the superior cervical ganglion of the cat. By using one ganglion to assay the resting level of ACh and by perfusing the contralateral organ, these investigators have determined the amount of transmitter synthesized and released under a variety of experimental conditions including electrical stimulation, the addition of an anticholinesterase to the perfusion fluid, and perfusion media of varying ionic composition. Their results may be summarized as follows:

1. During stimulation, ACh turns over at a rate of 8 to 10 percent of its resting content every minute (i.e., about 24 to 30 ng/min). At rest, the turnover rate is about 0.5 ng/min. Since there is no change in the ACh content of the ganglion during stimulation at physiological frequencies, it is evident that electrical stimulation not only releases the transmitter but also stimulates its synthesis.
2. Choline is the rate-limiting factor in the synthesis of ACh.
3. In the perfused ganglion, Na^+ is necessary for optimum synthesis and storage, and Ca^{2+} is necessary for the release of the neurotransmitter.
4. Newly synthesized ACh appears to be more readily released on nerve stimulation than depot or stored ACh.
5. About half of the choline produced by cholinesterase activity is reutilized to make new ACh.
6. At least three separate stores of ACh in the ganglion are inferred from these studies: "surplus" ACh, considered to be intracellular, which accumulates only in an eserine-treated ganglion and which is not released by nerve stimulation but is released by K^+ depolarization; "depot" ACh, which is released by nerve impulses and accounts for about 85 percent of the original store; and "stationary" ACh, which constitutes the remaining 15 percent that is nonreleasable.
7. Choline analogs such as triethylcholine, homocholine, and pyrrolcholine are released by nerve stimulation only after they are acetylated in the ganglia.
8. Increasing the choline supply in the plasma during perfusion of the ganglion only transiently increased the amount of ACh that was releasable with electrical stimulation, despite an accumulation of the transmitter in the ganglion.
9. The compound AH5183 (2-(4-phenylpiperidino) cyclohex-

anol), shown by the Parsons' lab to inhibit ACh transport into synaptic vesicles, ultimately blocks release of ACh from the stimulated ganglia. This finding supports the contention that there is vesicular release of the transmitter in the periphery.

As stated above, this work on the superior cervical ganglion represents the most complete information on the turnover of ACh in the nervous system. Using a constant infusion of labeled phosphoryl choline into the rat tail vein, Costa and co-workers have developed a procedure for determining the turnover rate of ACh in whole brain, various brain areas, and lately in specific nuclei. These studies have shown that the turnover of ACh is not uniform throughout the brain. With respect to regulation of ACh turnover in cholinergic terminals of skeletal muscle, Vaca and Pilar using the chick iris have elaborated on the work of Potter with the rat phrenic nerve-diaphragm preparation. With the iris and the diaphragm preparation, the same relationship of high-affinity choline uptake, ACh synthesis, and regulation by endogenous ACh has been demonstrated as previously described with CNS and autonomic nervous system preparations.

Brain Slices, Nerve-Ending Particles (Synaptosomes), and Synaptic Vesicles

In 1939 Mann, Tennenbaum, and Quastel demonstrated the synthesis and release of ACh in cerebral cortical slices. In the succeeding forty-seven years these observations have been repeatedly confirmed but only moderately extended. The major finding of interest in all these studies is that in the usual incubation medium the level of ACh in the slices reaches a limit and cannot be raised. In a high K^+ medium the total ACh is increased substantially because much of it leaks into the medium from the slices. The experiments again suggest that the intracellular concentration of the neurotransmitter may play a role in regulating its rate of synthesis, in addition to the high-affinity uptake system for choline. This concept of a feedback mechanism is supported by the findings that the administration of drugs such as morphine, oxotremorine, or anticholinesterases only succeed in, at the most, a doubling of the original level of ACh in the brain. Regardless of the dose of the drug, no higher level can be obtained.

Much of the current neurochemical work on the release of ACh from brain involves the use of nerve-ending particles (synaptosomes). This preparation, independently developed by DeRobertis and by Whittaker, is derived from a sucrose density gradient centrifugation of a crude mitochondrial fraction of brain. Although synaptosomes represent presynaptic terminals with enclosed vesicles and mitochondria, some postsynaptic fragments are often attached to them. A disadvantage of the preparation is its heterogeneity; the usual synaptosome fraction is a mixture of cholinergic, noradrenergic, serotonergic, and other terminals. In addition, as judged by electron microscopy and enzyme markers, the purity is around 60 percent; the contaminants may include glial cells, ribosomes, and membrane fragments that may be axonal, mitochondrial, or perikaryal. On the other hand, the advantage of these preparations is that they can be isolated easily and that synaptic vesicles can be collected by hypoosmotically shocking the synaptosomes. With respect to the disposition of ACh in synaptosomes, roughly half of the transmitter is found in vesicles and the other half in synaptosomal cytoplasm.

Following the discovery of these presynaptically localized vesicles that contained ACh, the conclusion was almost unavoidable that these organelles were the source of the quantal release of transmitter as described in the neurophysiological experiments of Katz and his collaborators. Thus, the obvious interpretation has been that as the nerve is depolarized, vesicles in apposition to the terminal fuse with the presynaptic membrane and ACh is released into the synaptic cleft to interact with receptors on the postsynaptic cell to change ion permeability. The subsequent sequence of events is not clearly pictured but in some fashion presynaptic membrane is pinocytotically recaptured, vesicles are resynthesized, and simultaneously or subsequently repleted with ACh; the neuron is then ready for the next quantal release of transmitter.

Although the exocytotic release of hormones is well established, the release of neurotransmitters from the CNS via this mechanism has been questioned by several investigators (see reviews by Cooper and Meyer, Dunant and Israel, MacIntosh, and Silensky). Clearly, neither purely morphological nor neurochemical experimentation can unequivocally answer this question with current techniques; what is needed is an innovative approach that will satisfy the objections of the proponents of vesicular release and of gated channel release.

Cholinergic Pathways

The identification of cholinergic synapses in the peripheral nervous system has been relatively easy, and we have known for a long time now that ACh is the transmitter at autonomic ganglia, at parasympathetic postganglionic synapses, and at the neuromuscular junction. In the CNS, however, until the past few years, technical difficulties have limited our knowledge of cholinergic tracts to the motoneuron collaterals to Renshaw cells in the spinal cord. With respect to the aforementioned technical difficulties, the traditional approach has been to lesion a suspected tract and then assay for ACh, choline acetyltransferase, or high-affinity choline uptake at the presumed terminal area. Problems with lesioning include the difficulty in making discrete, well-defined lesions and in interrupting fibers of passage. This latter problem is illustrated by the discovery that a habenula-interpeduncular nucleus projection which, based on lesioning of the habenula, was always quoted as a cholinergic pathway is not: it turned out that what was lesioned were cholinergic fibers which passed through the habenula. Thus, although the interpeduncular nucleus has the highest choline uptake and choline acetyltransferase activity of any area in the brain, the origin of this innervation remains largely unknown. A quantum leap in technology for tracing tracts in the CNS has occurred in the past few years. Through the use of histochemical techniques originally developed

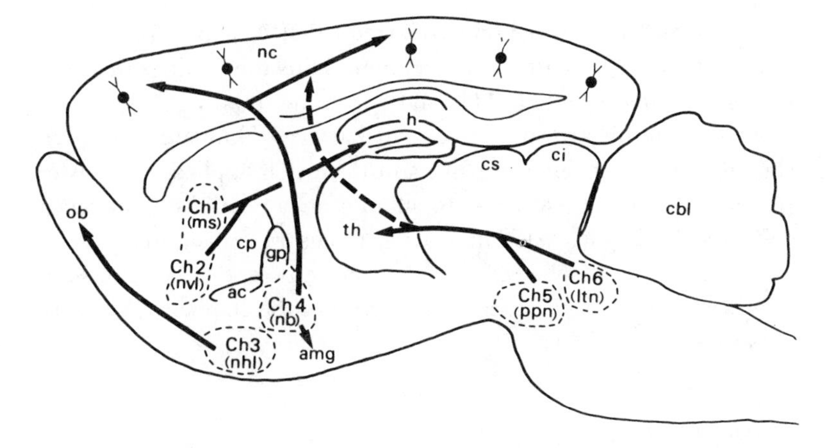

FIGURE 8-3. Schematic representation of some ascending cholinergic pathways. The traditional nuclear groups which most closely correspond to the Ch subdivisions are indicated in parentheses. Abbreviations: ac, anterior commissure; amg, amygdala; cbl, cerebellum; ci, inferior colliculus; cp, caudate-putamen; cs, superior colliculus; gp, globus pallidus; h. hippocampus; ltn, laterodorsal tegmental nucleus; ms, medial septal nucleus; nb. nucleus basalis; nc, neocortex; nhl, nucleus of the horizontal limb complex; nvl, nucleus of the vertical limb complex; ob, olfactory bulb; ppn, pedunculopontine tegmental nucleus; th, thalamus. (From Wainer *et al.*, (1984)

by Koelle and co-workers that stain for regenerated acetylcholinesterase after DFP treatment (Butcher, Fibiger), autoradiography with muscarinic receptor antagonists (Rotter, Kuhar), and immunohistochemical procedures with antibodies to choline acetyltransferase (McGeer, Salvaterra, Cuello, Wainer), a clear picture of cholinergic tracts in the CNS is now emerging. The well-documented tracts are depicted in Figure 8–3. There is additional information at hand that in the striatum and the nucleus accumbens septi, only cholinergic interneurons are found. Also, intrinsic cholinergic neurons recently have been reported to exist in the cerbral cortex, colocalized with VIP (vasoactive intestinal polypeptide) and often in close proximity to blood vessels.

With respect to other neurotransmitter functions of ACh, there are a few bits of information that suggest that ACh may participate in circuits involved with pain reception. Thus the findings that nettles *(Urtica urens)* contain ACh and histamine, that high concentrations of ACh injected into the brachial artery of humans have been shown to result in intense pain, and that ACh applied to a blister produced a brief but severe pain, all indicate a relationship between ACh and pain. That ACh may act as a sensory transmitter in thermal receptors, taste fiber endings, and chemoreceptors has also been suggested, based on the excitatory activity of the compound on these sensory nerve endings.

Cellular Effects

A variety of actions of ACh that may be viewed as cellular effects rather than neurotransmitter activity have been described. These include ciliary movement in the gill plates of Mytilis edulis, ciliary motility of mammalian respiratory and esophageal tracts, a hyperpolarizing effect on atrial muscle, limb regeneration in salamanders, protein production in the silk gland of spiders, the induction of sporulation in the fungus *Trichoderma*, protoplasmic streaming in slime molds, and photic control of circadian rhythms and seasonal reproductive cycles.

These situations describe an activity of ACh. There is also a variety of tissues and organisms, such as human placenta, *Lactobacillus plantarum*, *Trypanosoma rhodesience*, and a fungus *(Claviceps purpurea)* in which ACh is found but where nothing is known of its action. One of the most interesting situations is the corneal epithelium, which contains the highest concentrations of ACh of any tissue in the body and yet contains neither nicotinic nor muscarinic receptors. Some evidence suggests that ACh may be involved in sodium transport in this tissue.

All the activities of ACh and its localization in nonnervous tissue which we have noted above suggest that this agent may be a hormone as well as a neurotransmitter. It is worthwhile for the student to keep in mind that all known neurotransmitters may possess this dual function. These two activities have already been shown to occur with the biogenic amines. Even when a neurotransmitter is found in nervous tissue, its action may satisfy the criteria for defining a modulator rather than the currently strict criteria for a "classical" neurotransmitter.

Cholinergic Receptors

As noted in Chapter 5, cholinergic receptors fall into two classes, muscarinic and nicotinic. Based primarily on ligand affinities, and secondarily on tissue localization, muscarinic receptors have been classified as either M_1 of M_2 types. Whether these are distinct molecular entities or whether they represent interconvertible forms depending on the ionic milieu is still unclear. A functional discrimination in enteric neurons has been described by North and colleagues where the activation of an M_1 receptor decreases K conductance while activation of the M_2 receptor (thought to be presynaptic) increases K conductance. The muscarinic receptor from porcine atria has recently been purified to homogeneity and found to have a molecular weight of 78000 daltons; it exhibits both a high- and low-affinity site with carbamoylcholine as the agonist. Kuhar, and other laboratories have now succeeded in quantitatively mapping muscarinic receptors in the brain via autoradiography using radioactive muscarinic antagonists.

Unfortunately, the task of identifying nicotinic cholinergic receptors in the brain is fraught with uncertainties. Using labeled α-bungarotoxin, nicotine, mecamylamine, or dihydro-β-erythroidine, each investigation has yielded mystifying results in which the antagonist cannot be easily displaced by ACh or by unlabeled ganglionic or neuromuscular antagonists but on occasion is displaced by muscarinic agonists and antagonists. Among possible interpretations of these anomalies, the labeled probes may be binding to different recognition sites or different subunits of the nicotinic cholinergic receptor or they may be binding to nicotinic receptors but not to nicotinic cholinergic receptors. It is clear that an intensive electrophysiologic/pharmacological approach is needed to map these curious nonmuscarinic cholinergic synapses.

Conversely, a wealth of information is at hand on the properties of the nicotinic cholinergic receptor of *Torpedo* and *Electrophorus* electric organs. This reflects the abundance of the receptor in this tissue and the availability of two snake toxins, α-bungarotoxin and *naja naja siamensis*, that specifically bind to the receptor and have facilitated its isolation and purification.

As currently envisaged, the nicotinic ACh receptor is a pentameric integral membrane protein composed of four glycosylated polypeptide chains designated alpha, beta, gammma, and delta with a stoichiometry of two alpha subunits to one each of the other three. All subunits traverse the membrane and, when viewed face on, resemble a five-petal rosette with a central pit. ACh binds to the alpha subunits and produces a conformational change in the channel that selectively allows cations rather than anions with a diameter of about 0.65 nm to pass through. The molecular weight of the receptor complex is about 250,000. Although the receptor can be phosphorylated by a cAMP-dependent protein kinase, the physiological effect of phosphorylation is not known.

An unexpected windfall resulting from the isolation of the nicotinic cholinergic receptor has been the demonstration that when this lipoprotein as isolated from *Electrophorus* is injected into rabbits, all the signs of myasthenia gravis appear. This finding, coupled with autoradiographic studies of muscle biopsies of myasthenics using labeled α-bungarotoxin as a tag where a marked deficiency of receptors is observed, has led to a better understanding of the disease at a molecular level. It is now clear that myasthenia gravis is an autoimmune disease in which a circulating antibody appears to be involved in an increased rate of degradation and damage, as well as antagonism of the ACh receptor. Since antibodies produced from *Electrophorus* or *Torpedo* ACh receptor result in myasthenic signs in rabbits, guinea pigs, and monkeys, this work also carries the implication that the ACh receptor is a phylogenetically conserved protein which can exhibit immunological cross-reactivity. For a complete account of the physicochemical properties of the ACh receptor and the kinetics of channel gating, the review by Aldrich *et al.* is recommended.

Drugs That Affect Central Cholinergic Systems

Many of the cholinomimetics, cholinolytics, neuromuscular blocking drugs, and anticholinesterases with which the student is familiar do not affect the central nervous system because of their inability to cross the blood–brain barrier at a significant rate. However, atropine, scopolamine, physostigmine (eserine), and DFP (diisopropyl phosphorofluoridate) all penetrate the central nervous system, and their main effects have been studied in some detail.

The systemic injection of atropine and scopolamine leads to a decrease in the ACh content of the brain. This reduction is confined to the cerebral hemispheres. At the same time it has been shown, using the cortical cup technique, that the administration of these drugs either intravenously, intraventricularly, or via the cup results in the release of ACh from the exposed cortex into the collecting cup. This releasing effect of ACh by the antimuscarinic agents has also been observed in brain cortical slices. Conversely, muscarinic agents such as oxotremorine inhibit the release of ACh from hippocampal as well as cortical slices. This inhibition is antagonized by atropine. The mechanism by which the antimuscarinic agents lower the ACh content of brain and stimulate the release is currently interpreted as a reflection of presynaptic, muscarinic receptors that are subjected to feedback inhibition by ACh. Thus, it is argued that atropine, by blocking these receptors, enhances the release of the transmitter. The amnesic effect of atropine and scopolamine does not correlate with the lowered levels of ACh in the brain since the intraventricular injection of hemicholinum, which markedly reduces the ACh concentration in brain, has no effect on maze performance by trained rats.

The administration of oxotremorine and arecoline to rats causes a rise in the ACh content in brain which roughly coincides with the tremor period. This is presumed to be a muscarinic action of these agents since the tremor can be aborted by atropine.

A number of years ago it was observed that the long-term administration of DFP to myasthenics produced nightmares, confusion, and hallucinations. Since then, several studies using different organophosphorus anticholinesterases have shown behavioral effects in humans following the administration of these agents. These effects include anxiety, hostility, depression, and psychomotor retardation, all of which can be antagonized by atropine. Since physostigmine relieves manic symptoms and causes depressed patients to become more depressed, and since some anticholinergic drugs are useful in treating depression, some investigators have attempted to link cholinergic drugs with affective disorders and with stress. The results of these studies have been conflicting, indicating that the cholinergic system is but one of several systems involved.

ACh in Disease States

Aside from myasthenia gravis, the role of ACh in nervous system dysfunction is unclear. Certainly a strong case can be made for familial dysautonomia, an autosomal recessive condition affecting Ashkenazi Jews that is in fact diagnosed by a supersensitivity of the iris to methacholine. Less clear are convulsive disorders where, experimentally, seizures can be induced by either applying ACh topically to the exposed cortex or by the administration of DFP but where, clinically, cholinergic drugs have no therapeutic value. Huntington's chorea, involving a degeneration of Golgi Type-2 interneurons in the striatum, is reported to be partially ameliorated by physostigmine. The administration of physostigmine to patients with tardive dyskinesia has produced mixed results.

In Alzheimer's disease, characterized behaviorally by a severe impairment in cognitive function and neuropathologically by the appearance of neuritic plaques and neurofibrillary tangles, a cholinergic dysfunction has been implicated. This is based on the findings that: (1) the administration of centrally acting muscarinic blocking agents to normal individuals induces a loss of recent memory, (2) in the cerebral cortex and hippocampus of patients with Alzheimer's disease there is a dramatic reduction of ACh, choline acetyltransferase, and high-affinity choline uptake, (3) in Alzheimer

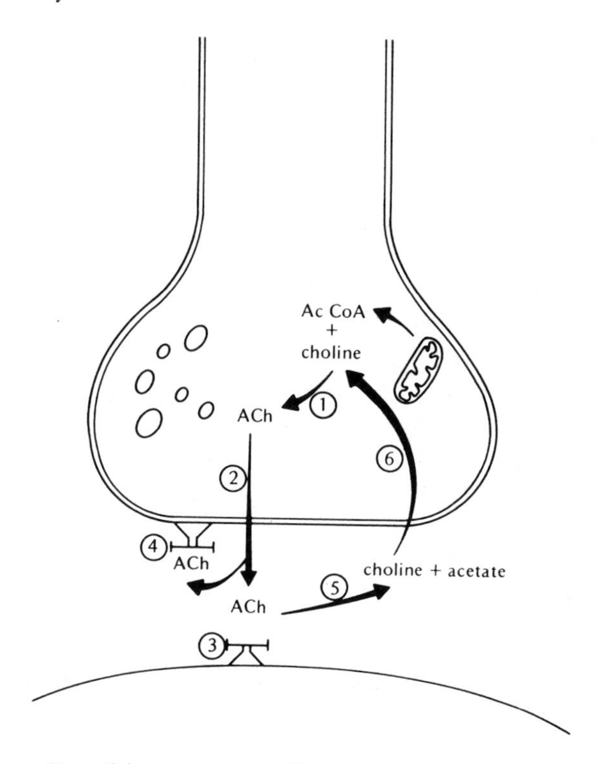

FIGURE 8-4. Sites of drug action at cholinergic synapses.

Site 1: ACh synthesis can be blocked by styryl pyridine derivatives such as NVP.

Site 2: Release is promoted by β-bungarotoxin, black-widow spider venom, and La^{3+}. Release is blocked by botulinum toxin, cytochalasin B, collagenase pretreatment, and Mg^{2+}.

Site 3: Postsynaptic receptors are activated by cholinometic drugs and anticholinesterases. Nicotinic receptors, at least in the peripheral nervous system, are blocked by α-bungarotoxin, rabies virus, curare, hexamethonium, or dihydro-β-erythroidine; muscarinic receptors are blocked by atropine, pirenzepine, and quinuclidinyl benzilate.

Site 4: Presynaptic muscarinic receptors may be blocked by atropine or quinuclidinyl benzilate. Muscarinic agonists (e.g., oxotremorine), will inhibit the evoked release of ACh by acting on these receptors.

Site 5: Acetylcholinesterase is inhibited reversibly by physostigmine (eserine) or irreversibly by DFP, or Soman.

Site 6: Choline uptake competitive blockers include hemicholinium-3, troxypyrrolium tosylate, or triethylcholine.

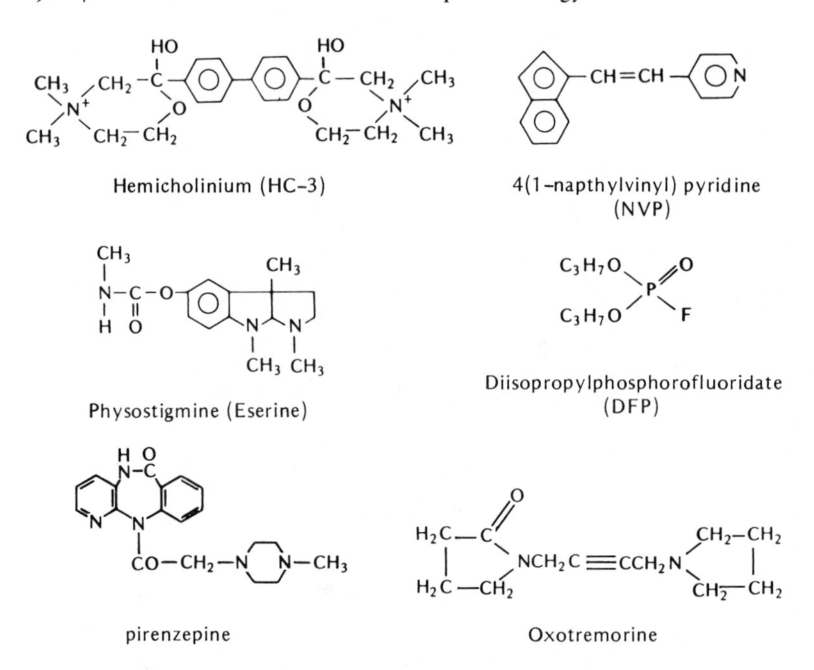

FIGURE 8-5. Structures of some drugs that affect the cholinergic nervous system.

patients there is a severe reduction of neurons in the nucleus basalis of Meynert, the primary cholinergic input to the cortex, and (4) in some, but not all, studies, a decrease in muscarinic receptors has been noted. It should also be noted, however, that patients with Alzheimer's disease have decreased levels of somatostatin as well as reduced numbers of locus coeruleus neurons, and that a decrease in neuronal content in the nucleus basalis of Meynert has been observed in some patients with Down's syndrome (trisomy 21), and with Parkinson's disease. Thus the marriage of Alzheimer's disease to a cholinergic dysfunction may involve some extramarital relationships. At any rate, little success has been achieved to date by treating patients with choline, lecithin, phsysostigmine, or the muscarinic agonist arecoline. Perhaps future therapy will require the development of a specific muscarinic agonist in combination with an adrenergic agonist or the transplantation of a peripheral cholinergic neuron to the brain with the hope that appropriate synaptic contacts can be restored.

Sites of drugs that affect the synthesis, release, and neurotransmitter activity of ACh are shown in Figure 8-4. Structures of drugs that affect the cholinergic nervous system are shown in Figure 8-5.

Selected References

Aldrich, R. W., V. E. Dionne, E. Hawrot, and C. F. Stevens (1985). Ion transport through ligand-gated channels. In *Physiology of Membrane Disorders* (T. E. Andreoli, J. F. Hoffman, D. D. Farestil, and S. G. Schultz, eds.). Plenum, New York.

Anderson, D. C., S. C. King, and S. M. Parsons (1983). Pharmacological characterization of the acetylcholine transport system in purified *Torpedo* electric organ synaptic vesicles. *Molec. Pharmacol. 24*, 48.

Benishin, C. G., and P. T. Carroll (1983). Multiple forms of choline-O-acetyltransferase in mouse and rat brain: Solubilization and characterization. *J. Neurochem. 41*, 1016.

Butcher, L. L. and N. J. Woolf (1984). Histochemical distribution of acetylcholinesterase in the central nervous system: Clues to the localization of cholinergic neurons. In *Handbook of Chemical Neuroanatomy*, Vol. 3, Part 2 (A. Björklund, T. Hökfelt, and M. J. Kuhar, eds.), p. 1. Elsevier, Amsterdam.

Cheney, D. L., and E. Costa (1977). Pharmacological implications of brain acetylcholine turnover measurements in rat brain nuclei. *Ann. Rev. Pharmacol. Toxicol. 17*, 369.

Cooper, J. R., and E. M. Meyer (1984). Possible mechanisms involved in the release and modulation of release of neuroactive agents. *Neurochem. Int. 6*, 419.

Conti-Tronconi, B. M., and M. A. Raftery (1982). The nicotinic acetylcholine receptor: Molecular structure and correlation with functional properties. *Ann. Rev. Biochem. 51*, 491.

Cuello, A. C., and M. V. Sofroniew (1984). The anatomy of the CNS cholinergic neurons. *Trends in Neurosci. 7*, 74.

Dolly, J. O., and E. A. Barnard (1984). Nicotinic acetylcholine receptors: An overview. *Biochem. Pharmacol. 33*, 841.

Dunant, Y., and M. Israel (1985). The release of acetylcholine. *Sci. Am. 252*, 58.

Fibiger, H. C. (1982). Organization and some projections of cholinergic neurons in the mammalian forebrain. *Brain Res. Rev. 4*, 327.

Hanin, I., and D. J. Jenden (1969). Estimation of choline esters in brain by a new gas chromatographic procedure. *Biochem. Pharmacol. 19*, 837.

Houser, C. R., and G. D. Crawford, R. P. Barber, P. Salvaterra, and J. E. Vaughn (1983). Organization and morphological characteristics of cholinergic neurons: An immunohistochemical study with monoclonal antibody to choline acetyltransferase. *Brain Res. 266*, 97.

Israel, M., and B. Lesbats (1981). Chemiluminescent determination of acetylcholine and continuous detection of its release from torpedo electric organ synapses and synaptosomes. *Neurochem. Int. 3*, 81.

Janowsky, D. S., and Risch, S. C. (1984). Cholinomimetic and anticholinergic drugs used to investigate an acetylcholine hypothesis of affective disorders and stress. *Drug Develop. Res. 4*, 125.

Kawashima, K., H. Ishikawa, and M. Mochizuki (1980). Radioimmunoassay for acetylcholine in the rat brain. *J. Pharmacol. Methods 3*, 115.

Kuhar, M. J. (1981). Autoradiographic localization of drug and neurotransmitter receptors in the brain. *Trends in Neurosci. 4*, 10.

MacIntosh, F. C. (1980). The role of vesicles in cholinergic systems. In *Proceedings of the Third Meeting of the European Society for Neurochemistry* (M. Brzin, D. Stut, and H. Bachelard, eds.), p. 11. Pergamon Press, Oxford.

MacIntosh, F. C., and B. Collier (1976). Neurochemistry of cholinergic nerve terminals. *Handbook of Experimental Pharmacology. Neuromuscular Junction*, Vol. 42, p. 991. Springer-Verlag, New York.

Massoulié, J., and S. Bon (1982). The molecular forms of cholinesterase and acetylcholinesterase in vertebrates. *Ann. Rev. Neurosci. 5*, 57.

McCaman, R. E., and J. Stetzler (1977). Radiochemical assay for ACh: Modifications for sub-picomole measurements. *J. Neurochem. 28*, 669.

McGeer, P. L., E. G. McGeer, and J. H. Peng (1984). Choline acetyltransferase: Purification and immunohistochemical localization. *Life Sci. 34*, 2319.

Meyer, E. M., and J. R. Cooper (1981). Correlations between Na^+-K^+ATPase activity and acetylcholine release in rat cortical synaptosomes. *J. Neurochem. 36*, 467.

Nonaka, R., and T. Moroji (1984). Quantitative autoradiography of muscarinic cholinergic receptors in the rat brain. *Brain Res. 296*, 295.

North, R. A., B. E. Slack, and A. Surprenant (1985). Muscarinic M_1 and M_2 receptors mediate depolarization and presynaptic inhibition in guinea enteric nervous system. *J. Phyiol. 368*, 435.

Olsen, J. S., and A. H. Neufeld (1979). The rabbit cornea lacks cholinergic receptors. *Invest. Opthalmol. 18*, 1216.

Peper, K., R. J. Bradley, and F. Dreyer (1982). The acetylcholine receptor at the neuromuscular junction. *Physiol. Rev. 62*, 1271.

Peterson, G.F L., G. S. Herron, M. Yamaki, D. S. Fullerton, and M. I. Schimerlik (1984). Purification of the muscarinic acetylcholine receptor from porcine atria. *Proc. Nat. Acad. Sci. 81*, 4993.

Potter, L. T. (1970). Synthesis, storage and release of [^{14}C]acetylcholine in isolated rat diaphragm muscles. *J. Physiol. 206*, 145.

Potter, P. E., M. Hadjiconstantinou, J. L. Meek, and N. H. Neff (1984). Measurement of acetylcholine turnover rate in brain: An adjunct to a simple HPLC method for choline and acetylcholine. *J. Neurochem. 43*, 288.

Rosenberry, T. L. (1975). Acetylcholinesterase. In *Advances in Enzymology*, Vol. 43, p. 103.

Rosenberry, T. L., P. Barnett, and C. Mays (1980). The collagenlike subunits of acetylcholinesterase from the eel Electrophorus electricus. *Neurochem. Int. 2*, 135.

Rotter, A., N. J. Birdsall, A. S. V. Burgen, P. M. Field, E. C. Hulme, and G. Raisman (1979). Muscarinic receptors in the central neurons system of the rat. *Brain Res. Rev. 1*, 141.

Rotundo, R. L. (1984). Purification and properties of the membrane bound form of acetylcholinesterase from chicken brain. *J. Biol. Chem. 259*, 13186.

Sastry, B. V. R., and C. Sadavongvivad (1979). Cholinergic systems in nonnervous tissue. *Pharmacol. Rev. 30*, 65.

Silensky, E. M. (1985). The biophysical pharmacology of calcium dependent acetylcholine secretion. *Pharmacol. Rev. 37*, 81.

Simon, J. R., S. Atweh, and M. J. Kuhar (1976). Sodium-dependent high affinity choline uptake: A regulatory step in the synthesis of acetylcholine. *J. Neurochem. 26*, 909.

Sokolovsky, M. (1984). Muscarinic receptors in the central nervous system. *Int. Rev. Neurobiol. 25*, 139.

Tucek, S. (1984). Problems in the organization and control of acetylcholine synthesis in brain neurons. *Prog. Biophys. Molec. Biol. 44*, 1.

Vaca, K., and G. Pilar (1979). Mechanisms controlling choline transport and acetylcholine synthesis in motor nerve terminals during electrical stimulation. *J. Gen. Physiol. 73*, 605.

Vickroy, T. W., M. Watson, H. I. Yamamura, and W. R. Roeske (1984). Agonist binding to multiple muscarinic receptors. *Fed. Proc. 43*, 2785.

Wainer, B. H., A. I. Levey, E. J. Mutson, and M.-M. Mesulam (1984). Cholinergic systems in mammalian brain identified with antibodies against choline acetyltransferase. *Neurochem. Int. 6*, 163.

Wilson, W. S., R. A. Schulz, and J. R. Cooper (1973). The isolation of cholinergic synaptic vesicles from bovine superior cervical ganglion and estimation of their acetylcholine content. *J. Neurochem. 20*, 659.

9 | Catecholamines I: General Aspects

The term "catecholamine" refers, generically, to all organic compounds that contain a catechol nucleus (a benzene ring with two adjacent hydroxyl substituents) and an amine group (Fig. 9-1). In practice the term "catecholamine" usually implies dihydroxyphenylethylamine (dopamine, DA) and its metabolic products, norepinephrine (NE) and epinephrine (E).

The great advances in the understanding of the biochemistry, physiology, and pharmacology of norepinephrine and related compounds during the last several decades have been made possible mainly through the development of new and sensitive assay techniques, and through methods for visualizing catecholamines and their metabolic enzymes or receptors in *in vivo* and *in vitro* preparations.

Methodology

Bioassay procedures used to be one of the most sensitive methods for the estimation of epinephrine and norepinephrine in tissue extracts or biological fluids. However, the more recently developed radioenzymatic, high performance liquid chromatographic and mass fragmentographic methods are more sensitive and specific and have virtually displaced all other assay methods.

Gas Chromatography–Mass Fragmentography (GCMF)

The technique of gas chromatography–mass fragmentography (GCMF) involves the combined instrumentation of gas chromatography and mass spectrometry. The advantages of mass fragmentography are derived from merging two useful analytic techniques, gas

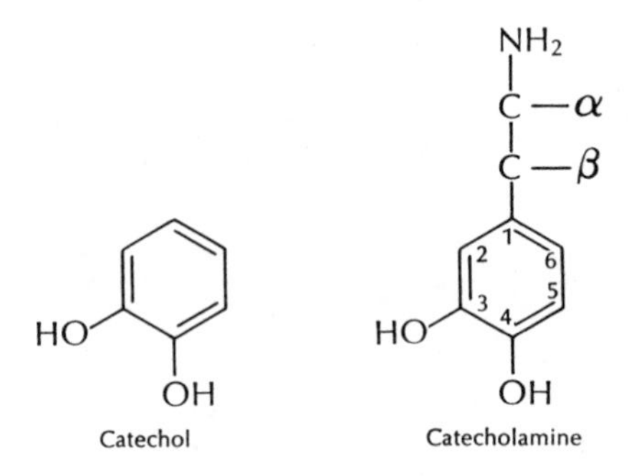

FIGURE 9-1. Catechol and catecholamine structure.

chromatography, which has great power in the separation of complex mixtures of compounds, and mass spectrometry, which has a high capability for structural identification. In GCMF the mass spectrometer is used as the detection system, and mass specific recordings are made of preselected ions. This provides one of the most sensitive of all gas chromatographic detection systems known. For studies that require methods capable of accurately and specifically measuring catecholamines in a few microliters of plasma or in microgram quantities of tissue, the utility of GCMF is unsurpassed. At present, the limit of sensitivity of GCMF for catecholamines and their metabolites is in the range of 10 to 50 femtomoles. (10^{-15}M). However, routine assays are usually conducted on samples containing low picomole amounts of these substances.

Gas chromatography–mass fragmentography has another advantage. It is possible to use stable isotopes as internal standards to compensate for errors introduced by factors that are normally difficult to control such as variations in derivative formation or extraction efficiency. From a chemical standpoint, stable isotopes are ideal standards since they behave in a manner identical to the compounds to be analyzed. Mass fragmentography has the unique ability to discriminate between such closely related compounds and thus permit the simultaneous quantification of each substance.

Radioisotopic Methods

The enzymatic isotopic assays are in general highly specific and at least 50 to 100 times more sensitive than the fluorometric assays currently available for catecholamine analysis. These radioenzymatic methods have proven to be precise, reproducible, and for the most part relatively simple to perform if the investigator has some experience in working with enzyme preparations. The assays are also flexible in that they can be readily adapted to measure certain catecholamine metabolites and precursors in addition to the parent amines. It should be noted, however, that the high specificity of these assays is determined to a large degree by the characteristics of the enzymatic reaction and by the techniques used to isolate the labeled metabolites. Since the enzyme preparations used routinely in assay of the catecholamines, catechol-*O*-methyltransferase and phenylethanolamine-*N*-methyltransferase are less specific, additional separation steps are needed in order to obtain the required degree of specificity. Similarly, radiochemical assays require several purification steps in order to assure the specificity of the assay technique.

High Performance Liquid Chromatography (HPLC) Coupled with Electrochemical Detection (LCEC)

In recent years the technology of high performance liquid chromatography (HPLC) has reached a degree of maturity which permits its routine use in most laboratories at a modest cost. High performance liquid chromatography usually involves ordinary column chromatography with the employment of stationary phases of a very small particle size so that excellent separations may be achieved with relatively short columns. However, because of the small particle size of the stationary phase, the columns will not achieve adequate flow without operation under pressure; hence the coinage of the term high pressure liquid chromatography. In general, nevertheless, at least for the biogenic amines and metabolites, liquid chromatography is usually carried out under pressures of 2000 psi or less.

The coupling of electrochemical methodology as a detection system for liquid chromatographic systems is a rather recent acquisition to the technology available for measurement of biogenic amines and metabolites. Rather than utilizing a chemical to oxidize catecholamines or metabolites (as is normal in fluorometric analysis), these substances are oxidized at an inert electrode surface by an applied potential. Quantification is thus achieved by counting the electrons released in the oxidation process. This is accomplished by measuring the current. The sensitivity of this technique as applied to catecholamine analysis is in the low picogram range and is comparable to the sensitivity achieved by routine gas chromatography–mass spectrometry (GCMS).

Although the technique of high performance liquid chromatography coupled with electrochemical detection (LCEC) only originated in 1972, it is already well established and has replaced many of the classical biochemical procedures for measurement of monoamines and their metabolites. The rapid acceptance of this technique is in part explained by the fact that the basic equipment required for LCEC analysis of monoamines is relatively inexpensive. Furthermore, the sensitivity achieved in routine use is excellent, and it is possible to quantitate numerous amines and metabolites in a single sample.

In vivo *Voltammetry—An Electroanalytical Technique for the* in vivo *measurement of Neurotransmitters in Brain*

Voltammetry is an established technique for the quantitative analysis of electroactive substances *in vitro* and, as indicated above, has been used extensively in conjunction with HPLC to measure monoamines in biological tissues and fluids. Adams and co-workers have initiated a new potentially powerful application of voltammetry to the *in vivo* detection of electroactive substances in brain. This is accomplished by measurement of oxidation currents generated at the carbon fiber electrode implanted in selected regions of the central nervous system. Most studies to date have been carried out in rats, but there is no reason to believe that this technique can not be extended to primates.

One major disadvantage of this *in vivo* method is the lack of specificity and the interference by endogenous electroactive substances like ascorbic acid. This problem has recently been circumvented in part by the application of differential pulse voltammetry to a pyrolytic carbon fiber electrode. This technique gives distinct oxidation current peaks in recordings from the rat neostriatum which are attributable to dopamine and ascorbic acid. Although it is not possible to differentiate between individual catecholamines or their metabolites employing this technique, the specificity of these electrochemical measures can be greatly improved by pharmacological manipulations and by careful experimental design. Thus, this technique of *in vivo* electrochemistry provides, for the first time, a means to follow the time course of transmitter or metabolite release in the central nervous system with a time resolution on the order of tens of seconds and a sensitivity sufficient to measure the minute amounts of transmitter released from central monoamine neurons.

Histochemical Fluorescence Microscopy and Microspectrofluorometry

In 1955 Eränko, and also Hillarp and Hökfelt, independently observed that flourescence microscopic examination of adrenal medullary tissue fixed in formaldehyde revealed a bright yellow fluorescence, which they attributed to catecholamines. However, attempts to visualize catecholamines in sympathetically innervated tissue remained unsuccessful for a number of years, probably because of the lower amine concentrations and of the diffusion of the fluorophores into the aqueous fixative. In fact, it was not until 1961, when Falck and his co-workers described the application of freeze drying and the treatment of tissue sections with dry formaldehyde vapor, that a sufficiently sensitive technique became available for the purpose of demonstrating the very small amounts of catecholamines present in adrenergic nerves.

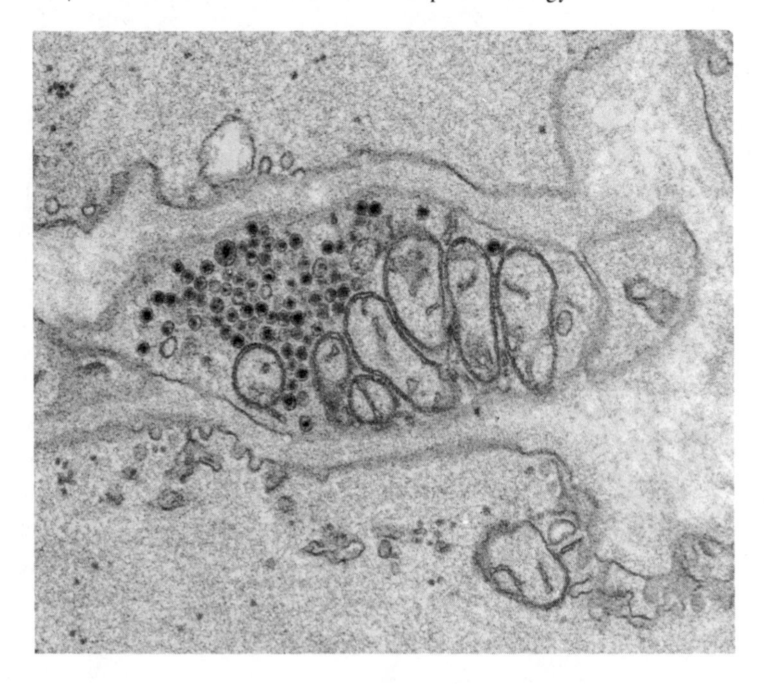

FIGURE 9-2. Electron micrograph of rat vas deferens. Potassium permanganate fixation. A cross section of an adrenergic nerve terminal or "varicosity" containing many synaptic vesicles, most with electron-opaque core (i.e., granular vesicles). (Supplied through the courtesy of L. S. Van Orden, III) × 37,000

The localization of catecholamines and their precursors within morphologically recognizable microscopic structures has provided a great advantage to those investigators interested in studying and understanding adrenergic mechanisms. This technique has also been employed in conjunction with electron miscroscopy quite successfully in the peripheral nervous system where it is possible to obtain a correlation between morphological changes (fluorescence intensity, content of granular vesicles) and monoamine content (Fig. 9-2).

The fluorescence histochemical technique applied to the central nervous system was the first to reveal the specific catecholaminergic neuronal pathways and cell bodies that were previously unrecognized by the conventional methods of neuroanatomy and has made possible an extensive mapping of monoaminergic pathways in brain. (See Chapter 10.) The anterograde and retrograde changes occurring after nerve sectioning have made it possible to map the ascending and descending monoamine systems within the central nervous system.

In the central nervous system it appears that practically the entire monoamine store is contained within enlargements of the nerve terminals. These varicosities or nerve terminal regions have been estimated by fluorescence microscopy to contain very high concentrations of catecholamines, approx. 1000 to 10,000 μg/gm. On the other hand, the monoamine cell bodies and axons have a much lower content (10–100 μg/gm). It is thought by most investigators that these varicosities, because of their relatively high amine content are the presynaptic structures involved in transmitter storage, release, re-uptake, and synthesis.

Electron Microscopy

In neuropharmacological research, special attention has been devoted to the problem of high resolution cytochemical methods for the identification of synaptic transmitter substances in intact nervous tissue, since this essential criterion must be satisfied in order to identify a substance as the likely transmitter for a particular defined synaptic complex. Thus electron miscroscopy coupled with the proper cytochemical techniques has provided the investigator in the catecholamine field with a technique for visualization of granular vesicles within the neuron suspected but not proven to contain catecholamines. Figures 9-2 and 9-3, electron micrographs of rat vas deferens, and rat hippocampus, illustrate the various types of vesicular structures found in these tissues. In peripheral noradrenergic neurons, there is a good correlation between norepinephrine depletion and the disappearance of the dense core from vesicles.

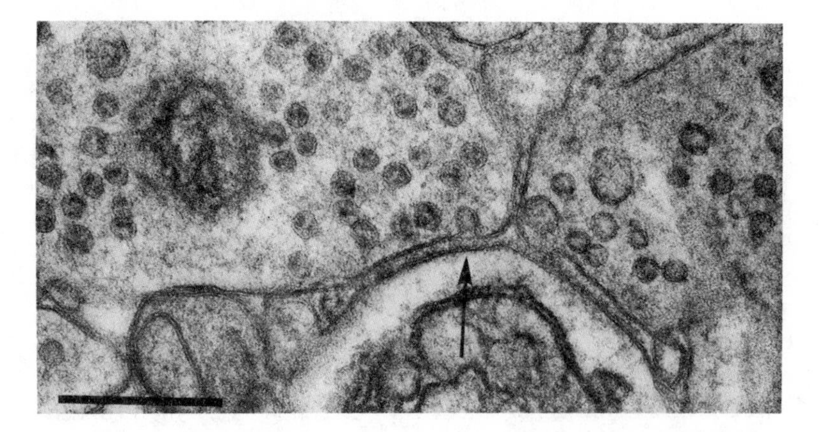

FIGURE 9-3. Presumed noradrenergic synaptic bouton in rat hippocampus, prepared with permanganate fixation after perfusion with ice-cold formaldehyde/glyoxylic acid. Note many small granular vesicles within bouton on left, one of which (arrow) resembles the oft-sought "omega" figure suggestive of an activevly releasing vesicle. Calibration bar = 250 nM. (From Koda and Bloom, unpublished)

Neurotoxins

The discovery of a rather unusual pharmacological agent has also aided in the study of adrenergic mechanisms. This compound is 6-hydroxydopamine (3,4,6-trihydroxyphenylethylamine). When administered intravenously or intraventricularly, 6-hydroxydopamine is rapidly and efficiently taken up into the catecholamine nerve endings. Perhaps because of its selective distribution and concentration as well as its extreme susceptibility to oxidation, this agent appears to selectively destroy sympathetic neurons by causing acute degeneration of adrenergic terminals. In effect this compound produces a "chemical sympathectomy." Thus, when the compound is given intravenously to animals, it causes almost total depletion of norepinephrine in sympathetically innervated tissues and also greatly reduces tyrosine hydroxylase activity. The only exception appears to be the adrenal medulla where 6-hydroxydopamine is not effectively taken up and the resultant demands placed on this organ to maintain homeostasis lead to a significant increase in tyrosine hydroxylase activity in medullary tissue. In most aspects, "chemical sympathectomy" produces effects comparable to those resulting from surgical denervation or from immunosympathectomy produced by the administration of antisera to nerve growth factor (NGF). However, this technique has the additional advantage that it may be applied to the central nervous system. When 6-hydroxydopamine is given intraventricularly or intracerebrally, it produces an extensive and lasting depletion of brain norepinephrine and dopamine and a degeneration of central dopamine- and norepinephrine-containing neurons. The depletion and degeneration are the result of what appears to be quite selective destruction of catecholamine-containing neurons by 6-hydroxydopamine. Since it has been shown that following appropriate administration this agent acts specifically to destroy catecholaminergic neurons, it has provided a useful tool for the evaluation of some of the functional roles ascribed to catecholamines in the central nervous system.

DSP4 [*N*-(2-chloroethyl)-*N*-ethyl-2-bromobenzylamine] is another neurotoxin which, unlike 6-OHDA, can cross the blood–brain barrier. This neurotoxin, when administered to adult rats in a dosage of 50 mg/kg, produces a permanent depletion of cortical and spinal cord norepinephrine without causing permanent alterations in brain dopamine and epinephrine. However, the action of DSP4 is not totally selective since a moderate neurotoxic effect on central serotonin neurons is observed as well as peripheral sympathetic neurons. This central 5-HT effect can be prevented by pretreatment with a serotonin uptake blocker.

In 1983 researchers at Stanford University identified a contaminant in a locally produced "synthetic heroin" that induced a Parkinson-like syndrome in some of the individuals who self-administered this preparation. The contaminant identified in this preparation, 1-methyl-4-phenyl-1,2,3,6-tetrahydropyridine (MPTP), exhibits a high degree of anatomical and species specificity. MPTP administered systemically in low doses to nonhuman primates produces Parkinsonian symptoms associated with a selective destruction of nigrostriatal dopamine neurons while sparing other brain dopamine systems. Thus, this neurotoxin appears to selectively destroy the same small group of brain cells in the substantia nigra that degenerate in naturally occurring Parkinson's disease. The discovery of the selective neurotoxic properties of MPTP and the development of a primate model of Parkinsonism has stimulated a tremendous resurgence of inquiry into Parkinson's disease both at the basic level of its possible cause and into potential new therapeutic treatments. The mechanism responsible for selective dopamine neurotoxic features of MPTP has not been conclusively established, although it has been shown that pretreatment with inhibitors of MAO (particularly MAO-B) protects against the MPTP-induced neurotoxicity, presumably because it prevents the formation of reactive oxidation products.

Distribution

In 1946 Euler in Sweden and shortly thereafter Holtz in Germany independently identified the presence of norepinephrine in adrenergic nerves. These findings laid the foundation for a new era of research in the field of catecholamines. At that time it was not technically possible to study the content of transmitter in the terminal parts of the adrenergic fiber. Despite this, Euler predicted that norepinephrine was, in fact, highly concentrated in the nerve terminal region from which it was released to act as a neurotransmitter. This prediction was conclusively documented some ten years later with the development of the fluorescence histochemical technique.

A systematic study of extracts of various mammalian nerves by Euler and his colleagues demonstrated that there was a very close correlation between the content of norepinephrine present and the proportion of nonmyelinated to myelinated nerve fibers. The correlation between the norepinephrine content of a nerve or an organ and its content of adrenergic fibers is now so well established that the occurrence of norepinephrine in a given organ or nerve can, in general, be taken as evidence for the presence of adrenergic fibers provided the presence of chromaffin tissue can be excluded. Table 9-1 illustrates the distribution of norepinephrine in various organs

TABLE 9-1. Tissue distribution of norepinephrine (μg/gm)

Organ	Species			
	Bovine	*Cat*	*Rabbit*	*Rat*
Artery	0.2–1.0			
Heart	0.48	0.5–1.0	1.4[b]	0.65
Liver	0.25	0.005–0.20		0.06
Lung	0.05		0.05–0.07[b]	
Skeletal muscle	0.04	0.03		
Spleen	1.5–3.5	0.8–1.4	0.3–0.5	0.40
Splenic nerve	8.5–18.5			
Vas deferens	9.3[a]	4.4[a]	6.7[a]	7.9[a]
Vein	0.1–0.5		1.0–2.5[b]	

[a]W. O. Sjostrand, *Acta Physiol. Scand.* 65, Suppl. 257, 1965.
[b]R. H. Roth and J. Hughes, unpublished data.

SOURCE: Unless indicated otherwise, data taken from Table 18, U. S. von Euler, *Noradrenaline* (1956).

and tissues. Tissues with a high density of sympathetic innervation have a relatively high concentration of sympathetic transmitter, norepinephrine, and vice versa. The ubiquitous distribution of norepinephrine in tissue is consistent with the presence of adrenergic vasomotor fibers in almost all peripheral tissue. The absence of norepinephrine in placenta or bone marrow is in agreement with histochemical and physiological evidence, which suggests that these tissues also lack any adrenergic vasomotor innervation.

Shortly after it was established that norepinephrine was the neurotransmitter substance of adrenergic nerves in the peripheral nervous system, norepinephrine was identified by Holtz as a normal constituent of mammalian brain. However, for some years it was believed that the presence of norepinephrine in mammalian brain only reflected vasomotor innervation to the cerebral blood vessels. In 1954 Vogt demonstrated that norepinephrine was not uniformly distributed in the central nervous system and that this nonuniform distribution did not in any way coincide with the density of blood vessels found in a given brain area. This characteristic regional localization of norepinephrine within mammalian brain suggested that norepinephrine might subserve some specialized function perhaps as a central neurotransmitter. In fact, this observation undoubtedly supplied the impetus for many investigators to pursue actively the functional reasons for the nonuniform presence of this active substance in the central nervous system. The relative distribution of norepinephrine is quite similar in most mammalian species. The highest concentration is usually found in the hypothalamus and other areas of central sympathetic representation. More norepinephrine is generally found in gray matter than in white matter.

Dopamine is also found in the mammalian central nervous system, and its distribution differs markedly from that of norepinephrine, suggesting that dopamine is present not only as a precursor of norepinephrine. In fact, it represents more than 50 percent of the total catecholamine content of the central nervous system of most mammals. The highest levels of dopamine are found in the neostriatum, nucleus accumbens, and tuberculum olfactorium. The presence of abundant dopamine in the basal ganglia has stimulated intensive research on the functional aspects of this compound. There is a growing wealth of evidence to suggest that this agent may have an important role in extrapyramidal function. (See Chapter 10.) Dopamine is also present in the carotid body and superior cervical ganglion, where it may play a role other than as a precursor of norepinephrine. The superior cervical ganglion appears to have at least three distinct populations of neurons: cholinergic neurons, noradrenergic neurons, and small intensely fluorescent (SIF) cells. The SIF cells are believed to be the small interneurons that contain dopamine. The actual functional significance of these SIF cells is at the present time unclear although it has been suggested that dopamine released from these interneurons is responsible for hyperpolarization of the ganglion.

Epinephrine concentration in the mammalian central nervous system is relatively low, approximately 5 to 17 percent (by bioassay) of the norepinephrine content. Many investigators have suggested that these original estimates are subject to error and have in the past discounted the importance of the occurrence of epinephrine in mammalian brain. However, the presence of epinephrine in mammalian brain has now been documented by more sophisticated analytical techniques such as GCMS and LCEC and the presence of epinephrine—containing neurons confirmed by immunohistochemical techniques. (See Chapter 10.)

A detailed topographical survey of brain catecholamines at different levels of organization within the central nervous system has helped to give us some sort of basic framework from which to organize and conduct logical experiments concerning the possible functions of these amines.

Intraneuronal Localization

With the refinement of physical and chemical techniques it became possible to further localize monoamines both in isolated subcellular particles obtained by differential and density gradient centrifugation and by fluorescence histochemistry and electron microscopy in axonal varicosities. These powerful techniques have provided us with our current concepts concerning the location of amines within the neuron in a structure referred to as synaptic vesicles. The concept of storage of monoamines will be covered in more detail when we consider the life cycle of the catecholamines.

Life Cycle of the Catecholamines

Biosynthesis

Catecholamines are formed in brain, chromaffin cells, sympathetic nerves, and sympathetic ganglia from their amino-acid precursor tyrosine by a sequence of enzymatic steps first postulated by Blaschko in 1939 and finally confirmed by Nagatsu and co-workers in 1964 when they demonstrated that an enzyme (tyrosine hydroxylase) is involved in the conversion of L-tyrosine to 3,4-dihydroxyphenylalanine (DOPA). This amino-acid precursor, tyrosine, is normally

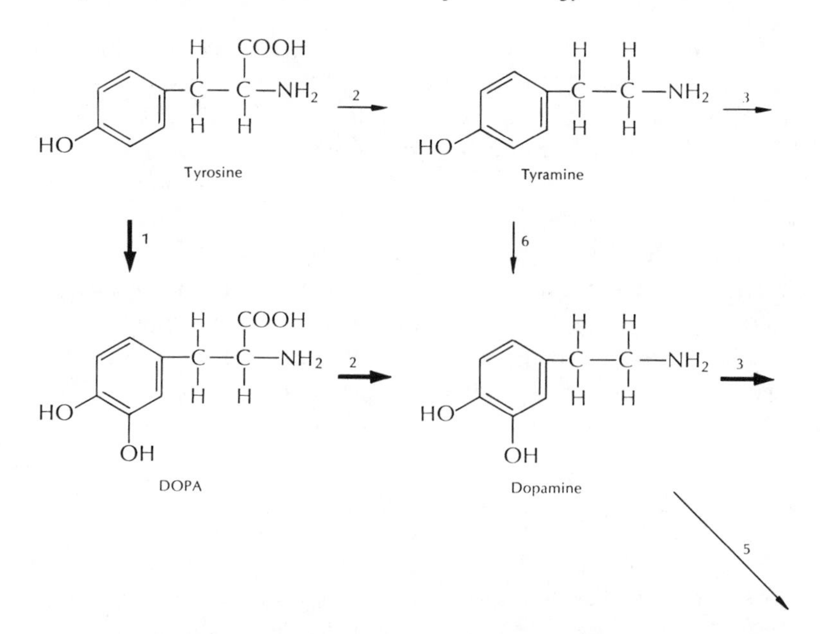

FIGURE 9-4. Primary and alternative pathways in the formation of catecholamine: (1) tyrosine hydroxylase; (2) aromatic amino-acid decarboxylase; (3) dopamine-β-hydroxylase; (4) phenylethanolamine-*N*-methyl transferase; (5) nonspecific *N*-methyl transferase in lung and folate-dependent *N*-methyl transferase in brain; (6) catechol-forming enzyme.

present in the circulation in a concentration of about 5 to 8 $\times$ 10^{-5} M. It is taken up from the bloodstream and concentrated within the brain and presumably also in other sympathetically innervated tissue by an active transport mechanism. Once inside the peripheral neuron tyrosine undergoes a series of chemical transformations resulting in the ultimate formation of norepinephrine or in brain, norepinephrine, dopamine, or epinephrine, depending upon the availability of phenylethanolamine-*N*-methyl transferase and dopamine-β-hydroxylase. This biosynthetic pathway for the formation of catecholamines is illustrated in Figure 9-4. The conversion of tyrosine to norepinephrine and epinephrine was first demonstrated in

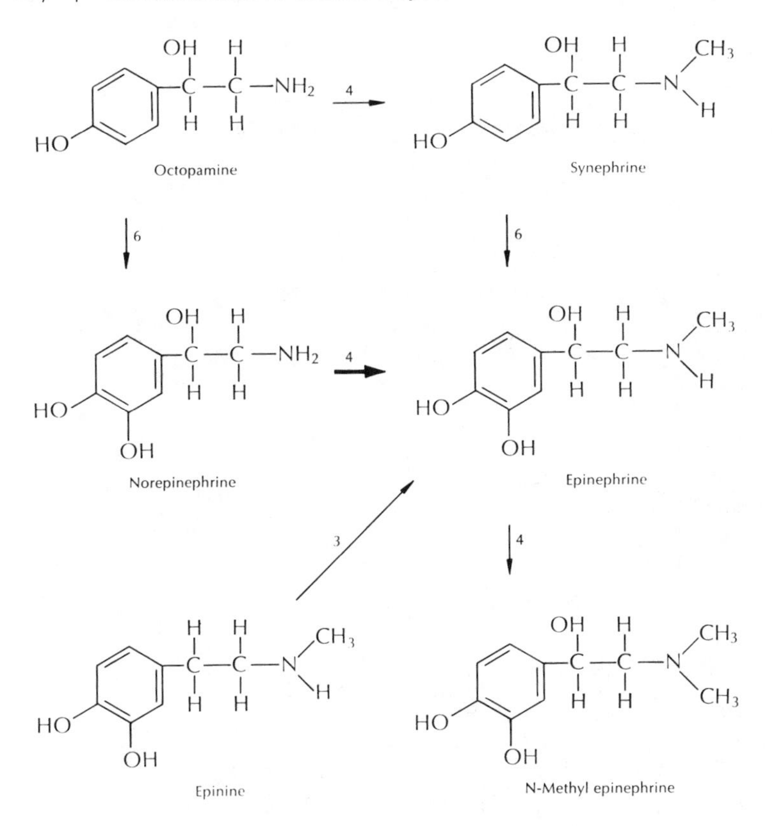

the adrenal medulla. The availability of radioactive precursors of high specific activity and chromatographic separation techniques has allowed the confirmation of the above-mentioned pathway to norepinephrine in sympathetic nerves, ganglia, heart, arterial and venous tissue, and brain. In mammals, tyrosine can be derived from dietary phenylalanine by a hydroxylase (phenylalanine hydroxylase) found primarily in liver. Both phenylalanine and tyrosine are normal constituents of mammalian brain, present in a free form in a concentration of about 5×10^{-5} M. However, norepinephrine biosynthesis is usually considered to begin with tyrosine, which represents a branch point for many important biosynthetic pro-

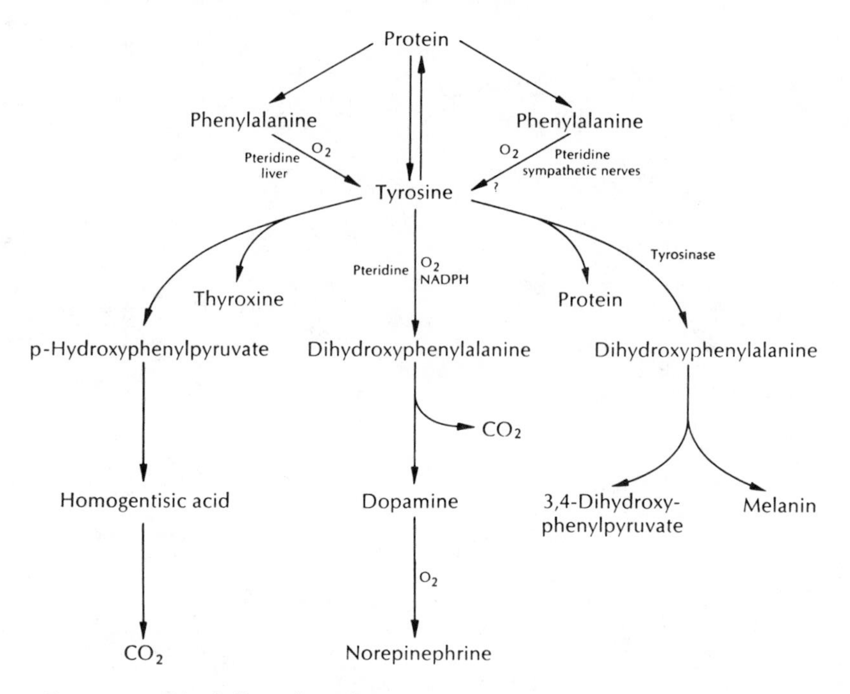

FIGURE 9-5. Metabolism of tyrosine.

cesses in animal tissues (Fig. 9-5). It should be emphasized that the percentage of tyrosine utilized for catecholamine biosynthesis as opposed to other biochemical pathways is very minimal (<2%).

Tyrosine Hydroxylase

The first enzyme in the biosynthetic pathway, tyrosine hydroxylase, was the last enzyme in this series of reactions to be identified. It was demonstrated by Udenfriend and his colleagues in 1964, and its properties have been reviewed repeatedly. It is present in the adrenal medulla, in brain, and in all sympathetically innervated tissue studied to date. This enzyme appears to be a unique constituent of catecholamine-containing neurons and chromaffin cells; it completely disappears from renal, salivary gland, vas deferens, and cardiac tissue upon chronic sympathetic denervation. The enzyme is sterospecific, requires molecular O_2, Fe^{2+}, and a tetrahydropteridine cofactor, and shows a fairly high degree of substrate specificity. Thus, this enzyme, in contrast to tyrosinase, oxidizes only the naturally occurring amino acid, L-tyrosine, and to a smaller extent L-phenylalanine. D-tyrosine, tyramine, or L-tryptophan will not serve as substrates for the enzyme. Phenylalanine hydroxylase and tyrosine hydroxylase appear to be distinct enzymes since phenylalanine hydroxylase does not hydroxylate tyrosine and is not inhibited by some potent tyrosine hydroxylase inhibitors.

The K_m for the enzymatic conversion of tyrosine to DOPA by purified adrenal tyrosine hydroxylase is about 2×10^{-5}M, and in a preparation of brain synaptosomes about 0.4×10^{-5}M. Tyrosine hydroxylation appears to be the rate-limiting step in the biosynthesis of norepinephrine in the peripheral nervous system and is likely to be the rate-limiting step in the formation of norepinephrine and dopamine in the brain as well. In most sympathetically innervated tissues including the brain, the activity of DOPA decarboxylase and that of dopamine-β-oxidase have a magnitude 100 to 1000 times that of tyrosine hydroxylase. This lower activity of tyrosine hydroxylase may be due either to the presence of less enzyme or to a lower turnover number of the enzyme. Since this enzyme has been demonstrated to be the rate-limiting step in catecholamine biosynthesis, it is logical that pharmacological intervention at this step would cause a reduction of norepinephrine biosynthesis. Many earlier attempts to produce chemical sympathectomy by blockade of the last two steps in the synthesis of norepinephrine have proved largely unsuccessful. On the other hand, studies with inhibitors of tyrosine hydroxylase have proved to be much more successful producing a marked reduction in endogenous norepinephrine and dopamine in brain and norepinephrine in heart, spleen, and other sympathetically innervated tissues. Effective inhibitors of this enzymatic step can be categorized into four main groups: (1) amino-acid analogues, (2) catechol derivatives, (3) tropolones, and (4) selective iron chelators. Some effective amino-acid analogues include α-methyl-p-tyrosine and its ester, α-methyl-3-iodotyrosine, 3-iodotyrosine, and α-methyl-5-hydroxytryptophan. In general, α-methyl-amino acids are more potent than the unmethylated analogues, and a marked increase in activity in the case of the tyrosine analogues can also be produced by substituting a halogen at the 3 position of the benzene ring. Most of the agents in this category act as competitive inhibitors of the substrate tyrosine. In this respect α-methyl-5-hydroxytryptophan appears to be unique, since it does not appear to compete with substrate or pteridine cofactor. Its actual mechanism at the present time remains unknown. The potent halogenated tyrosine analogues such as 3-iodotyrosine are about 100 times as active as α-methyl-ρ-tyrosine *in vitro*, but they are substantially less active *in vivo*. This is probably due to the very rapid deiodination of these compounds to tyrosine or tyrosine analogues which occurs *in vivo*. α-Methyl-ρ-tyrosine and its methyl ester have been the inhibitors most widely used to demonstrate the effects of exercise, stress, and various drugs on the turnover of catecholamines and also to lower norepinephrine formation in patients with pheochromocytoma and malignant hypertension.

Dihydropteridine Reductase

Although not directly involved in catecholamine biosynthesis, dihydropteridine reductase is intimately linked to the tyrosine hydroxylase step. This enzyme catalyzes the reduction of the quinonoid dihydropterin that has been oxidized during the hydroxylation of tyrosine to DOPA. Since reduced pteridines are essential for tyrosine hydroxylation, alterations in the activity of dihydropteridine reductase would effectively influence the activity of tyrosine hydroxylase. Thus, this might be a potential site for drug intervention in catecholamine biosynthesis. Dihydropteridines with amine substitution in positions 2 and 4 are effective inhibitors of this enzyme, while folic acid antagonists such as aminopterin and methotrexate are relatively ineffective. The distribution of dihydropteridine reductase is quite widespread, the highest activity being found in liver, brain, and adrenal gland. The distribution of this enzyme activity in brain does not appear to parallel the catecholamine or serotonin content of brain tissue, suggesting that reduced pterins most likely participate in other reactions besides the hydroxylation of tyrosine and tryptophan.

Dihydroxyphenylalanine Decarboxylase

The second enzyme involved in catecholamine biosynthesis is DOPA-decarboxylase, which was actually the first catecholamine synthetic enzyme to be discovered. Although originally believed to remove carboxyl groups only from L-DOPA, a study of purified enzyme preparations and specific inhibitors has subsequently demonstrated that this DOPA-decarboxylase acts on all naturally occurring aromatic L-amino acids, including histidine, tyrosine, tryptophan, and phenylalanine as well as both DOPA and 5-hydroxytryptophan. Therefore, this enzyme is more appropriately referred to as "L-aromatic amino acid decarboxylase." There is no appreciable binding of this enzyme to particles within the cell, since when tissues are disrupted and the resultant homogenates centrifuged at high speeds, the decarboxylase activity remains associated largely with the supernatant fraction. The exception to this is in brain, where some of the decarboxylase activity is associated with synaptosomes. However, since synaptosomes are in essence pinched-off nerve endings, they would be expected to retain entrapped cytoplasm as well as other intracellular organelles. The DOPA-decarboxylase found in synaptosomal preparations is thought to be present in the entrapped cytoplasm. DOPA-decarboxylase is, relative to other enzymes in the biosynthetic pathway for norepinephrine formation, very active and requires pyridoxal phosphate (vitamin B_6) as a cofactor. The apparent K_m value for this enzyme is 4×10^{-4} M. The high activity of this enzyme may explain why it has been difficult to detect endogenous DOPA in sympathetically innervated tissue and brain. It is rather ubiquitous in nature, occurring in the cytoplasm of most tissue including the liver, stomach, brain, and kidney in high levels, suggesting that its function in metabolism is not limited solely to catecholamine biosynthesis. Although decarboxylase activity can be reduced by production of vitamin B_6 deficiency in animals, this does not usually result in a significant reduction of tissue catecholamines although it appears to interfere with the rate of repletion of adrenal catecholamines after insulin depletion. In addition, potent decarboxylase inhibitors also have very little effect on endogenous levels of norepinephrine in tissue. However, these inhibitors have been useful as pharmacological tools (e.g., DOPA accumulation following administration of a decarboxylase inhibitor as an *in vivo* index of tyrosine hydroxylation).

Dopamine-β-Hydroxylase

Although it has been known for many years that brain, sympathetically innervated tissue, sympathetic ganglia, and adrenal medulla could transform dopamine into norepinephrine, it was not until 1960 that the enzyme responsible for this conversion was isolated from the adrenal medulla. This enzyme, called dopamine-β-hydroxylase, is, like tyrosine hydroxylase, a mixed function oxidase. It requires molecular oxygen and utilizes ascorbic acid as a cofactor. The K_m of this enzyme for its substrate dopamine is about 5×10^{-3}M. Dicarboxylic acids such as fumaric acid are not absolute requirements, but they stimulate the reaction. Dopamine-β-hydroxylase is a Cu^{2+}-containing protein, with about two moles of cupric ion/mole of enzyme. It appears to be associated with the particulate fraction from heart, brain, sympathetic nerve, and adrenal medulla, and it is currently believed that this enzyme is localized primarily in the membrane of the amine storage granules. This enzyme usually disappears after chronic sympathetic denervation and therefore is believed present largely only in adrenergic neurons or adrenal chromaffin tissue. Dopamine-β-hydroxylase does not show a high degree of substrate specificity and acts *in vitro* on a variety of substrates besides dopamine, oxidizing almost any phenylethylamine to its corresponding phenylethanolamine (i.e., tyramine → octopamine, α-methyldopamine → α-methylnorepinephrine). A number of the resultant, structurally analogous metabolites can replace norepinephrine at the noradrenergic nerve endings and function as "false neurotransmitters."

Dopamine-β-hydroxylase can be inhibited by a variety of compounds. The most effective are compounds which chelate copper: D-cysteine and L-cysteine, glutathione, mercaptoethanol, and coenzyme A. The inhibition can be reversed by addition of *N*-ethylmaleimide, which reacts with the sulfhydryl groups and interferes with the chelating properties of these substances. Copper chelating agents such as diethyldithiocarbamate and FLA-63 [bis-(l-methyl-4-homopiperazinyl-thiocarbonyl)-disulfide] have proved to be effective inhibitors both *in vivo* and *in vitro.* Thus it has been possible to treat animals with disulfuram or FLA-63 and produce a reduction in brain norepinephrine and an elevation of brain dopamine. This manipulation might serve as a useful tool in the assessment of the relative roles of norepinephrine and dopamine in the central nervous system. However, in most cases, especially in the periphery, the dopamine which is not β-hydroxylated to norepinephrine does not accumulate to any great extent since it appears to be very rapidly deaminated by monoamine oxidase (MAO). The major increase, corresponding to the missing norepinephrine, is usually found in dopamine metabolites. Some recent work has indicated that endogenous inhibitors are found in tissue homogenates which, when removed, lead to a significant enhancement of dopamine-β-hydroxylase activity. It has been suggested that these inhibitors may have a role in the regulation of norepinephrine biosynthesis *in vivo* but it is also possible that they are artifacts of homogenization.

Since dopamine-β-hydroxylase obtained from the bovine adrenal medulla can be prepared in a relatively pure form, it has been possible to produce a specific antibody to the enzyme. This antibody inactivates bovine dopamine-β-hydroxylase but does not appear to cross react with either DOPA-decarboxylase or tyrosine hydroxylase. However, cross-reactivity between dopamine-β-hydroxylase from human, guinea pig, and dog with the antibody against the bovine enzyme was observed and indicates that the enzymes from these various sources are probably structurally related. By coupling immunochemical techniques with fluorescence and electron microscopy, this antibody has already proved useful in localization of this enzyme in intact tissue.

*Phenylethanolamine-*N*-Methyl Transferase*

In the adrenal medulla norepinephrine is *N*-methylated by the enzyme phenylethanolamine-*N*-methyl transferase to form epinephrine. This enzyme is largely restricted to the adrenal medulla although low levels of activity have been reported in heart and mammalian brain (see Chapter 10). Like the decarboxylase, this enzyme also appears in the supernatant of homogenates. Demonstration of activity requires the presence of the methyl donor, *S*-adenosyl methionine. The adrenal medullary enzyme shows poor substrate specificity and will transfer methyl groups to the nitrogen atom on a variety of β-hydroxylated amines. However, this adrenal enzyme is distinct from the nonspecific *N*-methyl transferase of rabbit lung, which in addition to *N*-methylating phenylethanolamine derivatives will also react with many normally occurring indoleamines and such diverse structures as phenylisopropylamine, aromatic amines, and phenanthrenes.

In the early 1970s an *N*-methyl transferase was reported in brain that purportedly used 5-methyl-tetrahydrofolate as the methyl donor and dopamine and other monoamines as substrate. Although the idea that methyl-tetrahydrofolate serves as a methyl donor in brain captured the imagination of many investigators in the field, it was subsequently proven to be incorrect. It turned out that methyl-tetrahydrofolate was not a methyl donor at all but instead was degraded enzymatically to formaldehyde, which then condensed with amines such as dopamine and tryptamine to form tetrahydroisoquinolines and tetrahydro-β-carbolines, respectively (Fig. 9-6). Confusion arose since these ring closure compounds have many properties that are similar to the *N*-methylated amines and comigrated in many of the chromatographic systems used in the earlier studies.

Synthesis Regulation

Knowledge concerning the various enzymes involved in the synthesis of dopamine, norepinephrine, and epinephrine have provided an

FIGURE 9-6. Products formed when dopamine and tryptamine condense with formaldehyde. The structure at the left, 6,7-dihydroxy-1,2,3,4-tetrahydroisoquinoline is formed when dopamine condenses with formaldehyde. The structure at the right, 1,2,3,4,-tetrahydro-β-carboline is formed when formaldehyde condenses with tryptamine.

excellent example of how extensive information of the individual components of a given biological system can contribute to a better understanding of its overall function. Thus, once it was established that tyrosine hydroxylase was the rate-limiting step in the conversion of tyrosine to norepinephrine, it did not take long to realize and to demonstrate experimentally that potent inhibitors of this initial step would effectively reduce tissue levels of norepinephrine. However, even as it becomes possible to understand more fully the chemistry and functional aspects of these enzymes under isolated conditions, this information still proves to be inadequate for a complete understanding of how, in fact, these enzymes function *in vivo.* It is necessary to have an appreciation of the cytological arrangement of the enzymes and cofactors within the neuron as well as some insight as to the external factors which may influence the activity of the neuron. Therefore it is becoming more and more important for neuropharmacologists to have a broad basic training and to take the time to try to interrelate the *in vivo* and *in vitro* data obtained experimentally to see if they can be correlated to provide a better understanding of the physiological control mechanisms involved.

It has been known for a long time that the degree of sympathetic activity does not influence the endogenous levels of tissue norepinephrine; and it has been speculated that there must be some homeostatic mechanism whereby the level of transmitter is maintained at a relatively constant level in the sympathetic nerve endings despite the additional losses assumed to occur during enhanced sympathetic activity.

More than twenty years ago, Euler hypothesized on the basis of experiments carried out in the adrenal medulla that during periods of increased functional activity, the sympathetic neuron must also increase the synthesis of its transmitter substance, norepinephrine, to meet the increased demands placed upon the neuron. If the sympathetic neuron had the ability to increase transmitter synthesis, this would enable the neuron to maintain a constant steady-state level of transmitter despite substantial changes in transmitter utilization. Some years later, experiments carried out by several laboratories on different sympathetically innervated tissues directly demonstrated that this was, in fact, the case. Electrical stimulation of sympathetic nerves both *in vivo* or *in vitro* resulted in an increased formation of norepinephrine in the tissues innervated by these nerves. Further studies demonstrated that the observed acceleration of norepinephrine biosynthesis produced by enhanced sympathetic activity was due to an increase in the activity of the rate-limiting enzyme involved in catecholamine biosynthesis, tyrosine hydroxylase. Emphasis then shifted toward attempting to determine the mechanism by which an alteration in impulse flow within the sympathetic neuron could result in a change in the activity of this potential regulatory enzyme.

It was appreciated at an early stage that the observed enhancement of tyrosine hydroxylase activity that accompanied an increase in sympathetic activity was not the result of new enzyme synthesis but rather was related to an increase in the activity of existing enzyme molecules. The classical *in vitro* studies by Udenfriend and co-workers, demonstrating that catecholamines could act as feed-back inhibitors of tyrosine hydroxylase, provided a reasonable theoretical mechanism by which impulse activity might regulate tyrosine hydroxylase. Since tyrosine hydroxylase is inhibited by catechols and catecholamines, presumably due to their ability to antagonize competitively the binding of the pteridine cofactor to the apoenzyme, it was proposed that free intraneuronal norepinephrine may control its own synthesis by negative feedback inhibition of tyrosine hydroxylase. This concept was supported by many pharmacological studies, including the observation that an increase in the endogenous norepinephrine content of peripheral sympathetically innervated tissue or of brain catecholamine content, which is produced by treatment with monoamine oxidase inhibitors (MAOI), results in a marked reduction of catecholamine biosynthesis. In addition, the increase in synthesis observed *in vitro* during neuronal depolarization is dependent upon the presence of Ca^{2+} and appears inextricably coupled to catecholamine release. This latter observation suggested that the release-induced depletion of transmitter might be a prerequisite for the observed acceleration of catecholamine synthesis.

In the late 1960s the following conceptual model emerged. During periods of increased impulse flow when more transmitter is released and metabolized, a strategic regulatory pool of norepinephrine normally accessible to tyrosine hydroxylase is depleted, end-product inhibition is removed, and tyrosine hydroxylase activity is increased. Likewise, during periods of quiescence when transmitter utilization is decreased, norepinephrine accumulates and tyrosine hydroxylase activity is decreased. Over the past decade, much indirect evidence has amassed to suggest that the release of tyrosine hydroxylase from feedback inhibition is a mechanism that is operative in the sympathetic neuron and in central catecholamine neurons for increasing neurotransmitter synthesis in response to enhanced impulse flow.

In more recent years, it has been apparent that even in peripheral noradrenergic neurons the regulation of tyrosine hydroxylase is a much more complex process and that in addition to feedback regulation other processes, intimately linked to neuronal activity, control norepinephrine synthesis. The concept that the frequency of neuronal depolarization might in some way influence the physical properties of tyrosine hydroxylase is one of the more exciting possibilities to have emerged.

A significant advancement in our comprehension of the way in which short-term changes in impulse flow may regulate transmitter synthesis was provided by several experiments. It was demonstrated that electrical stimulation of both central and peripheral catecholamine neurons results in an apparent allosteric activation of tyrosine hydroxylase which persists following the termination of the stimulation period. This poststimulation increase in the activity of tyrosine hydroxylase is maximal after about ten minutes of continuous stimulation and persists for a measurable period of time after the stimulation period ends. Kinetic studies demonstrated that this activation of tyrosine hydroxylase observed following depolarization is mediated in part by an increased affinity of the enzyme for pteridine cofactor and a decreased affinity for the natural endproduct inhibitors, norepinephrine and dopamine. The increased affinity for pterin cofactor (whose concentration in tissue is thought to be subsaturating) and the decreased affinity of the enzyme for catecholamine may be important physiologically for the regulation of transmitter synthesis in response to increased impulse flow. These studies provided the first direct experimental evidence that some short-term mechanism in addition to depletion of endogenous catecholamine is operative in altering tyrosine hydroxylase activity during increased neuronal activity. The molecular mechanism(s) responsible for the impulse-induced alterations in the physical properties of tyrosine hydroxylase remains elusive, although it has been speculated that calcium and/or cAMP may be intimately involved. Numerous *in vitro* studies have demonstrated that tyrosine hydroxylase can be activated by phosphorylation, providing credence for the speculation that phosphorylation is a process by which catecholamine neurons may regulate tyrosine hydroxylase activity in response to altered neuronal activity. Although C-kinase, cAMP-dependent protein kinase, and calcium–calmodulin-dependent kinase all phosphorylate tyrosine hydroxylase producing a kinetic activation of the enzyme, the question of which of these mechanisms is operative physiologically is still unresolved. The most recent studies favor the involvement of a Ca^{++}-calmodulin-dependent phosphorylation in the impulse-dependent activation process.

In addition to the mechanisms for immediate adaptation to increased transmitter utilization discussed above, a second mechanism comes into play after prolonged increases in the activity of adrenergic neurons. This latter mechanism in contrast is reflected by an increase in the activity of tyrosine hydroxylase measured *in vitro* under saturating concentrations of substrate and cofactor. For a while, it was unclear whether this increase in activity observed *in vitro* was due to the disappearance of an inhibitor, the appearance of an activator, or an actual increase in the number of active enzyme sites. In recent years, however, much evidence has accumulated which indicates that there is a measurable increase in the formation of new enzyme molecules. Thus, prolonged neuronal activity is thought to result in an induction of tyrosine hydroxylase. This process of transsynaptic induction does not appear to be specific for tyrosine hydroxylase since an increase in the formation of dopamine-β-hydroxylase is also observed. No significant increase in the formation of DOPA-decarboxylase has been reported. It has been proposed by Joh and co-workers based on recent studies employing gene technology that the major enzymes involved in catecholamine biosynthesis—tyrosine hydroxylase, dopamine-β-hydroxylase, and phenylethanolamine-N-methyltransferase—may derive from a common gene and thus may be coregulated (i.e., coinduced by various treatments). However, whether the story of shared molecular sequences across this big family of amine synthetic enzymes will hold when the sequences of these enzymes are fully in hand remains to be determined.

It would be naïve to believe that catecholamine biosynthesis would not also be influenced by circulating hormones. In fact, steroid hormones will increase phenylethanolamine-*N*-transferase activity both in the adrenal gland and in the central nervous system. Furthermore, thyroidectomy increases the conversion of [^{14} C]-tyrosine to [^{14} C]-norepinephrine. This finding suggests that normal degradation products of thyroid hormone (i.e., iodinated tyrosines) may be circulating in the bloodstream and acting as endogenous inhibitors of catecholamine biosynthesis. This, however, seems unlikely in normal individuals owing to the presence of very active tissue dehalogenases. In certain patients with a defect in tissue dehalogenase, this could become a reality, since it is known that in this instance iodotyrosines accumulate in the body and may reach levels expected to significantly inhibit tyrosine hydroxylase.

Even though tyrosine hydroxylase has been demonstrated to be rate limiting under most circumstances, this reaction may not remain rate limiting under all physiological or pharmacological conditions. Since there is a sequence of reactions, any one of these steps could assume a rate-limiting role depending upon the given pathological or pharmacologically induced situations. Thus, reserpine can transform the dopamine-β-hydroxylase step into the rate-limiting step presumably by blocking access of the substrate dopamine to the site of its conversion to norepinephrine. In fact, experiments utilizing the bovine splenic nerve granule as a model system have indicated that extragranular norepinephrine can effectively inhibit the uptake of dopamine into the granules and thus also inhibit the conversion of dopamine to norepinephrine. However, whether or not this sort of mechanism is functional *in vivo* (i.e., if high concentrations of cytoplasmic norepinephrine can inhibit the conversion of dopamine to norepinephrine) remains to be demonstrated.

Although it is well known that alterations in brain tryptophan can influence serotonin biosynthesis, the failure of tyrosine administration to increase levels of norepinephrine in sympathetically innervated tissue or catecholamine levels in brain has led most investigators to assume that tyrosine hydroxylase in catecholamine-containing neurons is normally saturated with its amino-acid substrate tyrosine *in vivo* and therefore its activity not readily influenced by alterations in the availability of tyrosine. A number of recent studies by Wurtman, Fernstrom, and their co-workers have demonstrated, however, that under conditions which are believed to cause an increase in impulse flow in peripheral sympathetic or central noradrenergic or dopaminergic neurons that tyrosine hydroxylation and release (evaluated by measuring the accumulation of catecholamine metabolites) can be enhanced by systemic administration of tyrosine (Table 9-2). This conclusion is further supported by the finding that tyrosine hydroxylase in dopamine neurons with a high basal rate of firing (the mesoprefrontal dopamine neurons) is very susceptible to precursor regulation (see Chapter 10). These studies are indicative that under specialized conditions transmitter output and perhaps the functions of catecholamine neurons may be enhanced or maintained by administration of tyrosine.

TABLE 9-2. Influence of tyrosine administration on catecholamine synthesis and release[a] under conditions believed to increase impulse flow in catecholamine neurons

Tissue	*Treatment*	*Biochemical index*	*Tyrosine effect*
Striatum	Haloperidol	DOPA accumulation	15% inc.
Striatum	Haloperidol	DOPAC,HVA	60% inc.
Whole brain	Cold stress	$MHPG\text{-}SO_4$	70% inc.
Whole brain	Yohimbine	$MHPG\text{-}SO_4$	35% inc.
Hippocampus Hypothalamus	Tail shock	$MHPG\text{-}SO_4$	40% inc.
Whole brain, brain stem, or forebrain	Spontaneous Hypertensive Rats	$MHPG\text{-}SO_4$	15–40% inc.

[a]Estimated by measuring catecholamine metabolite levels.

SOURCE: Data taken from J. D. Milner and R. J. Wurtman, *Biochem. Pharm.*, 1986, in press.

Alternative Biosynthetic Pathways

The question has often been raised as to whether the sequence of enzymatic reactions described in detail above is in fact obligatory. Many attempts have been made to find alternative pathways *in vivo*, and to determine whether they are functionally important. Thus if labeled tyramine is administered to animals, both labeled norepinephrine and normetanephrine can be isolated in the urine. It has been very difficult to detect the conversion of tyramine to norepinephrine in sympathetically innervated tissue. However, the conversion of tyramine to dopamine has been demonstrated in liver microsomes. Therefore, it is possible that this conversion of tyramine to norepinephrine observed *in vivo* is a reflection of a metabolic reaction taking place in the liver rather than in the sympathetic neuron.

Turnover

The term "turnover" connotes the overall rate at which the whole amine store within a given tissue is replaced. Thus, turnover rates are not necessarily identical with rates of biosynthesis. However, estimation of turnover rate can be used as an index to the functional state of various populations of catecholamine neurons, although not as an index of the synthetic capacity of the neurons.

Many techniques are utilized for the assessment of catecholamine turnover in both peripheral and central sympathetic neurons.

1. Turnover of norepinephrine has been estimated by measurement of the rate of decline of the specific activity of norepinephrine after a tracer dose of labeled amine is introduced into the endogenous pool. This technique has been applied both to peripheral and central neurons. In the case of the central nervous system, the problem of penetration of the blood–brain barrier has been circumvented by administration of the labeled amine by intracisternal or intraventricular injections. This method for measurement of turnover, as one might expect, has a number of limitations. This method assumes that all the labeled norepinephrine found in the brain and in peripheral sympathetically innervated tissue is specifically retained by only the norepinephrine-containing neurons. This is not the case, since in addition to its retention in norepinephrine-containing neurons, labeled norepinephrine is also taken up and retained by a central network of dopamine-containing neurons in the central nervous system. In some instances this is not a serious limitation in the central nervous system since, except for the dopamine terminals in the striatum, olfactory tubercle, nucleus accumbens, median eminence, and certain specialized cortical areas (see Chapter 10), the remainder of the brain contains only a relatively small number of dopamine-containing terminals.

   The isotope method assumes that only tracer amounts of

labeled norepinephrine are introduced into the endogenous pool so that the size of the endogenous pool remains unaltered. This limitation has been largely overcome with the availability of norepinephrine of high specific activity and it is now possible to administer amounts of labeled norepinephrine which alter endogenous levels of amines by less than 5 percent. Perhaps the greatest disadvantage of this method, which also applies to many other methods used to measure turnover, is the assumption that norepinephrine is stored in a single homogeneous pool with which the labeled norepinephrine mixes completely. In view of much biochemical, histochemical, and pharmacological evidence indicating more than one "pool" of catecholamines in sympathetic neurons, this assumption appears somewhat unlikely.

2. Another technique for measuring norepinephrine turnover involves following the disappearance rate of endogenous norepinephrine after inhibition of catecholamine biosynthesis by α-methol-*p*-tyrosine or another potent inhibitor of tyrosine hydroxylase. This technique also suffers from several disadvantages. First, it involves measurement of an exponentially decreasing level of norepinephrine, which is difficult to measure accurately in small tissue samples. This difficulty can now be easily overcome with the availability of very sensitive assay techniques. Second, the required administration of large amounts of drugs (synthesis inhibitors) are almost certain to induce pharmacological actions aside from their "specific" action on tyrosine hydroxylase. Third, since turnover is being measured while endogenous levels of norepinephrine are being markedly depleted, this extensive depletion may in itself influence the normal control processes existing in the neuron.
3. An alternative isotopic method is to administer catecholamine precursors such as tyrosine or DOPA, which are ultimately converted to labeled dopamine and norepinephrine. Catecholamine turnover is then determined by applying principles of steady-state kinetics to the decline of norepi-

nephrine and/or dopamine specific activities. The disadvantage of this method is that the results are complicated by the persistence of labeled catecholamine precursors in the circulation and tissues for considerable lengths of time after administration. This complication can be minimized to some extent by "chasing" the label with unlabeled precursor in order effectively to dilute the specific activity of the original agent.

4. Another isotope method involves the administration of labeled tyrosine and calculation of the rate of catecholamine synthesis from the rate of conversion of labeled tyrosine to labeled catecholamine. Provided (a) that endogenous catecholamine tissue levels remain unchanged, (b) that total tissue norepinephrine is derived from synthesis, and (c) that newly synthesized norepinephrine is not utilized (released or metabolized) preferentially, this technique provides a useful estimate of turnover.
5. More indirect methods for estimation of amine turnover in the central nervous system involve the measurement of amine metabolites released into ventricular perfusion fluids or their concentration changes in the cerebrospinal fluid (CSF) or in brain tissue. The most recent extension of this technique has been the measurement of plasma monoamine metabolites. Preliminary studies have indicated that plasma levels of 3-methoxy-4-hydroxyphenethylene glycol and both 3,4-dihydroxyphenylacetic acid and homovanillic acid might provide an index of activity in central noradrenergic and dopaminergic neurons, respectively. An advantage of the CSF or plasma metabolite approach is its possible clinical application. It should be immediately pointed out that the degree of accumulation of a given metabolite in the central nervous system, cerebrospinal fluid, or blood is not necessarily an index of the quantity of metabolite formed, since most primary catecholamine metabolites can undergo further metabolism and be removed from the central nervous system, cerebrospinal fluid, or blood at different rates. Also drugs or various physiological conditions can alter routes of

> metabolism as well as rates of removal from the brain or cerebrospinal fluid independent of their effects on amine turnover. When estimation of metabolic end products of cerebral amine metabolism are used as indices to turnover, the interpretation remains extremely difficult, especially when alternative pathways for metabolism exist. Furthermore, many of the factors involved in the determination of steady-state concentrations of these metabolites in tissue, cerebrospinal fluid, and blood remain unknown.

In general the turnover and synthesis of brain monoamines is more rapid than that found in peripheral tissue. A major exception is the superior cervical ganglion where the synthesis and turnover of NE is quite rapid. In the brain the synthesis and turnover of dopamine and serotonin are more rapid than the synthesis of norepinephrine. This may explain why dopaminergic and serotonergic neurons, as opposed to noradrenergic neurons, can sustain high levels of physiological activity without producing substantial depletions of transmitter in their nerve terminals.

The dramatic difference between the rates of synthesis and turnover of brain catecholamines (i.e., NE and DA) may, in part, be explained by the fact that there appears to be a higher level of tyrosine hydroxylase associated with brain dopaminergic neurons. (See Table 9-3.) Brain areas enriched in dopaminergic cells or nerve terminals have higher tyrosine hydroxylase activity and turnover rates than noradrenergic-enriched areas. Similar to the observations made in the periphery, brain areas containing catecholamine cell bodies have higher tyrosine hydroxylase activity and turnover rates than the corresponding areas enriched in nerve terminals. However the synthesis and turnover rates measured in areas containing catecholamine cell bodies are not as closely coupled with synaptic function, like measures made in nerve terminal enriched areas.

Storage

A great conceptual advance made in the study of catecholamines more than twenty years ago was the recognition that in almost all

TABLE 9-3. Tyrosine hydroxylase activity and dopamine and norepinephrine turnover rate in rat brain areas

Region	*Tyrosine hydroxylase activity (nmoles DOPA/mg protein/hr)*	*Steady-state catecholamine content (μg/gm)*	*Catecholamine turnover rate (μg/gm/hr)*
DA Cell Bodies			
Substantia nigra	17.5 ± 2.7	1.7 ± .09	1.13
DA Terminals			
Caudate nucleus	12.0 ± 1.3	7.4 ± 0.4	2.12
Olfactory tubercle	11.1 ± 1.3	4.3 ± 0.5	1.54
Nucleus accumbens	7.7 ± 0.6	2.6 ± 0.2	.62
Amygdala	4.3 ± 0.8	0.9 ± 0.2	.25
NE Cell Bodies			
Locus ceruleus	4.6 ± 1.2	.99 ± 0.09	0.22
NE Terminals			
Pons-medulla	0.73 ± 0.06	.99 ± 0.03	0.13
Frontal cortex	0.27 ± 0.03	.33 ± 0.04	0.10
Cerebellum	0.20 ± 0.03	.40 ± 0.02	0.09
Hippocampus	0.17 ± 0.02	.43 ± 0.04	0.08

SOURCE: Data taken from N. J. Bacapoulos and R. K. Bhatnagar, *J. Neurochem.* *29*, 639, 1977.

tissues a large percentage of the norepinephrine present is located within highly specialized subcellular particles (colloquially referred to as "granules") in sympathetic nerve endings and chromaffin cells. Much of the norepinephrine in the central nervous system is also presumably located within similar particles. These granules contain adenosine triphosphate (ATP) in a molar ratio of catecholamine to ATP of about 4 : 1. Because of this perhaps fortuitous ratio, it is generally supposed that the anionic phosphate groups of ATP form a salt link with norepinephrine, which exists as a cation at physiological pH, and thereby serve as a means to bind the amines within the vesicles. Some such complex of the amines with ATP, with ATP associated with protein, or with protein directly is probable since the intravesicular concentration of amines, at least in the adrenal chromaffin granules and probably also in the splenic nerve granules (0.3 to 1.1 M), would be hypertonic if present in free solution and might be expected to lead to osmotic lysis of the vesicles.

The catecholamine storage vesicles in adrenal chromaffin and splenic nerve appear to have a number of general properties:

1. They possess an outer limiting membrane;
2. with appropriate fixation they possess an electron dense core when viewed in the electron microscope;
3. they contain the enzyme dopamine-β-hydroxylase;
4. they have a high concentration of catecholamine and ATP in a 4 : 1 ratio in adrenal and 6–8 : 1 in splenic nerve.
5. chromaffin and splenic nerve granules contain a characteristic soluble protein (chromogranin) also suggested to be involved in the storage process.

The various types of vesicles found in neural tissue or chromaffin cells are summarized in Figure 9-7, which indicates their approximate size and distribution. A number of possible functions that have been proposed for the catecholamine storage vesicles are summarized below:

1. They bind and store norepinephrine, thereby retarding its diffusion out of the neuron, and protect it from being destroyed by monoamine oxidase (MAO), which is considered to be intraneuronal;
2. they serve as a depot of transmitter that may be released upon the appropriate physiological stimulus;
3. they oxidize dopamine to norepinephrine;
4. they take up dopamine from the cytoplasm, protecting it from oxidation by monoamine oxidase.

It is generally believed that the vesicles are formed in the neuronal cell body and subsequently transported to the nerve terminal region. This assumption is based upon a number of findings:

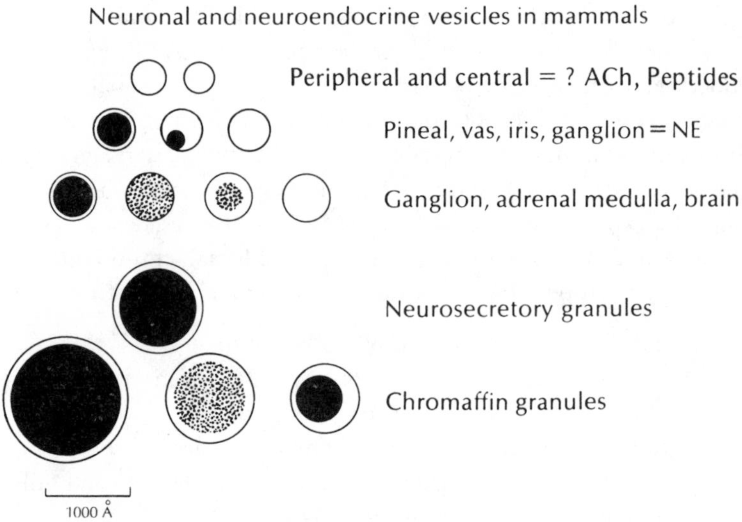

Neuronal and neuroendocrine vesicles in mammals

Peripheral and central = ? ACh, Peptides

Pineal, vas, iris, ganglion = NE

Ganglion, adrenal medulla, brain

Neurosecretory granules

Chromaffin granules

1000 Å

FIGURE 9-7. Size and electron microscopic appearance of the various types of synaptic vesicles, large granular vesicles, chromaffin granules, and neurosecretory granules found in various portions of the central and peripheral nervous system.

1. Only a minimal amount of protein biosynthesis occurs in axons or nerve terminals;
2. following the depletion of norepinephrine by reserpine, the amines first reappear within the perinuclear cytoplasm of the cell body;
3. when the axon of a sympathetic nerve is constricted, structures (mainly large granular vesicles containing norepinephrine) tend to accumulate at the proximal border.

Neuropeptides are also found in the adrenal medulla, and it has been demonstrated that enkephalins are stored in chromaffin vesicles. Just as it has been demonstrated that catecholamine synthesis increases during stimulation of chromaffin cells, enkephalin activity also increases with stimulation. The function of the released enkephalin, however, is still unclear.

Release

Much of the current knowledge regarding the release of catecholamines derives from the study of both adrenal medullary tissue and adrenergically innervated peripheral organs. Here it can be directly demonstrated that norepinephrine is released from the nerve terminals during periods of nerve stimulation and further that adrenergic fibers can sustain this output of transmitter during prolonged periods of stimulation, if synthesis and re-uptake of the transmitter are not impaired. In the periphery it is possible to isolate the innervated end-organs, to collect a perfusate from the vascular system during nerve stimulation, and to analyze this for the presence and quantity of a given putative transmitter. However, we really know little of the mechanism by which axonal nerve impulses arriving at the terminal cause the release of norepinephrine (excitation-secretion coupling). Our best appreciation of the events comes from the possibly analogous release of catecholamines from the adrenal medulla, which has been extensively studied. For a comprehensive review of this area the reader is referred to Castel *et al* (1984).

The mechanism of catecholamine release by the adrenal medulla is thought to take place as follows: with activity in the preganglionic fibers, acetylcholine is released and thought to combine with the plasma membrane of the chromaffin cells. This produces a change in membrane protein conformation altering the permeability of this membrane to Ca^{2+} and other ions, which then move inward. The influx of Ca^{+2} is believed to be the main stimulus responsible for the mobilization of the catecholamines and for their secretion. The current view is that the catecholamines are released from the chromaffin cell by a process of exocytosis, along with chromogranin, ATP, and a little dopamine-β-hydroxylase. Whether or not these cellular phenomena are applicable to the sympathetic nerve endings in general remains to be determined.

Teleologically, it appears rather unlikely that transmitter release from sympathetic nerve terminals does occur exclusively by a process of exocytosis. Exocytosis requires that the entire content of the granular vesicle be released (i.e., catecholamine, ATP, and soluble protein). Since the nerve terminal region, as far as we know now, cannot sustain any sort of protein biosynthesis, high rates of axonal flow from the nerve cell body would be required to replenish the protein lost during the process of exocytosis. Alternatively, one might propose a "protein re-uptake" mechanism in order to recapture that protein released during the process of synaptic transmission. In peripheral nerve, release of norepinephrine has been shown to be frequency dependent within a physiological range of frequency. This release of norepinephrine, similar to the release of ACh, is Ca^{2+}-dependent. Some evidence has also been presented which indicates that newly synthesized norepinephrine may be released preferentially. This preferential release is additional evidence to support the contention that norepinephrine exists in more than one pool within the sympathetic neuron.

It has been a great deal more difficult to demonstrate release of a putative transmitter from a given type of nerve ending in the central nervous system. Thus it is impossible to perfuse specific localized brain areas through their vascular system. In addition, it is also difficult but not necessarily impossible to stimulate selectively a well-defined neuronal pathway within the brain. (See Chapter 10.) However, with the independent development of the push-pull cannula and the chemitrode, and most recently electrochemical detectors, it has become possible to collect catecholamines from certain deep nuclear masses of the central nervous system or to measure release directly by *in vivo* voltammetry. In addition, many less-sophisticated techniques such as cortical cups, ventricular perfusions, brain slices, and isolated spinal cords have been used to demonstrate release of putative transmitters, including norepinephrine and dopamine from central nervous system tissue. All of these physiological techniques for studying release have some disadvantages and limitations, and many are somewhat gross and indirect. (See Chapter 2.) Nevertheless, they do represent increasing sophistication in research on release occurring in the central nervous system.

Regulation of Release

Much evidence is accumulating to suggest that endogenous humoral factors may regulate release by a direct local action on catecholamine nerve terminals. Not only can the local synaptic concentration of catecholamines modulate their own release by interacting with presynaptic autoreceptors but also prostaglandins, vasoactive amines, polypeptides such as angiotensin II, and acetylcholine have all been implicated in regulation of catecholamine release.

The most convincing evidence, however, has been generated in support of a role for presynaptic receptors in the modulation of impulse-induced catecholamine release in both central and peripheral catecholamine neurons. In most catecholamine-containing systems tested, administration of catecholamine agonists appears to attenuate stimulus-induced release while administration of catecholamine receptor blockers augments release. These pharmacological studies have established the concept that presynaptic receptors modulate release by responding to the concentration of catecholamine in the synapse (high concentrations inhibiting release and low concentrations augmenting release). Most recently presynaptic autoreceptors have also been implicated in the regulation of dopamine synthesis. (See Chapter 10.)

Prostaglandins of the E series are also potent inhibitors of neurally induced release of norepinephrine in a great number of tissues, and their action appears to be dissociated from any interaction with presynaptic receptors. These substances are released from sympathetically innervated tissues, and most evidence indicates that inhibition of local prostaglandin production is associated with an increase in the release of norepinephrine and subsequent effector responses induced by neuronal activity. The control of norepinephrine release by this prostaglandin-mediated feedback mechanism appears to operate through restriction of calcium availability for the norepinephrine release process and seems to be most efficient within the physiological frequency range of nerve impulses.

Metabolism

The metabolism of exogenously administered or endogenous catecholamines differs markedly from that of acetylcholine in that the speed of degradation of the amines is considerably slower than that of the ACh-ACh-esterase system. The major mammalian enzymes of importance in the metabolic degradation of catecholamines are monoamine oxidase and catechol-*O*-methyltransferase (COMT) (Fig. 9-8). Monoamine oxidase is an enzyme which converts catecholamines to their corresponding aldehydes. This aldehyde intermediate is rapidly metabolized, usually by oxidation by the enzyme aldehyde dehydrogenase to the corresponding acid. In some circumstances, the aldehyde is reduced to the alcohol or glycol by aldehyde reductase. In the case of brain norepinephrine, reduction of the aldehyde metabolite appears to be the favored route of metabolism. Neither of these latter enzymes has been extensively studied in neuronal tissue. Monoamine oxidase is a particle-bound protein, localized largely in the outer membrane of mitochondria, although a partial microsomal localization cannot be excluded. There is also some evidence for a riboflavin-like material in monoamine oxidase isolated from liver mitochondria. Monoamine oxidase is usually considered to be an intraneuronal enzyme, but it occurs in abundance extraneuronally. In fact, most experiments indicate that chronic denervation of a sympathetic end-organ leads only to a relatively small reduction in monoamine oxidase, suggesting that the greater proportion of this enzyme is, in fact, extraneuronal. However, it is the intraneuronal enzyme that seems to be important in catecholamine metabolism. Monoamine oxidase present in human and rat brain exists in at least two different forms designated type A and type B based on substrate specificity and sensitivity to inhibition by selected inhibitors. Clorgyline is a specific inhibitor of the A-type enzyme, which has a substrate preference for norepinephrine and serotonin. Deprenyl is a selective inhibitor of the B-type enzyme, which has a substrate preference for β-phenylethylamine and benzylamine as substrates. Dopamine, tyramine, and tryptamine appear to be equally good substrates for both forms of the enzyme.

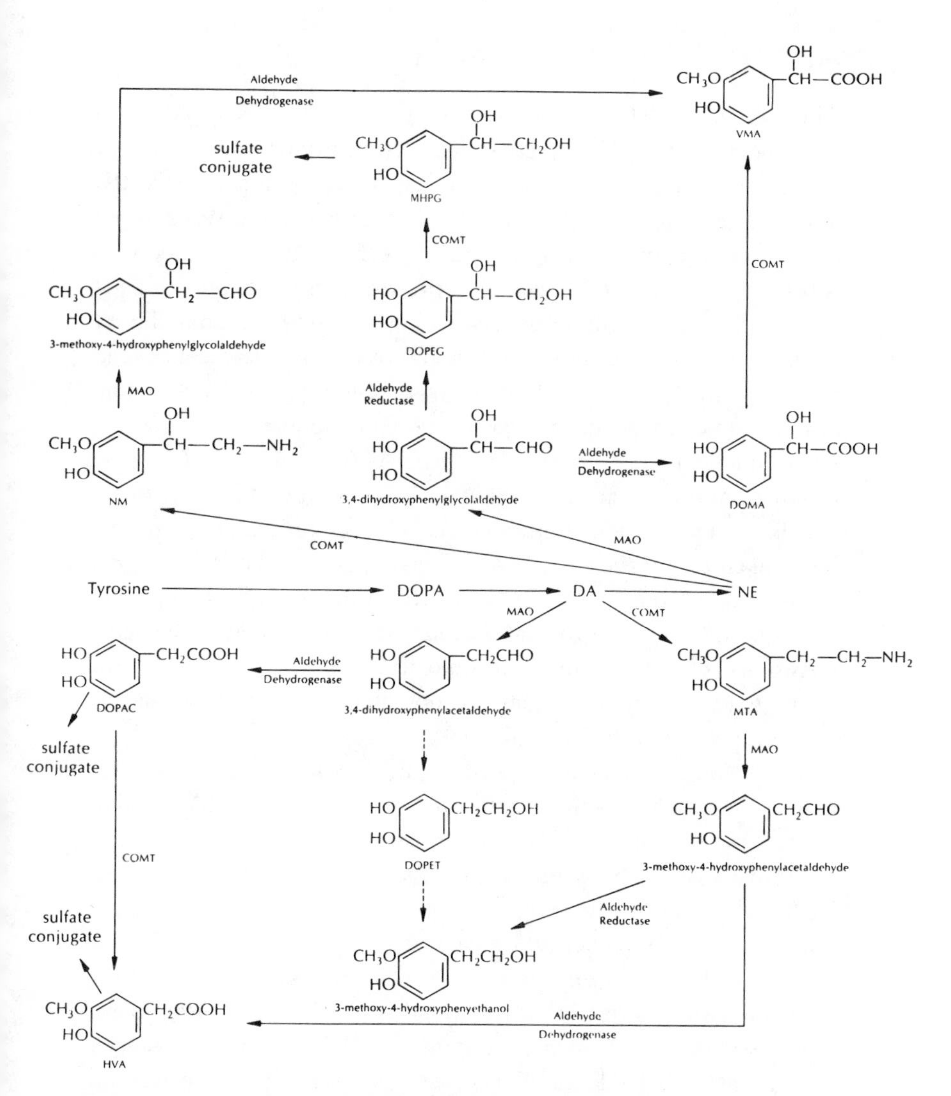

FIGURE 9-8. Dopamine and norepinephrine metabolism. The following abbreviations are used: DOPA, dihydroxyphenylalanine; DA, dopamine; NE, norepinephrine; DOMA, 3,4-dihydroxymandelic acid; DOPAC, 3,4-dihydroxyphenylacetic acid; DOPEG, 3,4-dihyroxyphenylglycol; DOPET, 3,4-dihydroxyphenylethanol; MOPET, 3-methoxy-4-hydroxyphenylethanol; MHPG, 3-methoxy-4-hydroxy-phenylglycol; HVA, homovanillic acid; VMA, 3-methoxy-4-hydroxy-mandelic acid; NM, normetanephrine; MTA, 3-methoxytyramine; MAO, monoamine oxidase; COMT, catechol-*O*-methyl transferase; Dashed arrows indicated steps that have not been firmly established.

The role and relative importance of these two types of MAO in physiological and pathological states is currently unknown, but this is an important area for further research. It has been speculated that under certain circumstances monoamine oxidase could serve to regulate norepinephrine biosynthesis by controlling the amount of substrate, dopamine, available to the enzyme dopamine-β-hydroxylase. Monoamine oxidase is not an exclusive catabolic enzyme for the catecholamines since it also oxidatively deaminates other biogenic amines such as 5-hydroxytryptamine, tryptamine, and tyramine. The intraneuronal localization of monoamine oxidase in mitochondria or other structures suggests that this would limit its action to amines that are present in a free (unbound) form in the axoplasm. Here, monoamine oxidase can act on amines that have been taken up by the axon before they are granule bound, or it can even act on amines that are released from the granules before they pass out through the axonal membrane. Interestingly, the latter possibility seems of minor physiological importance, since monoamine oxidase inhibition does not potentiate the effects of peripheral sympathetic nerve stimulation.

The second enzyme of importance in the catabolism of catecholamines is catechol-*O*-methyl transferase discovered by Axelrod in 1957. This enzyme is a relatively nonspecific enzyme that catalyzes the transfer of methyl groups from *S*-adenosyl methionine to the *m*-hydroxyl group of catecholamines and various other catechol compounds. Catechol-*O*-methyl transferase is found in the cytoplasm of most animal tissue, being particularly abundant in kidney and liver. A substantial amount of this enzyme is also found in the central nervous system and in various sympathetically innervated organs. The precise cellular localization of catechol-*O*-methyl transferase has not been determined although it has been suggested (with little foundation) to function extraneuronally. The purified enzyme requires *S*-adenosyl methionine and Mg^{2+} ions for activity. As with monoamine oxidase, inhibition of catechol-*O*-methyl transferase activity does not markedly potentiate the effects of sympathetic nerve stimulation, although in some tissue it tends to prolong the duration of the response to stimulation. Therefore, neither monoamine oxidase nor catechol-*O*-methyl transferase would seem to be the primary mechanism for terminating the action of norepinephrine liberated at sympathetic nerve terminals. It may be, however, that these enzymes play a more important role in terminating transmitter action and regulating catecholamine function in the central nervous system.

Uptake

When sympathetic postganglionic nerves are stimulated at frequencies low enough to be comparable to those encountered physiologically, very little intact norepinephrine overflows into the circulation, suggesting that local inactivation is very efficient. This local inactivation is not significantly blocked when catechol-*O*-methyl transferase or monoamine oxidase or both are inhibited, and it is believed to involve mainly re-uptake of the transmitter by sympathetic neurons.

Considerable attention has been directed to the role of tissue uptake mechanisms in the physiological inactivation of catecholamines. But only in recent years has this concept received direct experimental support, although a number of earlier findings had in fact suggested that catecholamines might be inactivated by some sort of nonmetabolic process. For example, more than forty years ago Burn suggested the possibility that exogenous norepinephrine might be taken up in storage sites in peripheral tissue. About thirty years ago it was demonstrated that an increase in the catecholamine content of cat and dog heart occurred after administration of norepinephrine and epinephrine *in vivo*. However, in view of the very high doses (mg) employed in these experiments, the significance of these findings remained questionable. It was really not until labeled catecholamines with high specific activity became available that similar experiments using doses of norepinephrine comparable to those likely to be encountered under physiological circumstances could be performed. Uptake studies of this nature carried out *in vivo* have indicated that approximately 40 to 60 percent of a relatively small intravenous dose of norepinephrine is metabolized enzymatically by catechol-*O*-methyl transferase and monoamine oxidase, while the remainder is inactivated by uptake into various tissues. This uptake appears to be extremely rapid, and in most cases the magnitude of the uptake is related to the density of sympathetic innervation and the proportion of cardiac output received by the tissue in question. Thus, after administration of ^{3}H-norepinephrine *in vivo*, the greatest uptake and binding occur in tissues such as spleen and heart. The brain is also capable of taking up norepinephrine but, since the uptake from the circulation is prevented by the blood–brain barrier, efficient uptake can only be observed when brain slices, minces, or brain synaptosomes are exposed to norepinephrine, or when the labeled amines are administered intraventricularly or intracisternally.

A large amount of data now suggests that the major site of this uptake (and subsequent binding) actually occurs in sympathetic nerves. This evidence can be summarized as follows:

1. In most cases norepinephrine uptake correlates directly with the density of sympathetic innervation (or the endogenous content of norepinephrine) providing that sufficient allowance is made for differences in regional blood flow to various tissues.
2. In tissues without a normal sympathetic nerve supply, the ability to take up exogenous catecholamines is severely impaired. This can be demonstrated by surgical sympathectomy, chemical sympathectomy, or immunosympathectomy.
3. Labeled norepinephrine taken up by heart, spleen, artery, vein, or other tissues can be subsequently released by sympathetic nerve stimulation.
4. Autoradiographic studies at the electron miscroscope level have directly localized labeled catecholamines within the neuronal elements of the tissue.
5. Fluorescence microscopy has also provided equally direct evidence for the localization of norepinephrine uptake to sympathetic nerves.

A number of studies demonstrate that the uptake of catecholamines is active as it proceeds against a concentration gradient. For example, slices or minces of heart, spleen, and brain can concentrate norepinephrine to levels five to eight times those in the incubation medium. In intact tissues even greater concentration ratios between tissue and medium may be obtained. In isolated rat heart perfused with Krebs solution containing 10 ng/ml of norepinephrine, Iversen finds concentrations of labeled norepinephrine rise thirty to forty times above that present in the perfusion medium. If we assume that norepinephrine uptake occurs almost exclusively into cardiac sympathetic neurons, the uptake process clearly has an exceptionally high affinity for norepinephrine. In fact, the actual concentration ratios between exogenous norepinephrine accumulated within the sympathetic neurons and that present in the medium could approach 10,000 : 1.

This uptake process is a saturable membrane transport process dependent upon temperature and requiring energy. The stereochemically preferred substrate is L-norepinephrine; furthermore, norepinephrine is taken up more efficiently than its *N*-substituted derivatives. The uptake process is sodium dependent and can be blocked by inhibition of Na^+, K^+-activated ATPase. In fact, most evidence suggests that catecholamine uptake is mediated by some sort of active membrane transport mechanism located in the presynaptic terminal membrane of postganglionic sympathetic neurons.

Axonal Catecholamine Transport

The early estimates by Weiss of the rate of axoplasmic transport based both on the rate of regeneration of axons and on nerve constriction experiments suggested that the total mass of axonal material is transported distally at a rate of approximately 1 mm/day. More recent studies employing the use of radioactive substances and autoradiography have evaluated the transport of individual components of the axoplasm. Numerous demonstrations indicate some proteins are transported at rates of approximately 1 mm/day; but, in addition, a heterogeneity of transport rates of some axoplasmic constituents, ranging up to 10 cm/day and even up to 1 meter/day, have also been reported. Many of these experiments have been concerned with the axonal transport of catecholamines, as described by Dahlström and her colleagues in Sweden. They found that ligated sympathetic nerves accumulate norepinephrine above the ligation indicating a proximodistal transport of amine-containing particles from the nerve cell body. Not only was this accumulation studied biochemically but also histochemical fluorescence observations of the zone of constriction were made at different times after ligation. It was found that when a sympathetic axon is crushed or ligated, there is an almost immediate accumulation of norepinephrine on the cell body side of the ligature or crush. After 12 to 24 hr there is a marked accumulation of norepinephrine in the segment proximal to the crush, with only a weak or small accumulation in the distal segment. The increase in norepinephrine on the proximal side proceeds linearly up to about 48 hr, at which time it tends to level off. Pharmacological treatment with reserpine indicated that the accumulated norepinephrine is stored within the granules. From the content of norepinephrine in 1 cm segments of nonligated axon and the rate of accumulation of norepinephrine in the 1 cm segment proximal to the ligature, the velocity of the transport of the norepinephrine-containing particles was calculated to be about 5 mm/hr in the rat, 10 mm/hr in the cat, and 2 to 3 mm/hr in the rabbit.

Dahlström also made tandem constrictions by using two ligatures and found that the accumulation of norepinephrine was considerably larger above the proximal ligation than above the distal ligation. In the part of the nerve between the two ligatures, the total norepinephrine did not exceed the amount in a normal unligated nerve segment of the identical length, suggesting that local synthesis was probably not responsible for the accumulation normally observed. Other experiments conducted in the central nervous system have suggested that the transport rate in this tissue is about 0.5 mm/hr.

Even in view of these fairly rapid rates of transport of catecholamine storage granules it seems unlikely that this transport process serves to replace transmitter normally lost during nerve activity. More probably this transport serves to replace the granular protein structure that houses or binds the catecholamine and turns over at a slower rate than that of the amines.

Neurotransmitter Role

On the basis of many experiments demonstrating that norepinephrine is synthesized and stored in sympathetic neurons, that it is released in significant quantities in response to sympathetic nerve stimulation, that it produces effector organ responses identical to those produced by sympathetic nerve stimulation, and that drugs which antagonize the action of exogenous norepinephrine also block the effect of sympathetic nerve stimulation of the organ in question, it has been accepted by most investigators that norepinephrine is in fact the sympathetic neurotransmitter in the mammalian peripheral sympathetic nervous system.

At the present time it is also well accepted that norepinephrine acts as a neurotransmitter in the central nervous system, although the most compelling evidence applies only to certain regions innervated by noradrenergic fibers. As described in Chapter 2, the central nervous system obviously presents an immediate obstacle to cellular investigations, namely, one of inaccessibility. Not only is the central nervous system of mammals relatively inaccessible but it does not have the clear-cut focal synaptic regions found in the peripheral nervous system. The cell bodies, dendrites, and axons of central neurons are usually studded with synaptic boutons. This, of course, provides the investigator with a totally heterogeneous system of chemical and perhaps electrical inputs in close proximity to each other, making it difficult to evaluate events taking place at a particular type of nerve ending. A number of criteria, of course, must be fulfilled before it is possible to accept a putative transmitter candidate as a true neurotransmitter at a given synaptic junction. These criteria have been reviewed in Chapter 2. Many of the proposed criteria for neurotransmitters have been satisfactorily fulfilled for norepinephrine and dopamine at some specific synaptic sites in the mammalian central nervous system. The major difficulties in transmitter identification in the central nervous system have been in demonstrating the release of the transmitter and in comparing the postsynaptic effects of presynaptic nerve stimulation and those obtained upon application of the putative transmitter to the postsynaptic receptor. If one imposes the most stringent criteria for evaluating release, it is necessary to demonstrate that the putative transmitter is released into the diffusible volume of the synaptic cleft in response to stimulation and depolarization of a given nerve terminal. At present no micromethod has yet been devised by which it is possible to study release at this fine level of refinement in the central nervous system. However, *in vivo* voltammetry shows great promise in this regard. If one is not so critical, release of monoamines from central nervous system tissue can be demonstrated by a number of experimental techniques. For example, norepinephrine is released from brain slices by gross nonspecific electrical stimulation. It has also been shown that stimulation of an isolated spinal cord produces a release of some norepinephrine and 5-hydroxytryptamine into the medium.

With even less stringent criteria for release, perfusion of the lateral ventricles of cats reveals detectable amounts of dopamine, norepinephrine, and their metabolites in the cerebrospinal fluid after periods of low-level stimulation of certain specific brain areas. Here, however, we must assume that the putative transmitter released from certain nerve endings within the brain escapes the recapture process and diffuses through the brain substance proper to be detected eventually in the cerebrospinal fluid. More "refined" techniques such as push-pull cannulae and chemitrodes have also provided evidence for norepinephrine and dopamine release in both acute and chronically implanted animals. This latter release is also enhanced by electrical stimulation or by certain drugs. A number of indirect techniques have also been used to demonstrate release *in vivo.* Thus, fluorescence histochemical experiments demonstrating depletion of nerve terminal fluorescence have been used as an index of release. Similarly, biochemical studies demonstrating depletion of catecholamines have been used as an index of release. However, other plausible explanations of this depletion such as acceleration of catabolism should also be considered.

A number of published reports have now demonstrated the feasibility of employing *in vivo* voltammetry to measure the release of monoamines in or from brain. For example, electrochemical methods have been used to measure *in vivo* changes in homovanillic acid (HVA) and 5-hydroxyindole acetic acid (5-HIAA) in the cerebrospinal fluid of anesthetized rats. More recently, this approach has been used successfully to monitor release of dopamine and/or 3,4-dihydroxyphenylacetic acid (DOPAC) from the striatum of anesthetized rats following stimulation of the substantia nigra or in unanesthetized rats following amphetamine administration. 5-Hydroxytryptamine release in the brain of the freely moving, unanesthetized rat has also been measured by voltammetry following administration of *p*-chloroamphetamine. Despite great potential, the technique of voltammetry is still in its infancy, and many technical problems must be overcome before it will be possible to measure synaptic release of monoamines in a selective, specific, and quantitative manner. However, this technique shows great promise for the future especially since it may be feasible to couple this technique with single-cell electrophysiological recordings.

It has also been difficult in the central nervous system to demonstrate that administration of an exogenous quantity of a putative transmitter produces the same effect as nerve terminal depolarization and that drugs which block the action of this natural transmitter at the receptor also exhibit the same degree of sensitivity toward the putative transmitter. Microiontophoresis has provided a powerful tool toward this end.

By means of this technique, these criteria have been satisfied for norepinephrine in the cerebellum, the hippocampus, cerebral cortex, and the olfactory bulb and for dopamine in the caudate nucleus. In the olfactory bulb norepinephrine may be a mediator of a recurrent inhibitory synaptic pathway since: (1) iontophoretically applied norepinephrine depresses spontaneous activity; (2) this effect as well as depression due to synaptic inhibition can be blocked by dibenamine, an α-adrenergic blocking agent; (3) synaptic inhibition is reduced when catecholamines are depleted by pretreating animals with reserpine and α-methyl-*p*-tyrosine; (4) norepinephrine-containing nerve endings appear to be restricted to the layer in the olfactory bulb where synaptic inhibition is most easily demonstrated.

In the caudate nucleus and spinal cord, areas rich in catecholamine-containing nerve endings, α-receptor blocking drugs are also capable of blocking the response to iontophoretically applied catecholamines. Further, in the caudate nucleus this response to the catecholamines is identical with that resulting from stimulation of the substantia nigra where dopamine-containing nerve endings are thought to arise. The effects of iontophoretically applied dopamine can also be blocked by chlorpromazine.

In almost all cases the response of single neurons to norepinephrine and dopamine is a depression of spontaneous activity. In the cerebellum, hippocampus, and cerebral cortex the norepinephrine response appears to be mediated by cyclic AMP.

It is clear that a great deal of evidence favors the possibility that both norepinephrine and dopamine do play such a transmitter role in the central nervous system.

Pharmacology of Catecholamine Neurons

In this section we will focus on the pharmacology of peripheral noradrenergic neurons and discuss possible sites of drug involvement in the life cycle of the catecholamines. Figure 9-9 depicts a schematic model of a noradrenergic nerve varicosity.

Site 1—Precursor transport. As noted earlier there appears to be an active uptake or transport of tyrosine as well as other aromatic amino acids into the central nervous system. In addition it is also probable that tyrosine is actively taken up by adrenergic neurons both in the periphery and in the central nervous system. At present, we have no pharmacologic agent that specifically antagonizes the uptake of tyrosine into the brain or into the catecholamine-containing neu-

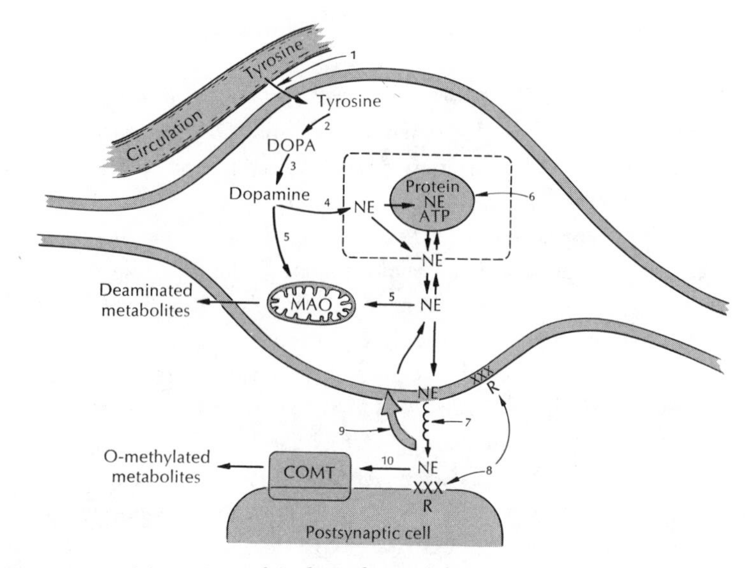

FIGURE 9-9. Schematic model of noradrenergic neuron.

ron. However, it is known that various aromatic amino acids will compete with each other for transport into the central nervous system. This competition might become important under a pathological situation in which blood aromatic amino acids are elevated. Thus in phenylketonuria, when plasma levels of phenylalanine are elevated to about 10^{-3} M, it might be expected that tyrosine and tryptophan uptake into the brain might be diminished.

Site 2—Tyrosine hydroxylase. The conversion of tyrosine to DOPA is the rate-limiting step in this biosynthetic pathway and thus is the step most susceptible to pharmacological manipulation. The various classes of tyrosine hydroxylase inhibitors were considered earlier in this chapter. α-methyl-ρ-tyrosine, an effective inhibitor of tryosine hydroxylase both *in vivo* and *in vitro*, is one of the few tyrosine hydroxylase inhibitors which has been employed clinically and found to be effective in the treatment of inoperable pheochromocytomas; it is of little or no value in the treatment of essential hypertension.

Sites 3 and 4—DOPA-decarboxylase and dopamine-β-hydroxylase. Potent inhibitors of DOPA-decarboxylase and dopamine-β-hydroxylase do exist, but they are generally not effective *in vivo* in reducing tissue concentrations of catecholamines.

Due to the nonspecificity of these two enzymes (see above) as well as the relatively nonspecific vesicular storage sites, certain drugs or even amines or amine precursors in foodstuff can be synthesized to phenylethylamine derivatives with a catechol or a β-hydroxyl group and taken up and stored in the storage vesicles. These compounds can subsequently be released by sympathetic nerves, in part replacing some of the norepinephrine normally released by sympathetic impulses. Since these "false neurotransmitters" are usually less potent in their action, replacement of some of the endogenous, releasable pool of norepinephrine with these agents results in a diminished sympathetic response. Thus a drug such as α-methyl-dihydroxphenylalanine, which is converted in the sympathetic neuron to α-methyl-norepinephrine, replaces a portion of the endogenous norepinephrine and itself can be subsequently released as a "false transmitter." The false transmitter theory also has been used to explain the hypotensive action of various monoamine oxidase inhibitors. Tyramine, an endogenous metabolite that is normally disposed of by monoamine oxidase in the liver and other tissues, is no longer rapidly deaminated in the presence of monoamine oxidase inhibitors and thus can be converted to the β-hydroxylated derivative, octopamine, which then accumulates in sympathetic neurons. Upon activation of these nerves, octopamine is released and acts as a false neurotransmitter, producing a resultant decrease in sympathetic tone (observed with monoamine oxidase inhibitors). "False neurotransmitters" can also be formed in the brain, but at the present time the action of these agents is unclear.

Site 5—Metabolic degradation by monoamine oxidase. The monoamine oxidase inhibitors comprise quite a heterogeneous group of drugs that have in common the ability to block the oxidative deamination of various biogenic amines. These agents all produce marked increases in tissue amine concentrations. These compounds are in most cases employed clinically for the treatment of depression. However, some of these agents are used for the treatment of hypertension and angina pectoris. In the case of monoamine oxidase inhibitors, it should be immediately pointed out that while these drugs inhibit monoamine oxidase, there is no firmly established correlation between monoamine oxidase inhibition and therapeutic effect. In fact, it is generally well appreciated that monoamine oxidase inhibitors inhibit not only enzymes involved in oxidative deamination of monoamines but in addition many other enzymes unrelated to monoamine oxidase such as succinic dehydrogenase, dopamine β-hydroxylase, 5-hydroxy-tryptophan decarboxylase, choline dehydrogenase, and diamine oxidase.

Site 6—Storage. A number of drugs, mostly notable the Rauwolfia alkaloids, interfere with the storage of monoamines by blocking the uptake of amine into the storage granules or by disrupting the binding of the amines. Thus drugs interfering with the storage mechanism will cause amines to be released intraneuronally. Amines released intraneuronally appear to be preferentially metabolized by monoamine oxidase.

Site 7—Release. The release of catecholamines is dependent upon Ca^{2+} ions. Thus, in an *in vitro* system it is possible to block release by lowering the Ca^{2+} concentration of the medium. Although drugs such as bretylium act to inhibit the release of norepinephrine in the peripheral nervous system, this drug does not easily penetrate the brain and thus has little or no effect on central monoaminergic neurons.

Site 8—Interaction with receptors. In the peripheral nervous system, α- and β-adrenergic blocking agents are effective. These agents block both the effects of sympathetic nerve stimulation and the action of exogenously administered norepinephrine. Some of these agents also have central activity. Evidence has indicated that presynaptic-α-receptors, termed α_2-receptors, are involved in the regulation of norepinephrine release. A blockade of these receptors facilitates release while stimulation leads to an inhibition of release. In certain tissues presynaptic β-receptors have also been identified. In contrast to α_2-receptors, activation of these presynaptic β-receptors facilitates rather than attenuates norepinephrine release.

Site 9—Re-uptake. Drugs such as cocaine, desipramine, amitryptyline, and other related tricyclic antidepressants are effective inhibitors of norepinephrine uptake both at peripheral and central sites. Pharmacological intervention at these sites makes more norepinephrine available to interact with postsynaptic receptors and thus tends to potentiate adrenergic transmission.

Site 10—Catechol-O-methyl transferase. Catechol-*O*-methyl transferase can be inhibited by pyrogallol or various tropolone derivatives. Inhibition of this enzyme in most sympathetically innervated tissue such as heart does not significantly potentiate the effects of nerve stimulation. In vascular tissue, however, inhibition of this enzyme does lead to a significant prolongation of the response to nerve stimulation. It appears possible that in this tissue catechol-*O*-methyl transferase may play some role in terminating transmitter action.

Selected References

Adams, R. N., and C. A. Marsden (1982). Electrochemical detection methods for monoamine measurements *in vitro* and *in vivo.* In *Handbook of Psychopharmacology*, Vol. 15, pp. 1–74. Plenum, New York. (L. L. Iverson, S. D. Iversen, and S. H. Snyder, eds.),

Axelrod, J. (1973). The fate of noradrenaline in the sympathetic neurone. *Harvey Lect. 67*, 175.

Björklund, A., and T. Hökfelt, eds. (1983). *Handbook of Chemical Neuroanatomy*, Vol. 1, Methods in Chemical Neuroanatomy. Elsevier, Amsterdam.

Bloom, F. E. (1972). Electromicroscopy of catecholamine-containing structures. In *Handbook of Experimental Pharmacology*, Vol. 33, Catecholamines, p. 46. Springer-Verlag, Berlin.

Burks, T. F., ed. (1985) Dopamine neurotoxic features of MPTP. *Life Sci. 36*, 201.

Castel, M., H. Gainer, and H.-D. Dellman (1984). Neuronal secretory systems. *International Rev. Cytology*, *88*, 303.

Costa, E., S. H. Koslow, and H. F. LeFevre (1975). Mass fragmentography: A tool for studying transmitter function at synaptic level. In *Handbook of Psychopharmacology*, Vol. 1 (L. L. Iversen, S. D. Iversen, and S. H. Snyder, eds.), p. 1. Plenum, New York.

Dunn, A. J., and N. R. Kramercy (1984). Neurochemical responses in stress: Relationships between the hypothalamic-pituitary-adrenal and catecholamine systems. In *Handbook of Psychopharmacology*, Vol. 18 (L. L. Iversen, S. D. Iversen, and S. Y. Snyder, eds.), p. 455. Plenum, New York.

El Mestikawy, S., H. Gozlan, J. Glowinski, and M. Hamon (1985). Characteristics of tyrosine hydroxylase activation by K^+-induced depolarization and/or forskolin in rat striatal slices. *J. Neurochem.* *45*, 173–184.

Euler, U. S. von (1956). *Noradrenaline*. Charles C Thomas, Springfield, Illinois.

Gibson, C. J. (1985). Control of monoamine synthesis by precursor availability. In *Handbook of Neurochemistry*, 2nd ed., Vol. 8 (A. Lajtha, ed.), p. 309. Plenum, New York.

Iversen, L. L., S. D. Iversen, and S. H. Snyder, eds. (1975a). Biochemical principles and techniques in neuropharmacology. In *Handbook of Psychopharmacology*, Vol. 1. Plenum, New York.

Iversen, L. L., S. D. Iversen, and S. H. Snyder, eds. (1975b). Biochemistry of biogenic amines. In *Handbook of Psychopharmacology*, Vol. 3. Plenum, New York.

Joh, T. H., E. E. Baetge, M. E. Ross, and D. J. Reiss (1983). Evidence for the existence of homolgous gene coding regions for the catecholamine biosynthetic enzymes. In *Cold Spring Harbor Symposia on Quantitative Biology*, Vol. XLVII, p. 327.

Jonsson, G., T. Malmfors, and C. Sachs, eds. (1975). 6-Hydroxydopamine as a denervation tool in catecholamine research. In *Clinical Tools in Catecholamine Research I*. Elsevier, New York.

Lindvall, O. and A. Björklund (1983). Dopamine and norepinephrine containing neuron systems: Their anatomy in the rat brain. In *Chemical Neuroanatomy* (P. C. Emson, ed.). Raven Press, New York.

Marsden, C. A., ed. (1984) Measurement of neurotransmitter release *in vivo*. In *Methods in Neuroscience*, Vol 6. Wiley, New York.

Milner, J. D., and R. J. Wurtman (1986). Catecholamine synthesis: Physiological coupling to precursor supply. *Biochemical Pharmacol.*, in press.

Usdin, E., N. Weiner, and E. Costa, eds. (1981). *Structure and Function of Monoamine Enzymes*, Vol. 2. Macmillan, New York.

Usdin, E., A. Carlsson, A. Dahlström, and A. Engel, eds. (1984). *Catecholamines* Part A, Basic and Peripheral Mechanisms. *Proceedings of the Fifth International Catecholamine Symposium*, Göteborg, Sweden. Alan R. Liss, New York.

10 | Catecholamines II: CNS Aspects

The cellular organization of the brain and spinal cord has been studied for many decades by classical histological and silver impregnation techniques. With the development in the last decade of completely different histological techniques based on the presence of a given type of transmitter substance or on specific synthetic enzymes involved in the formation of a given transmitter, it became possible to map chemically defined neuronal systems in the CNS of many species. With the formaldehyde fluorescence histochemical method of Falck and Hillarp (Chapter 9), it has been possible to delineate norepinephrine- dopamine- and 5-hydroxytryptamine-containing systems in mammalian brain. By the time such techniques had been employed for several years, it became clear that the distribution of these chemically defined neuronal systems did not necessarily correspond to that of systems described earlier with the classical techniques.

It was really not until the mid 1960s that the histochemical fluorescent technique was applied to brain tissue and the anatomy of the monoamine-containing neuronal systems was described. By a variety of pharmacological and chemical methods, it has subsequently been possible to discriminate between norepinephrine, epinephrine, and dopamine-containing neurons and to describe in detail the distribution of these catecholamine-containing neurons in the mammalian central nervous system. Several recent and thorough surveys of these systems are available.

Figure 10-1 provides a schematic model of a central monoamine-containing neuron as observed by fluorescence and electron microscopy. The cell bodies contain relatively low concentrations of amine (about 10–100 μg/gm), while the terminal varicosities contain a very

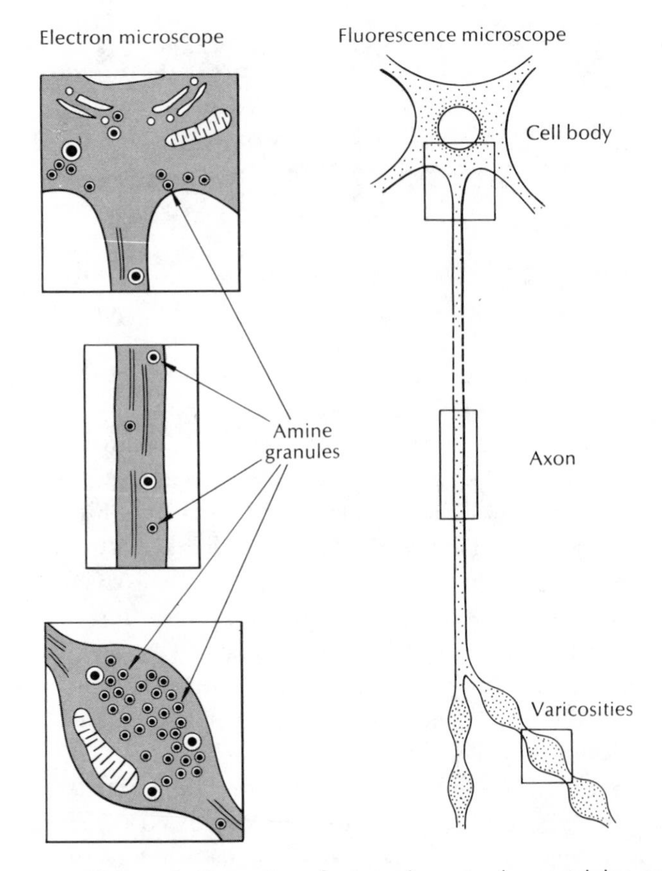

FIGURE 10-1. Schematic illustration of a central monoamine-containing neuron. The right side depicts the general appearance and intraneuronal distribution of monoamines based upon fluorescence microscopy. The cell bodies contain a relatively low concentration of catecholamine (about 10–100 μg/gm) while the terminal varicosities contain a very high concentration 1000–3000 μg/gm). The preterminal axons contain very low concentration of amine. At the electron microscopic level (left side), dense core granules can be observed in the cell bodies and axons but appear to be highly concentrated in the terminal varicosities. (Illustration adapted after K. Fuxe and T. Hökfelt, in *Triangle 30:* 73, 1971)

high concentration (about 1000–3000 μ/gm). The axons, on the other hand, consist of highly branched, largely unmyelinated fibers which have such a low concentration of amine that they are barely visible in untreated adult animals. With the electron microscope, depending on the type of fixation, it is possible to observe the presence of small granular vesicles that are thought to represent the subcellular storage sites containing the catecholamines. (See Chapter 9). These granular vesicles are concentrated in the terminal varicosities of the sympathetic neuron.

Systems of Catecholamine Pathways in the CNS

Detailed analysis of catecholamine pathways in the CNS has been greatly accelerated by recent improvements in the application of fluorescence histochemistry, such as the use of glyoxylic acid as the fluorogen, and by the application of numerous auxiliary mapping methods, such as anterograde and retrograde markers, immunocytochemical localization of catecholamine synthesizing enzymes, and direct methods for observing noradrenergic terminals with the electron microscope. Extensive progress in the functional analysis of these systems has also been made possible by evaluation with microelectrodes of the effects produced by selective electrical stimulation of the catecholamine (especially norepinephrine) cell-body groups. From such studies it is clear that the systems can be characterized in simple terms only by ignoring large amount of detailed cytological and functional data and that the earlier catecholamine wiring diagrams are no longer tenable. Furthermore, anatomic details for monoamine systems in rodents seem to bear only rudimentary homologies to their detailed selective anatomic configurations in human and nonhuman primates.

Noradrenergic Systems

There are two major clusterings of norepinephrine cell bodies from which axonal systems arise to innervate targets throughout the entire neuraxis:

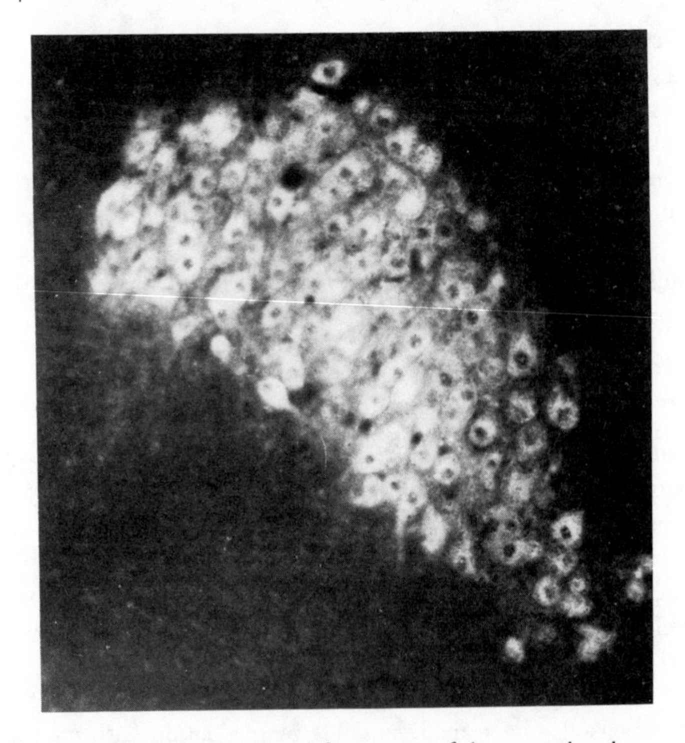

FIGURE 10-2. Formaldehyde-induced fluorescence of the rat nucleus locus coeruleus. In this frontal section through the principal portion of the nucleus, intensely fluorescent neurons can be seen clustered closely together. Within the neurons, the cell nucleus, which is not fluorescent after this treatment, appears dark except for the nucleolus. (Bloom, unpublished) × 650

LOCUS COERULEUS: This compact cell group within the caudal pontine central grey is named for the pigment the cells bear in humans and in some higher primates; in the rat the nucleus contains about 1500 neurons on each side of the brain (Fig. 10-2). The axons of these neurons form extensive collateral branches that project widely along well-defined tracts. At the electron microscope level, terminals of these axons exhibit—under appropriate fixation methods—the same type of small granular vesicles seen in the peripheral sympathetic nerves (Fig. 9-3).

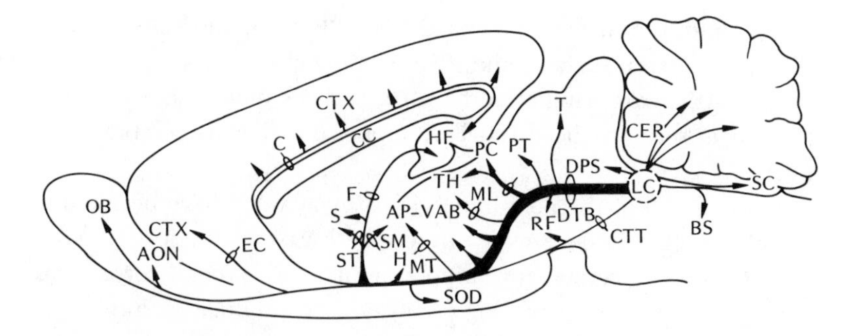

FIGURE 10-3. Diagram of the projections of the locus coeruleus viewed in the sagittal plane. See text for description. Abbreviations: AON, anterior olfactory nucleus; AP–VAB, ansa peduncularis-ventral amygdaloid bundle system; BS, brainstem nuclei; C, cingulum; CC, corpus callosum; CER, cerebellum; CTT, central tegmental tract; CTX, cerebral neocortex; DPS, dorsal periventricular system; DTB, dorsal catecholamine bundle; EC, external capsule; F, fornix; H, hypothalamus; HF, hippocampal formation; LC, locus coeruleus; ML, medial lemniscus; MT, mammillothalamic tract; OB, olfactory bulb; PC, posterior commissure; PT, pretectal area; RF, reticular formation; S, septal area; SC, spinal cord; SM, stria medullaris; SOD, supraoptic decussations; ST, stria terminalis; T, tectum; TH, thalamus. (Diagram compiled by R. Y. Moore. From the observations of Lindvall and Björklund, 1974b; Jones and Moore, 1977)

Fibers from the locus coeruleus form five major noradrenergic tracts (Fig. 10-3): the central tegmental tract (or dorsal bundle described by Ungerstedt), a central gray dorsal longitudinal facsiculus tract, and a ventral tegmental–medial forebrain bundle tract. These tracts remain largely ipsilateral, although there is a crossing over in some species of up to 25 percent of the fibers. These three ascending tracts then follow other major vascular and fascicular routes to innervate all cortices, specific thalamic and hypothalamic nuclei, and the olfactory bulb. Another major fascicle ascends in the superior cerebellar peduncle to innervate the cerebellar cortex, and the fifth major tract is that which descends into the mesencephalon and spinal cord, where the fibers course in the ventral–lateral column. At their terminals the locus coeruleus fibers form a plexiform network in which the incoming fibers gain access to a cortical region by passing through the major myelinated tracts, turning vertically toward the outer cortical surface, and then forming characteristic T-shaped branches that run parallel to the surface in the molecular layer; this pattern is seen in the cerebellar, hippocampal, and cerebral cortices.

Virtually all of the noradrenergic pathways that have been studied physiologically so far are efferent pathways of the locus coeruleus neurons; in cerebellum, hippocampus, and cerebral cortex, the major effect of activating this pathway is to produce an inhibition of spontaneous discharge. This effect has been associated with the slow type of synaptic transaction in which the hyperpolarizing response of the target cell is accompanied by increased membrane resistance. The mechanism of this action has been experimentally related to the second messenger scheme in which the noradrenergic receptor elicits its characteristic action on the target cells by activating the synthesis of cyclic AMP in or on the postsynaptic membrane. Pharmacologically and cytochemically, target cells responding to norepinephrine or the locus coeruleus projection in these cortical areas exhibit β-adrenergic receptors.

THE LATERAL TEGMENTAL NORADRENERGIC NEURONS: A large number of noradrenergic neurons lie outside of the locus coeruleus, where they are more loosely scattered throughout the lateral ventral tegmental fields. In general, the fibers from these neurons intermingle with those arising from the locus coeruleus, with those from the more posterior tegmental levels contributing mainly descending fibers within the mesencephalon and spinal cord, and those from the more anterior tegmental levels contributing to the innervation of the forebrain and diencephalon. Because of the complex intermingling of the fibers from the various noradrenergic cell-body sources, the physiological and pharmacological analysis of the effects of brain lesions becomes extremely difficult. The neurons of the lateral tegmental system may be the primary source of the noradrenergic fibers observed in the basal forebrain, such as amygdala and septum. No specific analysis of the physiology of these projections has yet been reported, and therefore it remains to be established whether the β-receptive cAMP mechanism associated with the cortical projections of the locus coeruleus group also apply to the synapses of the lateral tegmental neurons.

Dopaminergic Systems

The central dopamine-containing systems are considerably more complex in their organization than the noradrenergic systems. Not only are there many more dopamine cells, (the number of mesencephalic dopamine cells has been estimated to be about 15 to 20,000 on each side while the number of noradrenergic neurons in the entire brainstem is reported to be about 5000 on each side), there are also several major dopamine-containing nuclei as well as specialized dopamine neurons that make extremely localized connections within the retina and olfactory bulb. Recent anatomical studies support a much more topographic projection of the long-distance dopamine efferent fibers, such as the well-known nigrostriatal and so-called mesolimbic systems.

Based on this recent progress in the detailed anatomy of the dopamine systems (Fig. 10-4), it is convenient to consider them under three major categories in terms of the length of the efferent dopamine fibers.

1. *Ultrashort Systems.* Among the ultrashort systems are the *interplexiform amacrine-like neurons*, which link inner and outer plexiform layers of the retina, and the *periglomerular dopamine cells* of the olfactory bulb, which link together mitral cell dendrites in separated adjacent glomeruli.
2. *Intermediate-Length Systems.* The intermediate-length systems include (a) the *tuberohypophysial dopamine cells*, which project from arcuate and periventricular nuclei into the intermediate lobe of the pituitary and into the median eminence (often referred to as the tuberoinfundibular system); (b) the *incertohypothalamic neurons*, which link the dorsal and posterior hypothalamus with the dorsal anterior hypothalamus and lateral septal nuclei; and (c) the *medullary periventricular* group, which includes those dopamine cells in

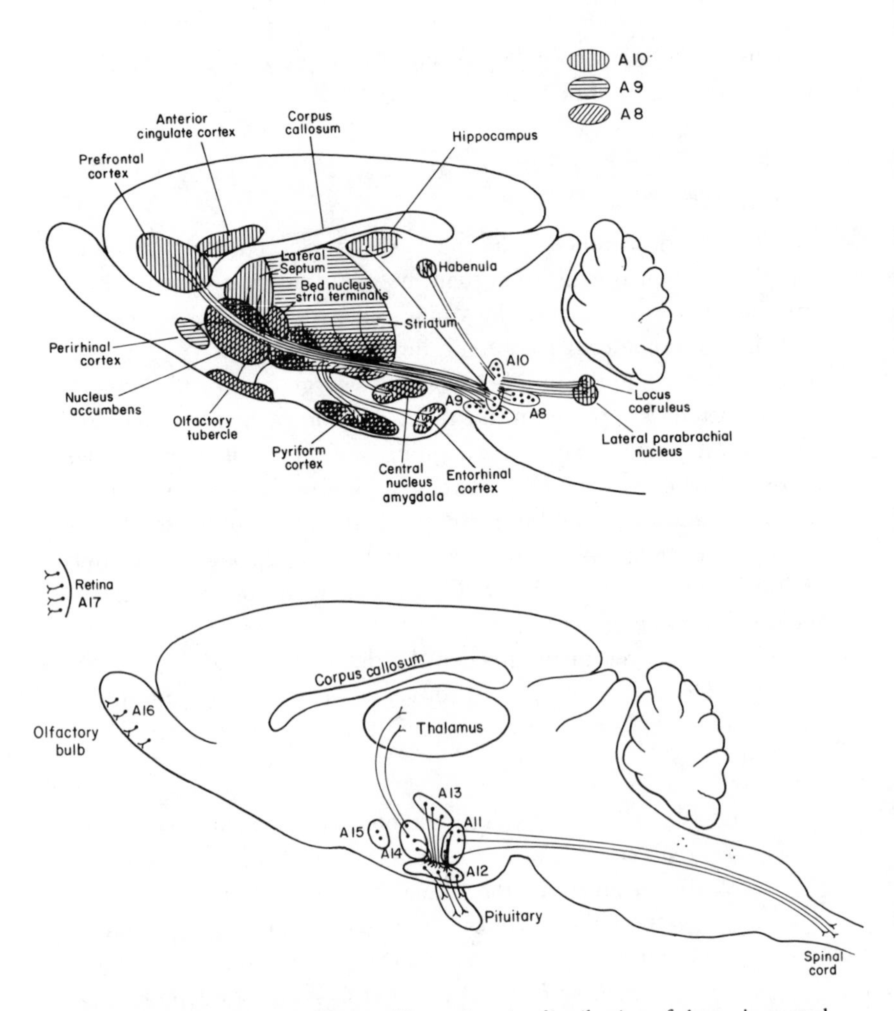

FIGURE 10-4. Schematic diagram illustrating the distribution of the main central neuronal pathways containing dopamine. The stippled regions indicate the major nerve terminal areas and their cell groups of origin. The cell groups in this figure are named according to the nomenclature of Dahlström and Fuxe (1965).

the perimeter of the dorsal motor nucleus of the vagus nerve, the nucleus tractus solitarius, and the cells scattered in the tegmental radiation of the periaqueductal grey matter.

3. *Long-Length Systems.* The long-length systems are the long projections linking the ventral tegmental and substantia nigra dopamine cells with three principal sets of targets: the *neostriatum* (principally the caudate and putamen); the *limbic cortex* (medial prefrontal, cingulate, and entorhinal areas); and *other limbic structures* (the regions of the *septum*, *olfactory tubercle*, *nucleus accumbens septi*, *amygdaloid complex*, and *piriform cortex*). These latter two groups have frequently been termed the *mesocortical* and *mesolimbic dopamine* projections, respectively. Under certain conditions these limbic target systems, when compared to the nigrostriatal system, exhibit some unique pharmacological properties, which are discussed in detail in the latter portion of this chapter.

With regard to cellular analysis of function, almost all reported data stem from studies of the nigrostriatal dopamine projection. Here the majority of studies on the iontophoretic administration of dopamine indicate that the predominant qualitative response is inhibition, and this effect, like that of apomorphine and cyclic AMP, has been found to be potentiated by phosphodiesterase inhibitors, providing the physiological counterpart to the second messenger hypothesis suggested by biochemical studies. On the other hand, the effects reported on the properties of caudate neurons when electrical stimulation is applied to the area of the ventral tegmentum and substantia nigra are considerably less homogeneous, with observations ranging from excitation to inhibition and with considerable variations in latency and sensitivity to dopamine antagonism. One study reported that neither the excitations nor the inhibitions elicited by nigral stimulation were prevented by 6-hydroxy-dopamine-induced destruction of the dopamine cell bodies. Unambiguous pharmacological and electrophysiological analysis of the dopamine pathway thus remains to be accomplished. Although at the present time the bulk of the evidence favors an inhibitory role for dopamine, inquisitive students will want to examine the arguments raised by both sides and, in particular, to develop their own evaluations of the effects reported. Important basic information is needed to rule out the spread of current to nearby nondopamine tracts and to determine accurately the latency of conduction reported for these extremely fine unmyelinated fibers.

Adrenergic Systems

In the sympathetic nervous system and the adrenal medulla, epinephrine shares with norepinephrine the role of final neurotransmitter; the proportion of this sharing being a species-dependent, hormonally modified arrangement. However, until recently, little evidence could be gathered to document the existence of epinephrine in the central nervous system because the chemical methods for analyzing epinephrine levels or for detecting activity attributable to the synthesizing enzyme, phenylethanolamine-N-methyl transferase were unable to provide unequivocal data. However, with the development of sensitive immunoassays for phenylethanolamine-N-methyl transferase, and their application in immunohistochemistry, and with the application of gas chromatography-mass fragmentography (GCMF) and high performance liquid chromatography (HPLC) with electrochemical detection to brain neurotransmitter measurements, the necessary data were rapidly acquired, and the existence of epinephrine-containing neurons in the CNS confirmed. By immunohistochemistry, epinephrine-containing neurons are defined as those that are positively stained (in serial sections) with antibodies to tyrosine hydroxylase, dopamine-β-hydroxylase, and with antibodies to phenylethanolamine-N-methyl transferase. These cells are found largely in two groups: one of which—called C-1 by Hökfelt—is intermingled with noradrenergic cells of the lateral tegmental system; the other C-2, is found in the regions in which the noradrenergic cells of the dorsal medulla are also found. The axons of these two epinephrine systems ascend to the hypothalamus with the central tegmental tract, then via the ventral periventricular system into the hypothalamus. Within the mesencephalon, the epinephrine-containing fibers innervate the nuclei of visceral efferent and afferent systems, especially the dorsal motor nucleus of the vagus nerve. In addition, epinephrine fibers also innervate the locus coeruleus, the intermediolateral cell columns of the spinal cord, and the periventricular regions of the fourth ventricle. Except for tests of epinephrine on the locus coeruleus (where it inhibits firing), no other tests of the cellular physiology of this system have thus far been reported. Our understanding concerning the function of epinephrine containing neurons in brain is still very limited, but based on the distribution of epinephrine in specific brain regions attention has been directed to their possible role in neuroendocrine mechanisms and blood pressure control.

Coexistence of Classical Transmitters and Peptides

The findings of numerous peptides in the CNS has raised the question of the relationship of the neurons containing these substances to neurons containing classical neurotransmitters. It was originally believed that the peptide–containing nerves represented separate identifiable systems. However, when studies in the peripheral nervous system clearly indicated the coexistence of peptides with the sympathetic and parasympathetic transmitters NE and ACh respectively, this initiated a series of systematic studies in the CNS to determine if classical neurotransmitter in the brain coexisted with one or more peptides. These studies revealed that coexistence of classical transmitter and peptides may represent a rule rather than an exception. Table 10-1 provides some examples of coexistence of classical neurotransmitter and peptides in selected brain regions. The functional significance of the occurrence of a classical transmitter and a peptide in the same neuron and their possible release from the same nerve ending is still unclear in the CNS. However, some insight may be obtained from studies in the periphery where two distinct types of interactions have been noted. In the salivary gland, the parasympathetic cholinergic neurons contain a VIP-like peptide while the sympathetic NE neurons contain a neuropeptide Y-like peptide. In this system, the peptides seem to augment the action of

TABLE 10-1. Coexistence of classical transmitters and peptides in the brain region. Species studied are shown in parentheses.

Classical transmitter	*Peptide*	*Brain region*
Dopamine	Neurotensin	Ventral tegmental area (rat)
	Cholecystokinin	Ventral tegmental area (rat, human)
Norepinephrine	Enkephalin	Locus ceruleus (cat)
	Neuropeptide Y	Medulla oblongata (human, rat)
		Locus ceruleus (rat)
Epinephrine	Neurotensin	Medulla oblongata (rat)
	Neuropeptide Y	Medulla oblongata (rat)
5-HT	Substance P	Medulla oblongata (rat, cat)
	TRH	Medulla oblongata (rat)
	Enkephalin	Medulla oblongata, pons (cat)
ACh	VIP	Cortex (rat)
	Enkephalin	Cochlear nerves (guinea pig)
	Substance P	Pons (rat)
GABA	Somatostatin	Thalamus and cortex (cat)
	Cholecystokinin	Cortex (cat, monkey)
	Neuropeptide Y	Cortex (cat)

SOURCE: Data taken in part from Hökfelt et al., *Science 225*, 1326, 1984.

the classical transmitters. Thus, VIP induces vasodilation and enhances the secretory effects of ACh while neuropeptide Y causes vasoconstriction like NE. In contrast, in the vas deferens where the NE neurons innervating this tissue also contain neuropeptide Y, the peptide seems to inhibit the release of norepinephrine via a presynaptic action. There is also some indication of preferential storage and release of NE and neuropeptide Y, with release of the peptide occurring preferentially at higher frequencies.

The physiological significance of multiple messengers at the synapse in the CNS is still unclear, but an appreciation of coexistence phenomena may influence our view on interneuronal communication and in a larger persepctive may be of importance in our efforts to treat or prevent various disease states or abnormalities of the nervous system.

Catecholamine Metabolism

Catecholamine metabolism has been dealt with in detail in Chapter 9, and only a few aspects pertinent to central catecholamine metabolism will be covered here.

In the peripheral sympathetic nervous system the aldehyde intermediate produced by the action of MAO on norepinephrine and normetanephrine can be oxidized to the corresponding acid or reduced to the corresponding glycol. Oxidation usually exceeds reduction and quantitatively vanilylmandelic acid (VMA) is the major metabolite of norepinephrine and is readily detectable in the urine. In fact, the urinary levels of VMA are routinely measured in clinical laboratories in order to provide an index of peripheral sympathetic nerve function as well as to diagnose the presence of catecholamine-secreting tumors such as pheochromocytomas and neuroblastomas. In the central nervous system, however, reduction of the intermediate aldehyde formed by the action of MAO on catecholamines or catecholamine metabolites predominates, and a major metabolite of norepinephrine found in brain is the glycol derivative, 3-methoxy-4-hydroxy-phenethyleneglycol (MHPG). Very little, if any, VMA is found in brain. In many species a large fraction of the MHPG formed in brain is sulfate conjugated. However in primates, MHPG exists primarily in an unconjugated "free" form in the brain. Some normetanephrine is also found in the brain and spinal cord. Destruction of noradrenergic neurons in the brain or spinal cord cause a reduction in the endogenous levels of these metabolites although not as marked as the corresponding reduction in endogenous norepinephrine. Recently, it has been demonstrated that direct electrical stimulation of the locus coeruleus or severe stress produces an increase in the turnover of norepinephrine as well as an increase in the accumulation of the sulfate conjugate of MHPG in the rat cerebral cortex. These effects are completely abolished by ablation of the locus coeruleus or by transection of the dorsal pathway, which suggests that the accumulation of MHPG/sulfate in brain may reflect the functional activity of central noradrenergic neurons. Since MHPG-sulfate readily diffuses from the brain into the CSF or general circulation, estimates of its concentration in the CSF or in urine are thought to provide a possible reflection of activity of noradrenergic neurons in the brain. Even though MHPG is proportionately a minor metabolite of norepinephrine in the peripheral sympathetic nervous system, a fairly large portion of the MHPG in the urine still derives from the periphery. In the rodent it has been estimated that brain provides a minor contribution (25–30%) to urinary MHPG while in primates the brain contribution is quite large (60%). Thus, at least in the rodent, it is quite probable that relatively large changes in the formation of MHPG by brain are necessary to produce detectable changes in urinary MHPG. For example, in the rat destruction of the majority of norepinephrine-containing neurons in the brain by treatment with 6-hydroxydopamine leads to only about a 25 percent decrease in urinary MHPG levels. Nevertheless, measurement of urinary changes in MHPG is still a reasonable strategy for obtaining some insight into possible alterations of brain norepinephrine metabolism in patients with psychiatric illnesses. Measurement of CSF levels of MHPG also provide another possible approach to assessing central adrenergic function in human subjects. More recent studies have suggested that plasma levels of MHPG might provide a reflection of central noradrenergic activity. In these studies, stimulation of the locus coeruleus in the rat was shown to result in a significant increase in the levels of plasma MHPG; and administration of drugs that are known to alter noradrenergic activity in rodents have predictably changed plasma levels and venous-arterial differences in MHPG in nonhuman primates. Although changes in either urinary, plasma, or CSF levels of MHPG and their relationship to central noradrenergic function have to be interpreted with considerable caution, these measurements do provide a starting point for clinical investigation.

The primary metabolites of dopamine found in the CNS are homovanillic acid (HVA) and dihydroxyphenylacetic acid (DO-

PAC) and a small amount of 3-methoxytyramine. In this system the acidic rather than the neutral metabolites thus appear to predominate. Accumulation of HVA in brain or CSF has often been used to provide an index of the functional activity of dopaminergic neurons in the brain. Drugs such as the antipsychotic drugs, which increase the turnover of dopamine (in part because of their ability to increase the activity of dopaminergic neurons and to augment release), also increase the amount of HVA in the brain and CSF. In addition, electrical stimulation of the nigrostriatal pathway causes an increase in brain levels of HVA and an increase in the release of HVA into ventricular perfusates. In disease states such as Parkinsonism, where there are degenerative changes in the substantia nigra and partial destruction of the dopamine neurons, a reduction of HVA is observed in the CSF.

In rat brain it has been demonstrated that the short-term accumulation of DOPAC in the striatum may provide an even more accurate reflection of activity in dopaminergic neurons of the nigrostriatal pathway. A cessation of impulse flow that is induced by the placement of acute lesions in the nigrostriatal pathway leads to a rapid decrease in striatal DOPAC, and the electrical stimulation of the nigrostriatal pathway results in a frequency-dependent increase in DOPAC within the striatum. Drugs such as the antipsychotic phenothiazines and butyrophenones, as well as anesthetics and hypnotics, which increase impulse flow in the nigrostriatal pathway, also increase striatal DOPAC. Drugs that block impulse flow such as γ-hydroxybutyrate, apomorphine and amphetamine cause a reduction in DOPAC. Thus, there appears to be an excellent correlation between changes in impulse flow in dopaminergic neurons, which are induced either pharmacologically or mechanically, and changes in the steady state levels of DOPAC. Unfortunately, in primates, DOPAC is a minor brain metabolite. Not only is it difficult to measure DOPAC in CSF but in primate brain this metabolite is unresponsive to drug treatments which cause large changes in dopamine metabolism (i.e., homovanillic acid).

By means of the sensitive and specific technique of gas chromatography–mass fragmentography, it has been possible to measure accurately DOPAC, HVA, and their conjugates in rat and human plasma. Studies in rats have demonstrated that stimulation of the nigrostriatal pathway and administration of antipsychotc drugs produces an increase in plasma levels of DOPAC and HVA. Several studies in humans have also indicated that dopaminergic drugs can influence plasma levels of HVA although the effect is quite modest in comparison to the effects observed in rodents. Thus, it may be that in several species plasma levels of these metabolites will prove useful indices of central dopamine metabolism.

Biochemical Organization

In general, noradrenergic neurons in the central nervous system appear to behave in a fashion quite similar to the postganglionic sympathetic noradrenergic neurons in the periphery. Electrical stimulation of the noradrenergic neurons in the locus coeruleus causes both an activation of tyrosine hydroxylase and an increase in the turnover of norepinephrine as well as a frequency-dependent accumulation of MHPG-sulfate (a major metabolite of norepinephrine in the rat CNS) in the cortex. Interruption of impulse flow in these neurons decreases turnover of norepinephrine but has little or no effect on the steady-state level of transmitter. Similar effects are also obtained in the serotonin-containing neurons of the midbrain raphe (stimulation increasing synthesis and turnover, and interruption of impulse flow slowing turnover).

It is well known that acute severe stress also causes an increase in the turnover of norepinephrine in the CNS, which is believed to occur as a result of increase in impulse flow in noradrenergc neurons. Recent experiments have confirmed this speculation. If impulse flow in noradrenergic neurons projecting to the cortex is acutely interrupted by destruction of the locus coeruleus, the increase in norepinephrine turnover and accumulation of MHPG-sulfate induced by stress is completely blocked.

Dopaminergic neurons in the CNS do not behave in a fashion completely analogous to peripheral or central noradrenergic neurons. An increase in impulse flow in the nigroneostriatal or mesolimbic dopamine system does lead to both an increase in dopamine synthesis and turnover and a frequency-dependent increase in the accumulation of dopamine metabolites in the striatum and olfactory tubercle. These observations are similar to those made in other monoamine systems where an increase in impulse flow causes an increase in the synthesis and turnover of transmitter.

Similar mechanisms also appear to be involved in the observed activation of transmitter synthesis. Short-term stimulation of central dopaminergic neurons results in an increase in tyrosine hydroxylation, which is mediated in part by kinetic alterations in tyrosine hydroxylase. The tyrosine hydroxylase prepared from tissue containing the terminals of the stimulated dopamine neurons has an increased affinity for pteridine cofactor and a decreased affinity for the natural end-product inhibitor, dopamine. As in central noradrenergic systems, there appears to be a finite period of time necessary for this activation to occur, and once activated, the enzyme remains in this altered physical state for about ten minutes after the stimulation period ends.

However, if impulse flow is interrupted in the nigroneostriatal or mesolimbic dopamine system either mechanically or pharmacologically by treatment with γ-hydroxybutyrate, the neurons respond in a rather peculiar fashion by rapidly increasing the concentration of dopamine in the nerve terminals of the respective dopamine systems. Not only do the terminals accumulate dopamine but there is also a dramatic increase in the rate of dopamine synthesis despite the steadily increasing concentration of endogenous dopamine within the nerve terminals. This observation is explained by the fact that a cessation of impulse flow in central dopaminergic neurons results in a change in the properties of tyrosine hydroxylase, making this enzyme insensitive to feedback inhibition by endogenous dopamine. This change in the properties of tyrosine hydroxylase removes one of the normal regulatory influences (feedback regulation) and allows enhanced transmitter formation until endogenous concentrations of dopamine are elevated to a level at which they can again exert a significant inhibitory influence on tyrosine hydroxylase. The actual mechanism whereby a cessation of impulse flow initiates changes in the properties of tyrosine hydroxylase is unclear, although many experiments have suggested that this effect may be caused by a decrease in the availability of intracellular calcium. Similar changes in the activity or properties of tyrosine hydroxylase are not observed in central noradrenergic neurons or in dopamine neurons lacking autoreceptors following periods of quiescence. At present, the physiological significance of this paradoxical response of certain dopaminergic neurons to a cessation of impulse flow is unclear. However, it is conceivable that these neurons may achieve some operational advantage by being able to increase rapidly their supply of transmitter during periods of quiescence.

From the discussion above, it should be apparent that the identification of specific chemically defined neuronal systems in the brain sets the stage for investigation of the functional organization of these systems as well as for integrating biochemical and neurophysiological knowledge of the mechanism of action of drugs. Some specific examples of the effects of drugs on the activity of noradrenergic and dopaminergic neurons and their sites of action will be given in the section following.

Pharmacology of Central Catecholamine-Containing Neurons

The psychotropic drugs (drugs that alter behavior, mood, and perception in humans and behavior in animals) can be divided into two main categories, the psychotherapeutic drugs an the psychotomimetic agents. The psychotherapeutic drugs can be further divided on the basis of their activity in humans into at least four general classes: antipsychotics, antianxiety drugs, antidepressants, and stimulants.

A great deal more information has become available concerning the anatomy, biochemistry, and functional organization of catecholamine systems in recent years. This has fostered knowledge concerning the mechanisms and sites of action of many psychotropic drugs. It is now appreciated that drugs can influence the output of monoamine systems by interacting at several distinct sites.

For example, drugs can act to influence the output of catecholamine systems by (1) acting presynaptically to alter the life cycle of the transmitter (i.e., synthesis, storage and release, etc. [Figs. 10-7 and 10-8]); (2) acting postsynaptically to mimic or block the action of the transmitter at the level of postsynaptic receptors; and (3) acting at the level of cell-body autoreceptors to influence the physiological activity of catecholamine neurons. The activity of catecholamine neurons can also be influenced by postsynaptic receptors via negative neuronal feedback loops. In the latter two actions, drugs appear to exert their influence by interacting with catecholamine receptors. (See Fig. 10-5.) In general, catecholamine receptors can be subdivided into two broad categories: those which are localized directly on catecholamine neurons (often referred to as presynaptic receptors); and receptors on other cell types which are often termed simply postsynaptic receptors since they are postsynaptic to the catecholamine neurons, the source of the endogenous ligand. When it became appreciated that catecholamine neurons, in addition to possessing receptors on the nerve terminals, appear to have receptors distributed over all parts of the neurons (i.e., soma, dendrites, and preterminal axons), the term presynaptic receptors really became inappropriate as a description for all these receptors. Carlsson, in 1975, suggested that "autoreceptor" was a more appropriate term to describe these receptors since the sensitivity of these catecholamine receptors to the neurons' own transmitter seemed more significant than their location relative to the synapse. The term autoreceptor achieved rapid acceptance and pharmacological research in the succeeding years resulted in the detection and description of autoreceptors on neurons in almost all chemically defined neuronal systems.

The presence of autoreceptors on some neurons may turn out to be only of pharmacological interest since they may never encounter effective concentrations of the appropriate endogenous agonist *in vivo*. Others, however, in addition to their pharmacological responsiveness, may play a very important physiological role in the maintenance of presynaptic events. This certainly appears to be the case for catecholamine autoreceptors.

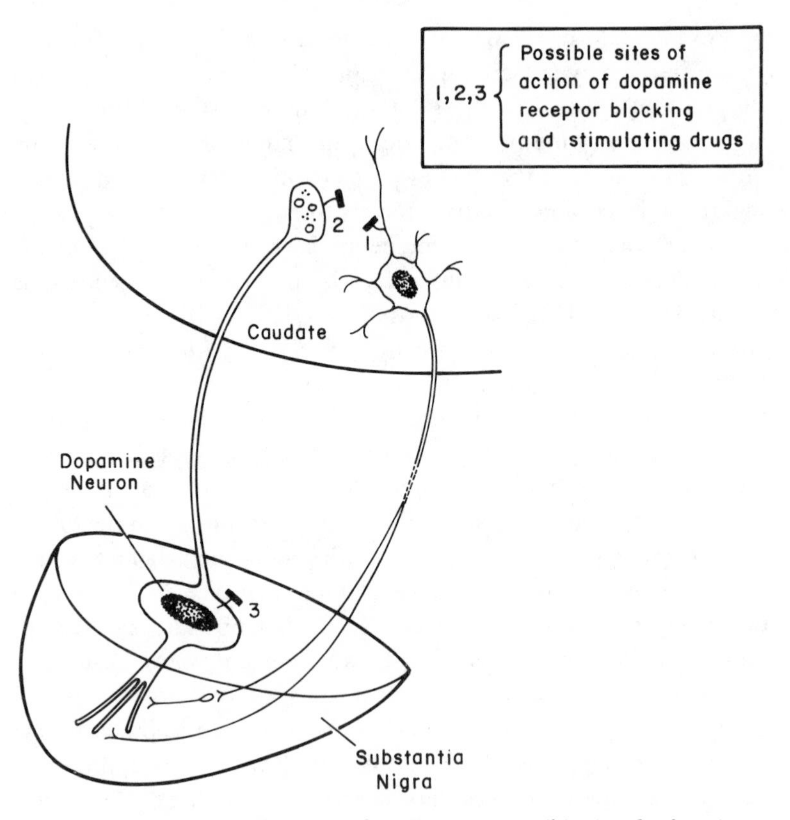

FIGURE 10-5. Schematic illustration depicting some possible sites for drug interaction with dopamine receptors.

Functional Significance of Autoreceptors

Autoreceptors localized on dopaminergic nerve terminals appear to modulate both the rate of dopamine biosynthesis and impulse-induced release of transmitter by a local negative feedback mechanism. In central noradrenergic systems, terminal autoreceptors are involved primarily in the modulation of impulse-induced transmitter release. Evidence has not yet been obtained to implicate these receptors in synthesis regulation. Cell-body autoreceptors in catecholamine systems appear to play a role in the modulation of the physiological activity of these neurons. Since transmitter release and synthesis are coupled to impulse flow, these cell-body autoreceptors influence transmitter synthesis and release indirectly by modifying the physiological activity of these neurons. In general, autoreceptors localized on different parts of the neuron appear to act in concert. Stimulation of these receptors diminishes the overall influence of catecholamine systems on postsynaptic follower cells, whereas blockade of these receptors seems to enhance the influence of catecholamine systems on postsynaptic follower cells.

Of the three well-studied mesotelencephalic dopamine systems (nigrostriatal, mesolimbic, and mesocortical), several mesocortical dopamine neurons appear to be unique: their nerve terminals and soma/dendrites are devoid of dopamine autoreceptors. In addition, the tuberoinfundibular neurons projecting to the median eminence appear to lack dopamine autoreceptors. Although these neurons retain the ability to regulate dopamine synthesis via end-product feedback inhibition, there is a complete absence of dopamine autoreceptor-mediated influence on synthesis. This lack of soma/dendritic and synthesis-modulating autoreceptors may explain in part the unique responsiveness of several mesocortical dopamine systems to certain types of drugs and the relative resistance of these neurons to the development of tolerance during long-term treatment with antipsychotic drugs (see below).

Pharmacology of Autoreceptors

Autoreceptors appear to be more sensitive to agonists than postsynaptic receptors of the same class. For example, clonidine, a norepinephrine agonist, in a dose of 5–10 μg/kg will produce noradrenergic autoreceptor stimulation whereas doses of 100 μg/kg or greater are required for postsynaptic effects. Likewise, apomorphine, a dopamine agonist, is ten to twenty times more potent in stimulating dopamine autoreceptors than postsynaptic dopamine receptors.

Adrenergic receptor blocking drugs such as idazoxane, piperoxane, and yohimbine at low doses appear to selectively block nor-

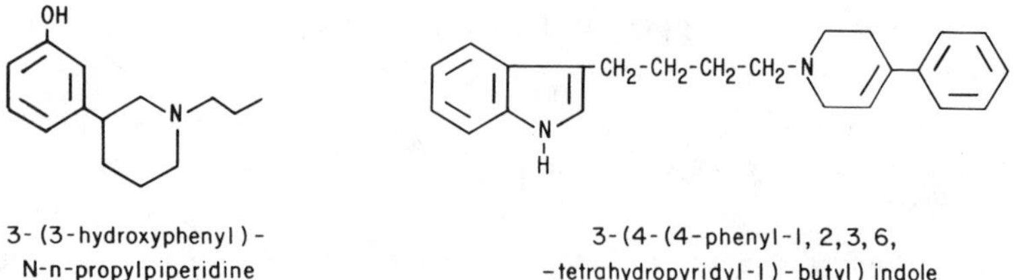

FIGURE 10-6. Structural formulas of several selective dopamine autoreceptor agonists.

adrenergic autoreceptors (α_2-receptors). However, if the dosage of these drugs is increased, the specificity of action is lost and the drugs exert significant α_1-effects. Dopamine autoreceptors in the nigrostriatal and mesolimbic systems also exhibit some subtle differences in response to dopamine receptor blocking drugs when compared to postsynaptic receptors. These observations are again indicative that autoreceptors differ in their pharmacological profile from postsynaptic receptors.

An important benefit which may ultimately derive from studying the properties and function of autoreceptors lies in the potential exploitation of differences between autoreceptors and postsynaptic receptors in the development of more selective and specific drugs. A step in this direction has already been achieved with the recent description of several new dopaminergic drugs which appear to have selective autoreceptor stimulating properties: 3-(3-hydroxyphenyl)-*N*-*n*-propylpiperidine (3-PPP); 3-(4-(4-phenyl-1,2,3,6-tetrahydropyridyl-1)-butyl) indole (EMD 23-448); see Fig. 10-6). It should be mentioned in this context that the enantiomers of 3-PPP have been extensively studied and found to differ in their pharmacological profile. +3-PPP is not a selective autoreceptor agonist and acts in a fashion similar to classic direct acting dopamine agonists, such as apomorphine, stimulating both dopamine autoreceptors and postsynaptic dopamine receptors. −3-PPP appears to act as a selective agonist at dopamine autoreceptors, but as an antagonist at postsynaptic dopamine receptors. Selective drugs of this nature may find some usefulness in decreasing excessive catecholamine transmission without the liability of inducing postsynaptic receptor supersensitivity as usually occurs with classical catecholamine receptor blocking agents.

Pharmacology of Dopaminergic Systems

Nigrostriatal and Mesolimbic Dopamine Systems

The nigrostriatal and mesolimbic dopamine neurons appear to respond in a similar manner to drug administration (Table 10-2). Acute administration of dopamine agonists (dopamine-receptor stimulators) decrease dopamine cell activity, decrease dopamine turnover, and decrease dopamine catabolism. Acute administration of antipsychotic drugs (dopamine-receptor blockers) increase dopaminergic cell activity, enhance dopamine turnover, increase dopamine catabolism, and accelerate dopamine biosynthesis. The increase in dopamine biosynthesis occurs at the tyrosine hydroxylase step and is explained in part as a result of the ability of antipsychotic drugs to block postsynaptic receptors and to increase dopaminergic activity via a neuronal feedback mechanism (Fig. 10-5). It is also apparent that some of the observed effects are enhanced as a result of interaction with nerve terminal autoreceptors. Blockade of nerve terminal autoreceptors increases both the synthesis and the release of dopamine. These systems respond to monoamine oxidase inhibitors (MAOI) with an increase in DA and decrease in DA synthesis, as do the other DA systems discussed below.

Recently it has been shown that chronic treatment with antipsychotic drugs produces a different spectrum of effects on central dopaminergic neurons. For example, following chronic treatment with haloperidol, *nigrostriatal* dopamine neurons in the rat become quiescent and dopamine metabolite levels and dopamine synthesis and turnover in the *striatum* return to normal limits. The kinetic activation of *striatal* tyrosine hydroxylase shown to occur following an acute challenge dose of an antipsychotic drug also subsides fol-

TABLE 10-2. Pharmacology of central dopaminergic systems

	System response					
Characteristics	*Nigrostriatal*	*Mesolimbic*	*Mesoprefrontal*	*Mesopiriform*	*Tuberoinfundibular*[a]	*Tuberohypophyseal*[b]
Respond to DA antagonist (increase in synthesis, catabolism, and turnover)	Yes	Yes	Yes (small)	Yes	No	Yes (small)
Respond to DA agonists (decrease in synthesis, catabolism, and turnover)	Yes	Yes	Yes (small)	Yes	No	Yes
Respond to monoamine oxidase inhibitors (increase in DA, decrease in synthesis)	Yes	Yes	Yes	Yes	Yes	Yes
Presence of nerve terminal synthesis modulating autoreceptors	Yes	Yes	No	Yes	No	Yes
High-affinity DA transport	Yes	Yes	Yes	Yes	No	No

Respond to mild stress (increase in DA synthesis and catabolism, blocked by diazepam)	Yes	Yes? (small)	Yes	No	No	No
Respond to anxiogenic β-carbolines (increase in DA catabolism)	No	No	Yes	No	—	—

[a]Cell bodies of this group of neurons are located in the arcuate and periventricular nuclei and their axons terminate in the external layer of the median eminence.

[b]Cell bodies of this group of neurons are located in the arcuate and periventricular nuclei and their axons terminate in the neurointermediate lobe of the pituitary (posterior pituitary).

lowing chronic treatment. These results are usually interpreted as indicative of the development of tolerance in the nigrostriatal DA system. In contrast (see below), tolerance to the biochemical effects observed following acute administration of antipsychotic drugs does not appear to develop in the mesoprefrontal and mesocingulate cortical dopamine pathways after chronic administration.

Mesocortical Dopamine System

It has only become possible with the advent of more sensitive analytical techniques (Chapter 9) to measure changes in the low levels of DA and related metabolites in this system, as well as in those discussed below. The response of the mesocortical dopamine systems to dopaminergic drugs is in most instances qualitatively similar to that of the nigrostriatal and mesolimbic systems (Table 10-2), although some notable exceptions have been recently observed. In fact, over the past few years the mesotelencephalic dopamine neurons which were once believed to be three relatively simple and homogenous systems have been found to be an anatomically, biochemically, and electrophysiologically heterogenous population of cells with differing pharmacological responsiveness. For example, although a great majority of midbrain dopamine neurons appear to possess autoreceptors on their cell bodies, dendrites, and nerve terminals, DA cells which project to the prefrontal and cingulate cortices appear to have either a greatly diminished number of these receptors or to lack them entirely. The absence (or insensitivity) of impulse-regulating somatodendritic as well as synthesis-modulating nerve terminal autoreceptors on the mesoprefrontal and mesocingulate cortical dopamine neurons may, in part, explain some of the unique biochemical, physiological and pharmacological properties of these two subpopulations of midbrain dopamine neurons (Table 10-3). For example, the mesoprefrontal and mesocingulate DA neurons appear to have a faster firing rate and a more rapid turnover of transmitter than the nigrostriatal, mesolimbic, and *mesopiriform* DA neurons. Transmitter synthesis is also more readily influenced by altered availability of precursor tyrosine in midbrain DA neu-

Table 10-3. Unique characteristics of mesotelencephalic dopamine systems lacking autoreceptors (mesoprefrontal and mesocingulate) compared to those possessing autoreceptors (mesopiriform, mesolimbic and nigrostriatal)

1. A higher rate of physiological activity (firing) and a different pattern of activity (more bursting).
2. A higher turnover rate and metabolism of transmitter dopamine.
3. Greatly diminished biochemical and electrophysiological responsiveness to dopamine agonists and antagonists.
4. Lack of biochemical tolerance development following chronic antipsychotic drug administration.
5. Resistance to the development of depolarization-induced inactivation following chronic treatment with antipsychotic drugs.
6. Transmitter synthesis more readily influenced by altered availability of precursor tyrosine.

rons lacking autoreceptors (mesoprefrontal and mesocingulate) than in those midbrain DA neurons possessing autoreceptors. This may be related to the enhanced rate of physiological activity in this subpopulation of midbrain DA neurons making them more responsive to precursor regulation. The mesoprefrontal and mesocingulate DA neurons also show a diminished biochemical and electrophysiological responsiveness to DA agonists and antagonists. Administration of low doses of apomorphine or autoreceptor selective dopamine agonists, in contrast to their inhibitory effects on other midbrain DA neurons, are ineffective in decreasing the activity or in lowering DA metabolite levels in these two cortical DA projections. DA receptor blocking drugs such as haloperidol produce large increases in synthesis and the accumulation of DA metabolites in the nigrostriatal, mesolimbic, and mesopiriform DA neurons, but have only a modest effect in the mesoprefrontal and mesocingulate DA neurons.

Heterogeneity among midbrain DA neurons is also found when

one studies the effects of chronic antipsychotic drug administration. When classical antipsychotic drugs are administered repeatedly over time, the great majority of dopamine cells cease to fire due to the development of a state of depolarization inactivation. However, some midbrain DA cells appear to be unaffected by repeated antipsychotic drug administration. These dopamine cells turn out to be those neurons projecting to the prefrontal and cingulate cortices. Parallel findings are observed biochemically. Following chronic administration of antipsychotic drugs, tolerance develops to the metabolite elevating effects of these agents in the midbrain dopamine systems possessing autoreceptors, but not in the systems lacking autoreceptors. When the atypical antipsychotic drug, clozapine (a drug which possesses therapeutic efficacy, but lacks Parkinson-like side effects and an ability to produce tardive dyskinesia) is administered repeatedly, dopamine neurons in the VTA develop depolarization inactivation, but neurons in the substantia nigra do not. The reason for this differential effect is unknown at present. Footshock, swim stress, or conditioned fear cause a selective (benzodiazepine reversible) metabolic activation of the mesoprefrontal DA neurons without causing a marked or consistent effect on other midbrain DA neurons including the mesocingulate DA neurons. Thus, this selective activation does not appear to be due solely to the absence of autoreceptors. The anxiogenic benzodiazepine receptor ligands such as the β-carbolines also produce a selective dose-dependent activation of mesoprefrontal DA neurons without increasing dopamine metabolism in other midbrain dopamine neurons.

In summary, certain mesotelencephalic DA systems, namely the mesoprefrontal and mesocingulate DA neurons, possess many unique characteristics compared to the nigrostriatal, mesolimbic, and mesopiriform DA systems (Table 10-3). Many of these unique characteristics may be the consequence of the lack of impulse-regulating somatodendritic and synthesis-modulating nerve terminal DA autoreceptors. However, some, such as the response to stress and anxiogenic β-carbolines, are clearly dependent upon other regulatory influences and not solely related to the absence of autorecep-

TABLE 10-4. Selective dopaminergic agents and their putative sites of action

Site and type of action	*Agent*
Dopamine autoreceptor agonist	3-PPP EMD 23-448
Selective agonist at dopamine receptors positively coupled to adenylate cyclase (D_1-receptor)	SKF-38393
Selective agonist at dopamine receptors noncoupled or negatively coupled to adenylate cyclase (D_2-dopamine receptors)	LY 141865
Antagonist at D_2-receptors	Sulpiride

tors. These findings suggest that DA action at autoreceptors may be one of the important ways that DA cells possessing these receptors modulate their function. Since autoreceptors appear to play such an important role in regulation of the systems that possess them, future studies will need to address how midbrain dopamine systems that lack autoreceptors are regulated (role of afferent neuronal systems). In fact, some recent studies have suggested that a Substance P/Substance K innervation of the ventral tegmental area (A10) may influence the functional activity of mesocortical and mesolimbic dopamine neurons.

The observation that central dopamine systems are quite heterogeneous from both a biochemical and functional point of view holds promise that in the near future it will be possible to develop drugs targeted to modify or restore function to selective dopamine systems which are abnormal in various behavioral or pathological states. Some progress has already been achieved in developing agents which appear to have an action at selective dopamine receptor sites (Table 10-4). However, whether these agents will be useful in selectively modifying the function of subsets of midbrain dopamine neurons remains to be determined.

Tuberoinfundibular and Tuberohypophysial Dopamine Systems

The tuberoinfundibular dopamine system responds to pharmacological and endocrinological manipulations in a manner that is qualitatively different from the other three dopamine systems—nigrostriatal, mesolimbic, and mesocortical—described above (Table 10-2). The tuberoinfundibular neurons appear to be regulated in part by circulating levels of prolactin. Prolactin increases the activity of these neurons by acting within the medial basal hypothalamus, possibly directly on the tuberoinfundibular neurons. These neurons in turn release DA, which then inhibits prolactin release from the anterior pituitary. Haloperiodol and other antipsychotic drugs have no effect on dopamine turnover in the tuberoinfundibular dopamine system until about sixteen hours after drug administration, whereas the biochemical effects in other systems are observed within minutes and are maximal in several hours. The absence of an acute response to DA antagonists and agonists may be related to the lack of autoreceptors in this system. While less is known about the pharmacology of the tuberohypophyseal DA neurons, this system seems to respond to drugs in a manner qualitatively similar to the better studied DA systems (Table 10-2).

Specific Drug Classes

Antipsychotic Drugs (Major Tranquilizers)

Although there are at least seven major classes of antipsychotic drugs, all of which have in common the ability to ameliorate psychosis and evoke extrapyramidal reactions, the most widely used are the phenothiazines, thioxanthenes, and butyrophenones. As might be expected from the name given to chlorpromazine by the French—Largactil—the phenothiazines are notorious for their wide spectrum of pharmacological and biochemical effects. Fortunately, however, it turns out that many actions attributable to the antipsychotic phenothiazines are nonspecific properties of the phenothiazine moiety itself and do not appear to be correlated with the antipsychotic potency of the therapeutically active subgroup of phenothiazines. Along with many of their other actions, the antipsychotic phenothiazines have been shown to interact with both noradrenergic and dopaminergic neurons. However, the antipsychotic potency of these compounds appears to be best correlated with their ability to interact with dopamine-containing neurons. Much evidence has accumulated to indicate that antipsychotic drugs are effective blockers of dopamine receptors. In fact, it is believed that this reduction in dopamine activity expressed postsynaptically is responsible for the extrapyramidal side effects observed with these drugs. The ability of antipsychotic phenothiazines to block dopamine receptors on as yet chemically unidentified postsynaptic neurons may in some way also be related to the antipsychotic potency of these agents.

For many years it has been appreciated that antipsychotic drugs of both the phenothiazine and butyrophenone classes can cause an increase in the turnover of dopamine in the CNS. Since these drugs appear to have potent dopamine-receptor blocking capabilities, it has been suggested that the increase observed in dopamine turnover results from blockade of both dopamine autoreceptors and postsynaptic dopamine receptors and a consequent feedback activation of the dopaminergic neurons, presumably by some sort of neuronal feedback loop. This speculation has been verified in part by direct extracellular recording techniques (Fig. 10-9).

Antianxiety Drugs (Minor Tranquilizers)

The antianxiety drugs appear to interact with both noradrenergic and selected dopaminergic neurons. They are very effective in blocking the stress-induced increases in the turnover of both cortical norepinephrine and mesoprefrontal cortical dopamine. The actual mechanism by which these drugs exert their effects on the norepinephrine system is at the present time unknown, although in general they are believed to facilitate GABAergic transmission. Electrophysiological studies have demonstrated that many of these

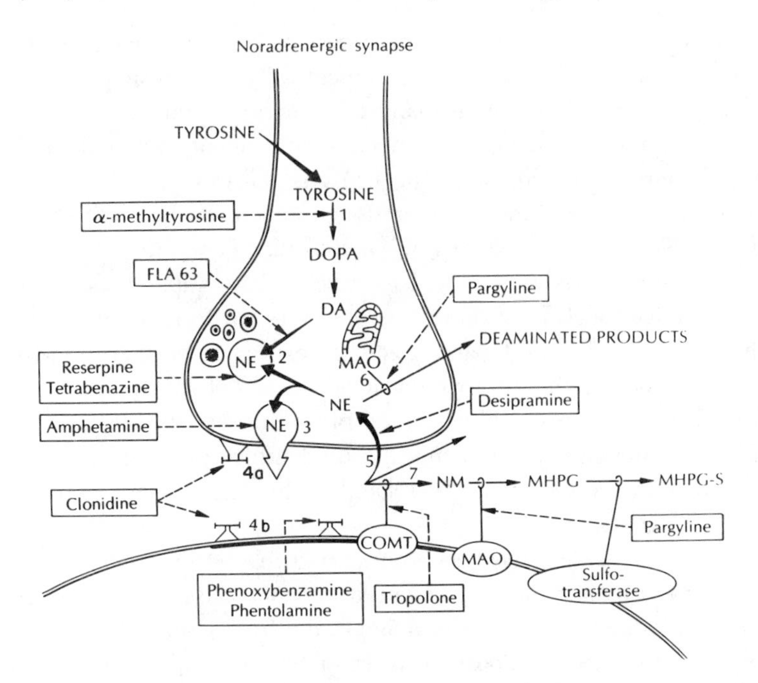

FIGURE 10-7. schematic model of central noradrenergic neuron indicating possible sites of drug action.

Site 1: *Enzymatic synthesis:*

a. Tyrosine hydroxylase reaction blocked by the competitive inhibitor, α-methyltyrosine.

b. Dopamine-β-hydroxylase reaction blocked by a dithiocarbamate derivative, Fla-63 bis-(1-methyl-4-homopiperazinyl-thiocarbonyl)-disulfide.

Site 2: *Storage:* Reserpine and tetrabenazine interfere with the uptake—storage mechanism of the amine granules. The depletion of norepinephrine produced by reserpine is long-lasting and the storage granules are irreversibly damaged. Tetrabenazine also interferes with the uptake–storage mechanism of the granules, except the effects of this drug are of a shorter duration and do not appear to be irreversible.

Site 3: *Release:* Amphetamine appears to cause an increase in the net release of norepinephrine. Probably the primary mechanism by which amphetamine causes release is by its ability to block effectively the re-uptake mechanism.

Site 4: *Receptor interaction:*

a. Presynaptic α_2-autoreceptors

b. Postsynaptic α-receptors

Clonidine appears to be a very potent autoreceptor-stimulating drug. At higher doses clonidine will also stimulate postsynaptic receptors.

agents can suppress unit activity in the locus coeruleus. It has often been speculated that the relief of stress symptoms in humans is mediated by "turning off" the central noradrenergic neurons. (Noradrenergic systems were believed to be the primary monoamine system where antianxiety agents interact.) However, in view of the recent studies demonstrating an effect of stress on the mesoprefrontal dopamine system and reversal of the stress-induced activation by antianxiety drugs, it is clear that the mode of action of antianxiety drugs is an open question. Furthermore, in tests of anxiolytic action in rats, complete destruction of the telencephalic projections of the locus coeruleus fails to alter the effectiveness of benzodiazepines.

Antidepressants

There are several major classes of antidepressant drugs, the monoamine oxidase inhibitors (MAOI), the tricyclic drugs, and the atyp-

---

Phenoxybenzamine and phenotolamine are effective α-receptor blocking agents. Recent experiments indicate that these drugs may also have some presynaptic $alpha_2$ receptor blocking action. However, yohimbine and piperoxane are more selective as α_2-receptor-blocking agents.

Site 5: *Re-uptake:* Norepinephrine has its action terminated by being taken up into the presynaptic terminal. The tricyclic drug desipramine is an example of a potent inhibitor of this uptake mechanism.

Site 6: *Monoamine Oxidase (MAO):* Norepinephrine or dopamine present in a free state within the presynaptic terminal can be degraded by the enzyme MAO, which appears to be located in the outer membrane of mitochondria. Pargyline is an effective inhibitor of MAO.

Site 7: *Catechol-O-methyl transferase (COMT):* Norepinephrine can be inactivated by the enzyme COMT, which is believed to be localized outside the presynaptic neuron. Tropolone is an inhibitor of COMT. The normetanephrine (NM) formed by the action of COMT on NE can be further metabolized by MAO and aldehyde reductase to 3-methoxy-4-hydroxyphenylglycol (MHPG). The MHPG formed can be further metabolized to MHPG-sulfate by the action of a sulfotransferase found in brain. Although MHPG-sulfate is the predominant form of this metabolite found in rodent brain, the free, unconjugated MHPG is the major form found in primate brain.

ical agents. The MAOIs interact with both noradrenergic and dopaminergic neurons. By definition these agents inhibit the enzyme MAO, which is involved in oxidatively deaminating catecholamines and thus cause an increase in the endogenous levels of both dopamine and norepinephrine in the brain. At least from a biochemical standpoint, the tricyclic antidepressants appear to interact primarily with noradrenergic rather than dopaminergic neurons. The secondary amine derivatives such as desipramine are exceptionally potent inhibitors of norepinephrine uptake but have only a minimal effect on dopamine uptake. The tertiary amine derivatives are less effective in inhibiting norepinephrine uptake but are very effective in blocking serotonin uptake. The atypical antidepressants do not appear to have a selective interaction with monoamine systems when given by acute administration.

For many years it was postulated that the therapeutic action of antidepressant drugs was directly related to their uptake-blocking capability. However, since the therapeutic action of antidepressant drugs is usually delayed in onset for a week or more, the hypothesis that the rapid, acute actions of these drugs in blocking norepinephrine and serotonin re-uptake are responsible for the long-term clinical antidepressant effects has been challenged. Recent investigations demonstrating that clinically effective antidepressant drugs such as iprindole and mianserin fail to significantly inhibit neuronal uptake of 5-HT and NE while effective uptake inhibitors such as amphetamine, femoxetine, and cocaine are not effective in the treatment of depression have further discredited this hypothesis. Current research has consequently shifted from the study of acute drug effects on amine metabolism to studies on the adaptive changes in NE and 5-HT receptor function induced by chronic antidepressant drug administration. In general, studies of this nature have revealed that chronic treatment with clinically effective antidepressant drugs causes a diminished responsiveness of central β-receptors and enhanced responsiveness of central serotonin and α_1-receptors. The time course of these receptor alterations more closely parallels the time course of the clinical antidepressant effects in humans. At present, however, the relationship between these receptor alterations and clinical antidepressant effects is unknown.

Stimulants

The therapeutic usage of this class of drugs is becoming less and less common as awareness of their abuse potential increases. At the present time, in fact, their use is largely restricted to the treatment of narcolepsy and hyperkinetic children and to their general use as anoretic agents. The principal drugs in this category are the various analogues and isomers of amphetamine and methylphenidate.

For many years it has been known that ingestion of large amounts of amphetamine often leads to a state of paranoid psychosis that may be hard to distinguish from the paranoia associated with schizophrenia. It now appears that this paranoid state can be readily and reproducibly induced in human volunteers given large amounts of amphetamine so that the drug may provide a convenient "model psychosis" for experimental study. It is of interest in this regard that antipsychotic drugs such as chlorpromazine can readily reverse the amphetamine-induced psychosis.

On a biochemical level it was no surprise to learn that amphetamine and related compounds interact with catecholamine-containing neurons, since amphetamine is a close structural analogue of the catecholamines. However, there was no clear evidence that amphetamine produced its CNS effects through a catecholamine mechanism until it was demonstrated that α-methyl-ρ-tyrosine (a potent inhibitor of tyrosine hydroxylase) prevented most of the behavioral effects of D-amphetamine. The question as to which catecholamine, norepinephrine, or dopamine, is involved in the behavioral effects of amphetamine is still an open question. But it is generally believed that the so-called stereotypic behavior in animals (i.e. compulsive gnawing, sniffing) induced by amphetamine is associated with a dopaminergic mechanism and that the increase in locomotor activity involves a noradrenergic mechanism or both.

Almost all classes of psychotropic drugs interact in one way or another with catecholamine-containing neurons. Figures 10-7 and 10-8 outline the life cycle of the transmitters of dopaminergic and noradrenergic neurons in the CNS and indicate possible sites at which drugs may intervene in this cycle. These schematic models also provide examples of drugs or chemical agents which interfere

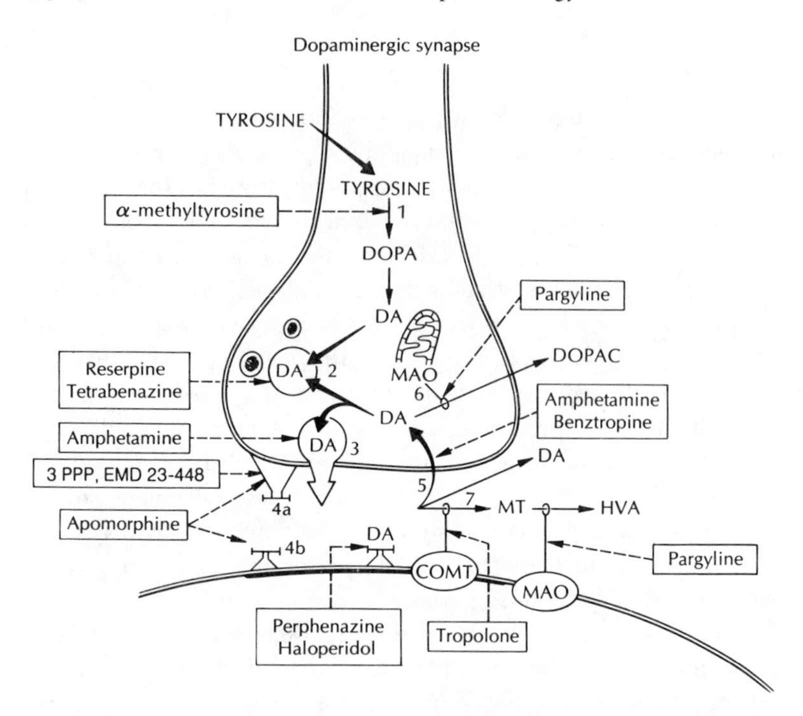

FIGURE 10-8. Schematic model of a central dopaminergic neuron indicating possible sites of drug action.

Site 1: *Enzymatic synthesis:* Tyrosine hydroxylase reaction blocked by the competitive inhibitor, α-methyltyrosine and other tyrosine hydroxylase inhibitors. (See Chapter 9.)

Site 2: *Storage:* Reserpine and tetrabenazine interfere with the uptake–storage mechanism of the amine granules. The depletion of dopamine produced by reserpine is long-lasting and the storage granules appear to be irreversibly damaged. Tetrabenazine also interferes with the uptake–storage mechanism of the granules except that the effects of this drug do not appear to be irreversible.

Site 3: *Release:* γ-Hydroxybutyrate effectively blocks the release of dopamine by blocking impulse flow in dopaminergic neurons. Amphetamine administered in high doses releases dopamine, but most of the releasing ability of amphetamine appears to be related to its ability to effectively block dopamine re-uptake.

Site 4: *Receptor interaction:* Apomorphine is an effective dopamine receptor-stimulating drug, with both pre- and postsynaptic sites of action. Both-3-PPP and EMD-23 (an indolebutylamine) are autoreceptor agonists. Perphenazine and haloperidol are effective dopamine receptor blocking drugs.

at these various sites within the life cycle of the transmitter substances. These numerous sorts of interactions ultimately result in an increase, decrease, or no change in the functional activity of the catecholamine neuron in question.

Only recently did it become clear exactly how various pharmacological agents alter activity in defined catecholamine neuronal systems in the brain. In most cases, the turnover of monoamines depends essentially upon impulse flow in the neuron. An increase in impulse flow usually causes an increase in turnover, and a decrease in impulse flow a reduction in turnover. As mentioned above, however, this is not always the case in the dopamine system if synthesis is used as an index of turnover. Measurement of turnover has been employed to gain some insight into the activity of various types of monoamine-containing neurons during different behavioral states or after administration of different psychotropic drugs. As might be predicted, psychotropic drugs can have a variety of effects, and these effects can alter the turnover of a given transmitter substance without necessarily altering impulse flow in the neuronal system under study. For example, a drug can have a direct effect on the synthesis, degradation, uptake, or release of a given transmitter which will in turn alter the turnover of the transmitter in question but will not necessarily lead to an increase or decrease in the activity of the neuronal system which utilizes that substance as a transmitter. So it becomes clear that an alteration in turnover of a transmitter is

---

Site 5: *Re-uptake:* Dopamine has its action terminated by being taken up into the presynaptic terminal. Amphetamine as well as the anticholinergic drug, benztropine, are potent inhibitors of this re-uptake mechanism.

Site 6: *Monoamine Oxidase (MAO):* Dopamine present in a free state within the presynaptic terminal can be degraded by the enzyme MAO, which appears to be located in the outer membrane of the mitochondria. Dihydroxyphenylacetic acid (DOPAC) is a product of the action of MAO and aldehyde oxidase on dopamine. Pargyline is an effective inhibitor of MAO. Some MAO is also present outside the dopaminergic neuron.

Site 7: *Catechol-O-methyl transferase (COMT):* Dopamine can be inactivated by the enzyme COMT, which is believed to be localized outside the presynaptic neuron. Tropolone is an inhibitor of COMT.

not necessarily a clear indication that there has been a change in impulse flow in a given neuronal pathway. Therefore, the most direct way to determine if a drug alters impulse flow in a chemically defined neuronal system is to measure the activity of that system while the animal is under the influence of the drug.

It now appears that there are also a number of ways in which drugs can alter impulse flow. For example, a drug can act directly on the nerve cell body; it can act on other neurons, which then influence impulse flow in the neuron under study; or it can act at the postsynaptic receptor to cause stimulation or blockade, which then results in some sort of feedback influence on the presynaptic neuron. This feedback information could either be neuronal or perhaps transsynaptically mediated by release of some local chemical from the postsynaptic membrane. The recent combined histochemical-neurophysiological identification of dopaminergic and noradrenergic neurons has made possible the direct study of the effects of various drugs on the firing of these chemically defined neurons.

Effect of Drugs on the Activity of Dopaminergic Neurons

More than a decade ago, entirely on the basis of indirect biochemical evidence, Swedish pharmacologists speculated that antipsychotic drugs, as a result of their ability to block dopaminergic receptors, caused a feedback activation of dopaminergic neurons. Recently, direct extracellular recording experiments have validated these earlier speculations. Antipsychotic drugs of both the phenothiazine and butyrophenone classes cause a dramatic increase in the firing rate of dopamine neurons in rat brain (Fig. 10-9), an action presumably mediated by a neuronal feedback loop. Phenothiazines without antipsychotic potency are completely inactive in this regard. Of course, it is still a big jump to determine whether this interesting ability of antipsychotic durgs to block dopamine receptors is in any way related to their antipsychotic properties. In fact, we already know that there is a temporal dissociation of the two actions.

A naturally occurring metabolite of mammalian brain, γ-hydrox-

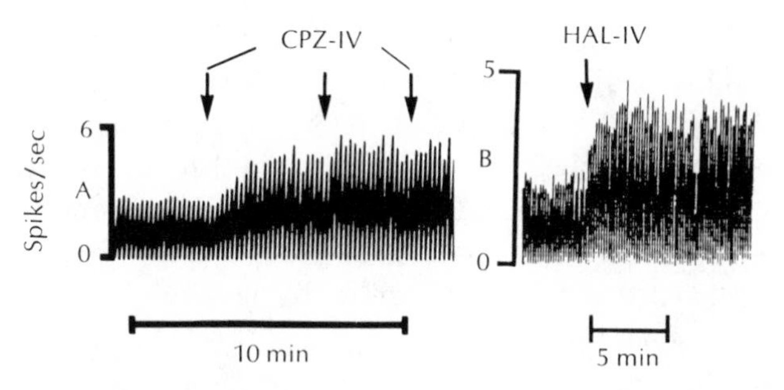

FIGURE 10-9. Effect of antipsychotic drugs on the firing rate of dopamine cells in the zona compacta of the rat.

A. Effect of chlorpromazine on the activity of a dopaminergic cell in the zona compacta of a rat anesthetized with chloral hydrate. Chlorpromazine (CPZ) administered intravenously (iv) in divided doses of 0.5, 0.5, and 1.0 mg/kg caused a marked acceleration of the firing of this dopaminergic neuron.
B. Effect of haloperidol on the activity of a dopaminergic cell in a nonanesthetized rat. An intravenous (iv) injection of haloperidol (HAL) 0.04 mg/kg increased basal activity about 100 percent. (Courtesy of G. K. Aghajanian and B. S. Bunney)

ybutyrate, also has an interesting interaction with dopaminergic neurons. When γ-hydroxybutyrate is administered to rats in anesthetic doses, this compound causes a complete cessation of impulse flow in the nigrostriatal pathway. This inhibition does not appear to be mediated via activation of a neuronal feedback loop. On the other hand, the activity of other monoamine-containing neurons such as the norepinephrine-containing neurons in the locus coeruleus or the 5-HT-containing neurons of the midbrain raphe are relatively unaffected. Thus, this drug appears to be capable of producing a reversible lesion in the nigrostriatal and mesolimbic dopamine systems. The ability of this drug to block impulse flow in dopamine neurons most likely explains its unique biochemical properties as well as why it can effectively antagonize the ability of the antipsychotic drugs to increase the accumulation of dopamine metabolites in the striatum. Whether the drug will prove useful in the treatment of certain disease or drug-induced states thought to be related to dopaminergic hyperactivity remains to be determined.

There is also compelling biochemical and behavioral evidence that amphetamine, L-dihydroxyphenylalanine, and apomorphine interact with dopamine neuronal systems in the CNS. Until recently, however, there has been no direct evidence that these drugs affect neuronal activity. Since all three drugs presumably lead either directly or indirectly to a stimulation of dopamine receptors, it has been predicted that they might slow the activity of dopaminergic neurons via some sort of neuronal feedback loop. Extracellular unit-recording experiments conducted on dopamine cells in the zona compacta and ventral tegmental area have borne out this speculation for all three drugs. In doses as low as 0.25 mg/kg, D-amphetamine inhibits firing of these dopamine neurons. It appears to act indirectly through the release of newly synthesized dopamine since α-methyl-ρ-tyrosine can prevent or reverse the depression of dopamine cell firing. L-Amphetamine will also inhibit the firing of dopaminergic neurons but is much less potent than D-amphetamine in producing this effect. L-Dihydroxyphenylalanine also inhibits firing of these neurons and appears to exert its effects indirectly by being converted to dopamine. Pretreatment with a decarboxylase inhibitor will antagonize completely the inhibitory effect of dihydroxyphenylalanine on these cells. Likewise, apomorphine appears to inhibit effectively the firing of dopamine cells, but this drug seems to act directly on dopamine receptors as its actions are not altered by inhibition of dopamine synthesis with α-methyl-ρ-tyrosine. It is also noteworthy that antipsychotic drugs have the ability to block and reverse the depressant effects of D-amphetamine, L-dihydroxyphenylalanine, and apomorphine on the firing of dopamine cells. This observation suggests that all three drugs may exert their effects on firing rate through some sort of postsynaptic feedback pathway. Other experiments indicate that apomorphine also has a direct effect on dopaminergic neurons of the nigrostriatal and mesolimbic systems. This effect is believed to be mediated by an interaction with autoreceptors on dopamine cell bodies. As indicated earlier, apomorphine acts on dopamine autoreceptors in dosages which are ineffective on postsynaptic dopamine receptors. Administered in autoreceptor selective doses apomorphine is ineffective in influencing the activity of dopamine neurons which lack autoreceptors.

Effect of Drugs on the Electrophysiological Activity of Noradrenergic Neurons

Within recent years a number of drugs have been found to influence the activity of the noradrenergic neurons in the locus ceruleus. Table 10-5 lists the drugs that have been studied and their influence on locus cell firing. Amphetamine inhibits locus cell firing, and it appears to do so by activating a neuronal feedback loop. The effects of amphetamine on locus firing are partially blocked by chlorpromazine. L-Amphetamine is equipotent with D-amphetamine in its inhibitory effects. Alpha-receptor agonists and antagonists also influence locus cell firing. This is due in part to the interaction of these adrenergic agents with autoreceptors on the NE cell bodies or dendrites in the locus. Alpha$_2$-agonists such as clonidine or guanfacine cause a suppression of locus cell firing and these inhibitory effects are reversed by α_2-antagonists. Piperoxane, yohimbine and idazoxane, α_2-antagonists, cause a marked increase in single-cell activity. Morphine and morphine-like peptides such as enkephalins or endorphins have been shown to exert an inhibitory influence on locus cell firing. The inhibitory effects of morphine or enkephalin are reversed by the opiate antagonist, naloxone but uninfluenced by α_2-antagonists. Methylxanthines such as isobutylmethylxanthines (IBMX), which induce a quasi-opiate withdrawal syndrome, cause an increase in locus cell firing, and this increase is antagonized by α_2-agonists such as clonidine. LC cells in chronic morphine treated rats increase their firing dramatically during naloxone-induced withdrawal, and this increase in firing is suppressed by clonidine. These animal studies, viewed in conjunction with the clinical data demonstrating the effectiveness of treating opiate withdrawal with clonidine, have suggested that NE hyperactivity may be an important component of the opiate withdrawal syndrome in humans. These studies have thus provided new in-

TABLE 10-5. Correlation between brain levels of MHPG and electrophysiological activity of noradrenergic neurons in the locus coeruleus

Drug or experimental condition	*Change in locus coeruleus unit activity*	*Change in brain MHPG*
Alpha$_2$-agonists	Decrease	Decrease
Clonidine		
Guanfacine		
Alpha$_2$-Antagonists	Increase	Increase
Yohimbine		
Piperoxane		
Idazoxane		
Tricyclic antidepressants	Decrease	Decrease
Desipramine		
Imipramine		
Amitriptyline		
Amphetamine	Decrease	Decrease, increase (dose dependent)

Morphine	Decrease	Increase
Methylxanthines Isobutylmethylxanthine	Increase	Increase
Naloxone	No change	No change
Naloxone-precipitated morphine withdrawal	Increase	Increase
Stress	Increase	Increase
Electrical stimulation LC	Increase	Increase
Transection of dorsal NE bundle	Decrease	Decrease

sight into the rational design of a new class of drugs (α_2-agonists) to treat opiate withdrawal, although the role of the locus ceruleus in this treatment effect has not actually been established.

Numerous pharmacological studies, most conducted in rodents, have demonstrated that there is a good correlation between drug-induced changes in the firing rate of locus coeruleus neurons monitored by extracellular single-unit recording and alterations in brain levels of MHPG. For example, drug-induced suppression of central noradrenergic activity produced by administration of clonidine or tricyclic antidepressants is accompanied by a reduction in MHPG. The α_2-antagonists (e.g., idazoxane, piperoxane or yohimbine) or experimental conditions (e.g., stress or naloxone-precipitated withdrawal), which cause an increase in noradrenergic activity, produce an increase in the brain levels of MHPG. The magnitude of the increase in MHPG produced by naloxone-precipitated morphine withdrawal exceeds that produced by administration of supramaximal doses of α_2-antagonists such as piperoxane or yohimbine. The biochemical observation is consistent with electrophysiological studies indicating that the increase in locus ceruleus cell activity produced during naloxone-precipitated withdrawal exceeds that achieved by administration of α_2-antagonists such as piperoxane and yohimbine. Since MHPG levels in brain measured under controlled experimental conditions provide an index of physiological activity in the locus coeruleus, measures of this metabolite in accessible body fluids may be useful for the assessment of changes in central NE function in intact animals or patients. Many clinical researchers are currently monitoring this metabolite in CSF, plasma, and urine in the hope that this measure will provide insight concerning alterations in central noradrenergic function.

Physiological Functions of Central Noradrenergic Neurons

Many functions have been proposed for the central norepinephrine neurons and their several sets of synaptic connections. Among the hypotheses that have the most supportive data are ideas concerning their role in affective psychoses (described below), in learning and memory, in reinforcement, in sleep–wake cycle regulation, and in the anxiety–nociception hypothesis. It has also been suggested that a major function of central noradrenergic neurons is not on neuronal activity and related behavioral phenomena at all, but rather a more general role in cerebral blood flow and metabolism. However, data available at present fit better a more general proposal: The main function of the locus coeruleus (LC) and its projections is to determine the global orientation of brain between events in the external world or within the viscera. Such an hypothesis of central noradrenergic neuron function has been generated by observations of locus coeruleus unit discharge in untreated, awake, behaviorally responsive rats and monkeys. These studies reveal the locus coeruleus units to be highly responsive to a variety of nonnoxious sensory stimuli, and further, that the responsiveness of these units varies as a function of the animal's level of behavioral vigilance. Increased neuronal activity in locus coeruleus is associated with unexpected sensory events in the subject's external environment, while decreased noradrenergic activity is associated with behaviors that mediate tonic vegetative behaviors. Such a global-orienting function can also incorporate other aspects of presumptive function expressed by earlier data, but none of those more discrete functions can be documented as necessary or sufficient explanations of locus coeruleus function.

Pharmacology of Adrenergic Neurons

Limited experiments have suggested that the pharmacology of central adrenergic neurons is similar to that of central noradrenergic neurons. Agents which block tyrosine hydroxylase, DOPA decarboxylase, and dopamine β-hydroxylase lead to a reduction of both norepinephrine and epinephrine in brain. Depleting agents such as reserpine which cause release of norepinephrine and dopamine also are effective in releasing epinephrine. MAO inhibitors cause an elevation of norepinephrine, dopamine, and epinephrine, but inhibitors of the A form of MAO are much more effective in elevation of epinephrine. In fact, most of the pharmacological data is consistent with the hypothesis that, at least in the rat hypothalamus, oxidative deamination is an important metabolic process by which epinephrine is degraded and type A MAO is predominantly involved in this degradation.

Similar to observations made in central noradrenergic neurons, α_2-agonists such as clonidine decrease epinephrine formation and α_2-antagonists increase epinephrine turnover. These data are also consistent with the possibility that α_2-receptors are involved in the regulation of synthesis and release of epinephrine and perhaps also in the control of the functional activity of adrenergic neurons.

Catecholamine Theory of Affective Disorder

The catecholamines play a fairly well-established role in the periphery with regard to stress and emotional behavior. By contrast, their suggested role in the central nervous system in affective disorders is still quite speculative although evidence for some involvement of these agents is becoming increasingly compelling.

A catecholamine hypothesis of affective disorder states that, in general, behavioral depression may be related to a deficiency of catecholamine (usually norepinephrine) at functionally important central adrenergic receptors, while mania results from excess catecholamine. While substantial experimental work bears on this proposal, it must be kept in mind that most of the experiments on which this hypothesis is based derive from "normal" animals. Second, from a nosological standpoint, depression itself is rather ill defined. In fact, Giarman has suggested that "nosologically it might be fair to compare the depressive syndrome with the anemias. Certainly, no self-respecting hematologist would subscribe to a unitary biochemical explanation for all of the anemias."

The original impetus for formulation of the catecholamine hypothesis was the finding that various monoamine oxidase inhibitors, notably iproniazid, acted clinically as mood elevators or antidepressants. Shortly thereafter, it was found that this class of compounds also produced marked increases in brain amine levels. By the same token, reserpine, a potent tranquilizer that causes depletion of brain amines, sometimes produces a serious depressed state (clinically indistinguishable from endogenous depression) and even suicidal attempts in humans. Although both classes of drugs altered brain levels of catecholamines and serotonin quite equally, the fact that a precursor of catecholamine biosynthesis, notably DOPA, could reverse most of the reserpine-induced symptomology in animals has tended to bias many in favor of the catecholamine theory. In general, many pharmacological studies appear to implicate the catecholamines as the amines involved in affective disorders. However, it must be realized that a great deal more work has been done on norepinephrine than on other transmitter candidates. In fact, by no means does the available evidence obtained for the involvement of norepinephrine rule out the participation of dopamine, 5-hydroxytryptamine epinephrine, or other putative transmitters in similar events.

The three general classes of drugs most commonly used to treat various depressive disorders are the monoamine oxidase inhibitors, the tricyclic antidepressants, and the psychomotor stimulants of which amphetamine was the prototype. All these pharmacological agents appear to interact with catecholamines in a way that is consistent with the catecholamine hypothesis.

The amphetamines, even if one considers all the suggested modes of action—(1) partial agonist, mimicking action of norepinephrine at receptor, (2) inhibition of catecholamine re-uptake, (3) inhibition of monoamine oxidase, (4) displacement of norepinephrine, releasing it onto receptors—appear to exert an action compatible with the hypothesis above since the net result of all these actions is a temporary increase of norepinephrine at the receptor or a direct stimulatory action at the receptor. Upon administration of chronic or high doses of amphetamine, it is possible to produce an eventual depletion of brain norepinephrine perhaps owing to a displacement of norepinephrine in the neuron and/or an inhibition of synthesis. Also, as indicated earlier, amphetamine causes an inhibition of neuronal activity in catecholamine neurons. This chronic depletion of transmitter or a prolonged inactivity of catecholamine neurons may relate to the clinical observation of amphetamine tolerance or to the well-known poststimulation depression or fatigue observed after chronic administration of this class of drugs.

Many of the tricyclic antidepressants influence catecholamine neurons by inhibiting the catecholamine re-uptake system (see above). While these findings are of course also consistent with the catecholamine hypothesis, the tricyclics do not affect this system exclusively. Thus, some of the tricyclic agents, notably imipramine, also potentiate peripheral and central effects of serotonin; they also have both central and peripheral anticholinergic actions.

The action of the monoamine oxidase inhibitors also supports the catecholamine hypothesis. All these agents inhibit an enzyme responsible for the metabolism of norepinephrine and various other amines (5-hydroxytryptamine, dopamine, tyramine, tryptamine). This inhibition results in a marked increase in the intraneuronal levels of norepinephrine. Presumably, according to the conceptual model of the adrenergic neuron presented above, this interneuronal norepinephrine might eventually diffuse out of the neuron and reach receptor cells, thus overcoming the presumed deficiency. A similar mechanism may also explain the antagonism of reserpine-induced sedation with monoamine oxidase inhibitors, since there will be a replenishment of the norepinephrine deficiency initially caused by reserpine.

Lithium, one of the main agents used in the treatment of mania has also been studied with regard to its action on the life cycle of the catecholamines. Interestingly, pretreatment with lithium blocks the stimulus-induced release of norepinephrine from rat brain slices. Other investigators have suggested that lithium may facilitate reuptake of norepinephrine. If the inhibition of release observed in stimulated brain slices is due to a facilitated recapture mechanism, then the mechanism of action of lithium is the exact opposite of that of the antidepressant drugs, as would be expected according to the catecholamine hypothesis. Although single injection of lithium may antagonize responses to norepinephrine, chronic treatment of rats leads to modest supersensitive norepinephrine responses.

As summarized in a review by Berger and Barchas, closer scrutiny of the pharmacological data cited in support of the catecholamine theory of affective disorder reveals some important discrepancies. For example, (1) cocaine is a very potent inhibitor of catecholamine re-uptake and thus, like the tricyclic antidepressants, it should increase the availability of norepinephrine at central synapses. But cocaine does not possess any significant antidepressant activity. (2) Iprandole, a tricyclic compound without any significant effect on catecholamine uptake or any influence on central noradrenergic neurons, is an effective antidepressant. Furthermore, desensitization of postsynaptic β-receptors after chronic administration of tricyclic antidepressant drugs is not entirely compatible with the proposed theory of a synaptic catecholamine deficit. (3) In laboratory studies the ability of tricyclic antidepressants to inhibit catecholamine uptake and the ability of MAO inhibitions to block MAO and elevate brain catecholamines are apparent soon after administration. Clinically, antidepressants require several weeks of continuous administration to produce therapeutic effects. (4) On the surface, the pharmacological and clinical effects of the tricyclics and lithium seem to fit nicely with the catecholamine theory of affective disorder. These agents produce opposite effects on norepinephrine disposition, and lithium is effective in treating mania while the tricyclics are useful in treating depression. The water is muddied, however, by the finding that lithium is also effective in the treatment of bipolar depressed patients.

The fact that several questions have been raised about the pharmacological data used to support the catecholamine theory of affective disorder has prompted many investigators to attempt to obtain some direct evidence of the involvement of norepinephrine systems in affective disorders. The most extensive studies have involved analysis of urinary excretion patterns of catecholamine metabolites in patients with affective disorders. The rationale has been that the urinary excretion of a particular metabolite like MHPG may be a useful reflection of central catecholaminergic processes. Despite the inherent problems in urinary catecholamine metabolite measurement, such as complications because of large contributions from peripheral sources, some interesting finds have emerged.

Findings from several clinical studies now indicate that: (1) depressed patients as a group excrete less than normal quantities of MHPG; (2) diagnostic subgroups of depressed patients are particularly likely to have low urinary MHPG values; (3) patients who switch from a depressive to a euthymic or hypomanic state show a corresponding increase in urinary MHPG; and (4) pretreatment urinary MHPG values are predictive of the type of therapeutic response obtained with amitriptyline or imipramine. Although provocative and in essence consistent with the norepinephrine theory of affective disorder, these clinical findings all hinge on the issue of the degree to which urinary MHPG reflects central norepinephrine metabolism.

For the above reasons, the original hypothesis that the mechanism of action of antidepressant drugs is to increase the availability of monoamines in brain has been updated to include the effects of long-term antidepressant treatment on monoamine receptor sensitivity. A wide array of effects of long-term treatment with antidepressants have been observed in several monoamine systems. These include long-term changes in receptor number and changes in physiological and behavioral sensitivity to monoamine agonists and antagonists. The most consistent findings observed following chronic administration of most of the clinically effective antidepressant drugs to experimental animals are (1) a reduction in the number of β-adrenoceptor recognition sites and a down regulation of β-adrenoceptor functioning, (2) an increase in the sensitivity of central α-adrenoceptors suggesting an upregulation in central α-adrenoceptor functioning, (3) similar to the observation on α-adrenoceptors, both behavioral and electrophysiological studies also point to an up regulation in the sensitivity of central 5-HT receptors. More direct studies will be necessary, however, before it will be possible to conclude whether any or all of these changes are related to the therapeutic action of antidepressant drugs. These studies await the development of appropriate models for monitoring central neurotransmitter functioning in humans.

Dopamine Hypothesis of Schizophrenia

The growing conviction that antipsychotic agents act therapeutically by decreasing central dopaminergic transmission led to the dopamine theory of schizophrenia. This hypothesis in its simplest form states that schizophrenia may be related to a relative excess of central dopaminergic neuronal activity. Although there is a lot of data that supports the dopamine hypothesis, at present we have no direct experimental evidence of an excess of dopamine-dependent neuronal activity or an elevation of dopamine levels at central synapses in schizophrenics. Thus, the cornerstone of this hypothesis is the idea that the therapeutic effects of antipsychotic medication are due exclusively to the ability of these drugs to block central dopamine receptors and thereby reduce dopaminergic transmission.

Over the past decade experiments in animals have generated much data in support of the contention that antipsychotic drugs are effective blockers of dopamine receptors. Indirect, clinical studies also suggest that antipsychotic drugs interact with dopamine neurons and effectively block central dopamine receptors. Nevertheless, it remains unclear whether dopamine receptor blockade is an essential component of the therapeutic effect of antipsychotic medication. Most behavioral, biochemical, and electrophysiological studies that demonstrate an interaction of antipsychotic drugs with dopaminergic neurons have been carried out following acute drug administration. This is a very serious drawback since it is well known that the clinical effects of antipsychotic drugs (both antipsychotic and neurological) take days, weeks, or even months to develop. Even if we are willing to agree that the pharmacological studies strongly suggest that antipsychotic agents work through blockade of central dopamine systems the evidence that schizophrenia is caused by a defect in one or more central dopamine systems is far weaker. In fact, at the present time there is really no direct evidence implicating a defect in any of the dopamine systems in schizophrenia.

Two reasonable approaches to testing the dopamine hypothesis of schizophrenia have been the measurement of HVA in the CSF and the examination of catecholamine metabolites and related enzymes in autopsied brain obtained from controls and schizophrenics. No significant differences have yet been demonstrated in brain or CSF levels of HVA in normals and unmedicated schizophrenics. However, a significant elevation of HVA levels in the CSF of patients on antipsychotic medications has been found, and this supports the idea that these drugs interact with central dopamine neurons in human patients.

In nonhuman primates chronic treatment with antipsychotic drugs causes tolerance to develop to the HVA increase normally observed in the putamen, caudate, and olfactory cortex after a challenge dose of an antipsychotic drug. However, in cingulate, dorsal frontal, and orbital frontal cortex, increased levels of HVA are maintained throughout the time course of chronic treatment. Similar observations have also been made in studies carried out on autopsied brain specimens obtained from human subjects. In schizophrenic patients chronically treated with antipsychotic drugs, a significant increase in HVA is found in the cingulate and perifalciform cortex but not in the putamen and nucleus accumbens, suggesting a possible locus for the therapeutic action of these drugs and providing the first direct experimental evidence that antipsychotic drug treatment increases the metabolism of dopamine in human brain in a regionally specific manner.

In studies examining both synthesis and degradation of dopamine in autopsied brain, no significant differences have been found between schizophrenic and control samples. There has been one report of a significant decrease in the activity of dopamine-β-oxidase in schizophrenic brain, but this observation has not been replicated in subsequent studies.

As neurochemical techniques of increasing sophistication are developed, studies of more discrete components of schizophrenic brain obtained postmortem will have greater potential for elucidating possible biochemical defects associated with schizophrenia. However the possibility of obtaining the appropriate postmortem samples, whatever they may be, seems quite remote, especially in view of the nearly universal use of antipsychotic medication in patients who are diagnosed as schizophrenic and who are likely to be available for autopsy soon enough after death to be useful for research purposes.

Even though the evidence that synthesis, metabolism, and turnover of brain dopamine is aberrant in schizophrenia remains inconclusive, with the identification of specific brain receptors for dopamine and neuroleptic drugs it became possible to examine whether brain dopamine receptors were abnormal in schizophrenics. Dopamine receptors in postmortem schizophrenic brains have been reported by several groups to be increased in numbers. Increased binding was found in both the nucleus accumbens and striatum using haloperidol as the ligand. However, neither the D_1 site labeled by *cis*-flupenthixol nor the so-called D_3-site labeled by (H_3)-dopamine reflected this increase. There also still remains a controversy as to whether the elevated density of dopamine receptors observed is drug induced (since most schizophrenic subjects have been drug treated) or is a cause or consequence of the disease process. This is despite the fact that two groups have independently demonstrated an increase in H^3-spiperone binding in brains of schizophrenics not exposed to neuroleptic drugs for over a year prior to coming to autopsy, although the observed increases were less than that usually recorded in drug-treated patients. Establishing with certainty that the elevation in dopamine receptor density is part of the disease process associated with schizophrenia would be very important for both etiological and diagnostic purposes as well as serving as a possible basis for treatment strategies. The application of new noninvasive techniques such as positron emission tomography to examine dopamine receptor distribution and density in schizophrenia hold promise that this may be accomplished in the near future.

Selected References

Aston-Jones, G., and F. E. Bloom (1981). Norepinephrine-containing locus coeruleus neurons in behaving rats exhibit pronounced responses to nonnoxious environmental stimuli. *J. Neurosci. 1*, 887.

Bacopoulos, N. G., S. E. Hattox, and R. H. Roth (1979). 3,4-Dihydroxyphenylacetic acid and homovanillic acid in rat plasma:

Possible indicators of central dopaminergic activity. *Eur. J. Pharmacol. 56*, 225.

Bannon, M. J., and R. H. Roth (1983). Pharmacology of mesocortical dopamine neurons. *Pharmacol. Rev. 35(1)*, 53.

Bannon, M. J., A. S. Freeman, L. A. Chiodo, B. S. Bunney, and R. H. Roth (1986). The pharmacology and electophysiology of mesolimbic dopamine neurons. In *Handbook of Psychopharmacology*, Vol. 19. Plenum, New York.

Billingsley, M. L., M. P. Galloway, and R. H. Roth (1984). Possible role of protein carboxymethylation in the autoreceptor-mediated regulation of dopamine neurons. In *Proceedings of the 5th International Catecholamine Symposium*, Göteborg, Sweden, p. 37. Alan R. Liss, Inc., New York.

Björklund, A., and T. Hökfelt, eds. (1984). *Handbook of Chemical Neuroanatomy*. Vol. 2, Classical Transmitters in the CNS, Part I., Elsevier, New York.

Bunney, B. S. (1979). the electrophysiological pharmacology of midbrain dopaminergic systems. In *The Neurobiology of Dopamine* (A. S. Horn, J. Korf, and B. H. C. Westerink, eds.), p. 417. Academic Press, New York.

Clark, D., S. Hjorth, and A. Carlsson (1985). Dopamine-receptor agonists: Mechanisms underlying autoreceptor selectivity. *J. Neural. Trans. 62*, 1.

Dahlström, A., and K. Fuxe (1965). Evidence for the existence of monoamine-containing neurons in the central nervous system. I. Demonstration of monoamines in the cell bodies on brain stem neurons. *Acta Physiol. Scand. 62*, Suppl. 232, 1.

Descarries, L., T. R. Reader, and H. H. Jasper (1984). Monoamine innervation of cerebral cortex. In *Neurology and Neurobiology*, Vol. 10, p. 1. Alan R. Liss, Inc., New York.

Elsworth, J. D., D. E. Redmond, Jr., and R. H. Roth (1982). Plasma and cerebrospinal fluid 3-methoxy-4-hydroxyphenylethylene glycol (MHPG) as indices of brain norepinephrine metabolism in primates. *Brain Res. 235*, 115.

Foote, S. L., F. E. Bloom, and G. Aston-Jones (1983). Nucleus locus ceruleus: New evidence of anatomical and physiological specificity. *Physiol. Rev. 63*, 844.

Freed, W. J., B. J. Hoffer, L. Olson, and R. J. Wyatt (1984). Transplantation of catecholamine-containing tissue to restore the functional capacity of the damaged nigrostriatal system. In *Neuronal Transplants* (J. R. Sladek, Jr. and D. M. Gash, eds.), pp. 373–402. Plenum, New York.

Fuller, R. W. (1982). Pharmacology of brain epinephrine neurons. *Ann. Rev. Pharmacol. Toxicol. 22*, 31.

Hoffer, B. J., G. R. Siggins, A. P. Oliver, and F. E. Bloom (1973). Activation of the pathway from locus coeruleus to rat cerebellar Purkinje neurons: Pharmacological evidence of noradrenergic central inhibition. *J. Pharmacol. Exp. Ther. 184*, 553.

Hoffmann, P. C., and A. Heller (1984). Embryonic dopaminergic neurons in culture and as transplants. In *Early Brain Damage*, Vol. 2, Neurobiology and Behavior. Academic Press, New York.

Hökfelt, T., O. Johansson, and M. Goldstein (1984). Chemical anatomy of the brain. *Science 225*, 1326.

Horn, A. S., J. Korf, and B. H. C. Westerink, eds. (1979). *The Neurobiology of Dopamine*. Academic Press, New York.

Kebabian, J. W., M. Beaulieu, and Y. Itoh (1984). Pharmacological and biochemical evidence for the existence of two categories of dopamine receptor. *Can. J. Neurol. Sci. 11*, 114.

Lindvall, O., and A. Björklund (1974). The organization of the ascending catecholamine neuron systems in the rat brain as revealed by the glyoxylic and fluorescence method. *Acta Physiol. Scand.*, Suppl. 412, 1.

Maas, J. W., S. E. Hattox, N. M. Greene, and D. H. Landis (1979). 3-Methoxy-4-hydroxyphenethyleneglycol (MHPG). Production by Human Brain *in vivo*. *Science 205*, 1025.

Maas, J. W. (ed.) (1983). *MHPG: Basic Mechanisms and Psychopathology*. Academic Press, New York.

Meltzer, H. Y. (1980). Relevance of dopamine autoreceptors for psychiatry: Preclinical and clinical studies. *Schizophrenia Bull. 6*, 456.

Meltzer, H. Y., and S. M. Stahl (1976). The dopamine hypothesis of schizophrenia: A review. *Schizophrenia Bull. 2*, 19.

Moore, R. Y., and F. E. Bloom (1978). Central catecholamine neuron systems: Anatomy and physiology of the dopamine systems. *Ann. Rev. Neurosci. 1*, 129.

Moore, R. Y., and F. E. Bloom (1979). Central catecholamine neuron systems: Anatomy and physiology of the norepinephrine and epinephrine systems. *Ann. Rev. Neurosci. 2*, 113.

Roth, R. H. (1984). CNS dopamine autoreceptors: Distribution, pharmacology and function. *Annals New York Acad. Sci. 430*, 27.

Salzman, P. M., and R. H. Roth (1979). Role of impulse flow in the short-term regulation of norepinephrine biosynthesis. *Prog. Neurobiol. 13*, 1.

Segal, M., and F. E. Bloom (1976). The acton of norepinephrine in the rat hippocampus: IV. The effects of locus coeruleus stimulation on evoked hippocampal activity. *Brain Res. 107*, 513.

Siggins, G. R. (1978). The electrophysiological role of dopamine in striatum: Excitatory or inhibitory? In *Psychopharmacology—A Generation of Progress* (M. E. Lipton, K. C. Killam, and A. DiMascio, eds.), p. 143. Raven Press, New York.

Snyder, S. H. (1981). Dopamine receptors, neuroleptics, and schizophrenia. *Am. J. Psychiat. 138*, 460.

Sugrue, M. F. (1984). Do antidepressants possess a common mechanism of action? *Biochemical Pharmacol. 32*, 1811.

Usdin, E., A. Carlsson, A. Dahlström, and A. Engels (eds.) (1984). *Catecholamines Part B, Neuropharmacology and Central Nervous Systems—Theoretical Aspects.* Proceedings of the Fifth International Catecholamine Symposium. Göteborg, Sweden, Alan R. Liss, New York.

Wagner, H. N., Jr., H. D. Burns, R. F. Dannals, D. F. Wong, B. Langström, T. Duelfer, J. J. Frost, H. T. Ravert, J. M. Links, S. Rosenbloom, S. E. Lukas, A. V. Kramer, and M. J. Kuhar (1985). Imaging of dopamine receptors in the human brain by positron tomography. In *The Metabolism of the Human Brain Studied with Positron Emission Tomography* (T. Greitz et al., eds.), p. 251. Raven Press, New York.

11 | Serotonin (5-hydroxytryptamine) and Histamine

Serotonin

Of all the neurotransmitters discussed in this book, serotonin remains historically the most intimately involved with neuropsychopharmacology. From the mid-nineteenth century, scientists had been aware that a substance found in serum caused powerful contraction of smooth muscle organs, but over a hundred years passed before scientists at the Cleveland Clinic succeeded in isolating this substance as a possible cause of high blood pressure. By this time, investigators in Italy were characterizing a substance found in high concentrations in chromaffin cells of the intestinal mucosa. This material also constricted smooth muscular elements, particularly those of the gut. The material isolated from the bloodstream was given the name "serotonin" while that from the intestinal tract was called "enteramine." Subsequently, both materials were purified, crystallized, and shown to be 5-hydroxytryptamine (5-HT), which could then be prepared synthetically and shown to possess all the biological features of the natural substance. The indole nature of this molecule bore many resemblances to the psychedelic drug LSD, with which it could be shown to interact on smooth muscle preparations *in vitro*. 5-HT is also structurally related to other psychotropic agents (Fig. 11-1).

When 5-HT was first found within the mammalian central nervous system, the theory arose that various forms of mental illness could be due to biochemical abnormalities in its synthesis. This line

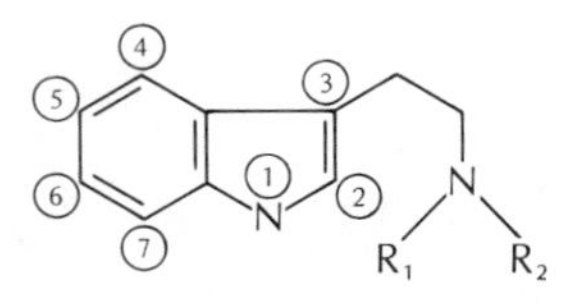

FIGURE 11-1. Structural relationships of the various indolealkyl amines.

Compound	*Substitutions*
Tryptamine	R_1 and R_2 = H
Serotonin	Tryptamine with 5 hydroxy
Melatonin	5 methoxy, *N*-acetyl
DMT*	R_1 and R_2 = methyl
DET*	R_1 and R_2 = ethyl
Bufotenine*	5 hydroxy, DMT
Szara psychotrope*	6 hydroxy, DET
Psilocin*	4 hydroxy, DMT
Harmaline*	6 methoxy; R_1 forms isopropyl link to C_2.
5-MT	5-methoxytryptamine
5,6 DHT	5,6 dihydroxytryptamine
5,7 DHT	5,7 dihydroxytryptamine

* = psychotropic or behavioral effects.

of thought was even further extended when the tranquilizing substance, reserpine, was observed to deplete brain 5-HT; throughout the duration of the depletion, profound behavioral depression was observed. As we shall see, many of these ideas and theories are still maintained although we now have much more ample evidence on which to evaluate them.

Biosynthesis and Metabolism of Serotonin

Serotonin is found in many cells that are not neurons, such as platelets, mast cells, and the enterochromaffin cells mentioned above. In fact, only about 1 to 2 percent of the serotonin in the whole body is found in the brain. Nevertheless, as 5-HT cannot cross the blood–brain barrier, it is clear that brain cells must synthesize their own.

For brain cells, the first important step is the uptake of the amino acid tryptophan, which is the primary substrate for the synthesis. Plasma tryptophan arises primarily from the diet, and elimination of dietary tryptophan can profoundly lower the levels of brain serotonin. In addition, an active uptake process is known to facilitate the entry of tryptophan into the brain and this carrier process is open to competition by large neutral amino acids including the aromatic amino acids (tyrosine and phenylalanine), the branch-chain amino acids (leucine, isoleucine, and valine), and others (i.e., methionine and histidine). The competitive nature of the large neutral amino-acid (LNAA) carrier means that brain levels of tryptophan will be determined not only by the plasma concentration of tryptophan but also by the plasma concentration of competing neutral amino acids. Thus, dietary protein and carbohydrate content can specifically influence brain tryptophan and serotonin levels by effects on plasma amino-acid patterns. Because plasma tryptophan has a daily rhythmic variation in its concentration, it seems likely that this concentration variation could also profoundly influence the rate and synthesis of brain serotonin.

The next step in the synthetic pathway is hydroxylation of tryptophan at the 5 position (Fig. 11-2) to form 5-hydroxytryptophan (5-HTP). The enzyme responsible for this reaction, tryptophan hydroxylase, occurs in low concentrations in most tissues, including the brain, and it was very difficult to isolate for study. After purifying the enzyme from mast cell tumors and determining the characteristic cofactors, however, it became possible to characterize this enzyme in the brain. (Students should investigate the ingenious methods used for the initial assays of this extremely minute enzyme activity.) As isolated from brain, the enzyme appears to have an absolute requirement for molecular oxygen, for reduced pteridine cofactor, and for a sulfhydryl-stabilizing substance, such as mercaptoethanol, to preserve activity *in vitro*. With this fortified system of assay, there is sufficient activity in the brain to synthesize one μg of 5-HTP per gram of brain stem in an hour. The pH optimum is approximately 7.2 and the K_m for tryptophan is $3 \times 10^{-4}M$. Additional research into the nature of the endogenous cofactor, tetrahydrobiopterin, yielded a K_m for tryptophan of

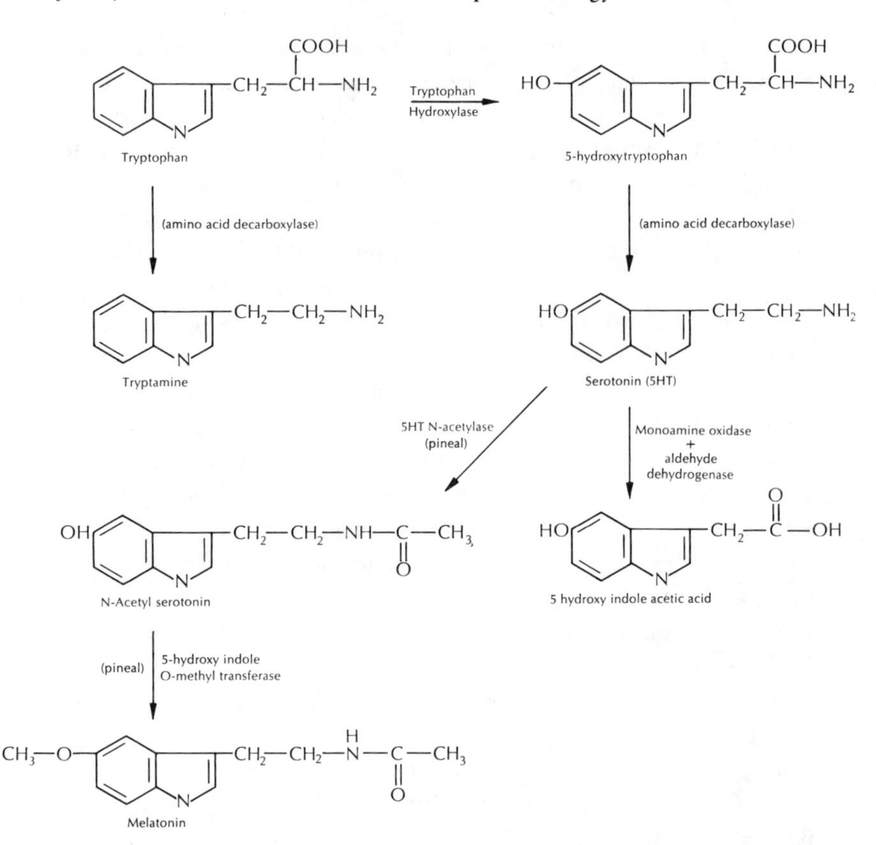

FIGURE 11-2. The metabolic pathways available for the synthesis and metabolism of serotonin.

5×10^{-5}M, which is still above normal tryptophan concentrations. Thus, the normal plasma tryptophan content and the resultant uptake into brain leave the enzyme normally "unsaturated" with available substrate. Tryptophan hydroxylase appears to be a soluble cytoplasmic enzyme, but the procedures used to extract it from the tissues may greatly alter the natural particle-binding capacity. Investigators examining the relative distribution of particulate and soluble tryptophan hydroxylase have reported that the particulate enzyme may be associated with 5-HT-containing synapses, while the soluble form is more likely to be associated with the perikaryal cytoplasm. The particulate form of the enzyme shows the lower K_m and bears an absolute requirement for tetrahydrobiopterin.

This step in the synthesis can be specifically blocked by *p*-chlorophenylaline, which competes directly with the tryptophan and also binds irreversibly to the enzyme. Therefore, recovery from tryptophan hydroxylase inhibition with *p*-chlorophenylalanine appears to require the synthesis of new enzyme molecules. In the rat, a single intraperitoneal injection of 300 mg/kg of this inhibitor lowers the brain serotonin content to less than 20 percent within three days, and complete recovery does not occur for almost two weeks.

Considerable attention has been directed recently to the overall regulation of this first step of serotonin synthesis, especially in animals and humans treated with psychoactive drugs alleged to affect the serotonin systems as a primary mode of their action. These studies have made an important general point that seems to apply, in fact, to the brain's response to drug exposure in many other cell systems as well as to serotonin. This point is that because transmitter synthesis, storage, release, and response are all dynamic processes; the acute imbalances produced initially by drug treatments are soon counteracted by the built-in feedback nature of synthesis regulation. Thus, if a drug acts to reduce tryptophan hydroxylase activity, the nerve cells may respond by increasing their synthesis of the enzyme and transporting increased amounts to the nerve terminals.

Mandell and colleagues have provided evidence, for example, that short-term treatment with Li will initially increase tryptophan uptake, resulting in increased amounts being converted to 5-HT. After fourteen to twenty-one days of chronic treatment, however, repetition of the measurements shows that while the tryptophan uptake is still increased, the activity of the enzyme is decreased, so that normal amounts of 5-HT are being made. In this new equilibrium state, the neurochemical actions of drugs like amphetamine and cocaine on 5-HT synthesis rates are greatly reduced, as are their behavioral actions (see below). In this way the 5-HT system, by shifting the relationship between uptake and synthesis during Li exposure, can be viewed as more "stable." This factor may be more fully appreciated when one considers that in the treatment of manic-depressive psychosis a minimum of seven to ten days is usually required before the therapeutic action of Li begins, a period during which the reequilibration of the 5-HT synthesizing process could be undergoing restabilization.

Decarboxylation

Once synthesized, 5-HTP is almost immediately decarboxylated to yield serotonin. The enzyme responsible for this was presumed to be identical with the enzyme that decarboxylates dihydroxyphenylalanine (DOPA). This information was gained mainly by the study of decarboxylase activities partially purified from liver or kidney and of the regional distribution of enzyme activity. More recent studies permit a reconsideration of this viewpoint. When the decarboxylating step is examined in brain homogenates, the activity for DOPA differs widely from 5-HTP with respect to pH, temperature, and substrate optima; and within brain homogenates, the two activities become enriched in different subfractions—5-HTP decarboxylase being mainly associated with the synaptosomal and DOPA-decarboxylase with the soluble or microsomal fraction. Finally, the case for two separate enzymes is supported by the observation that treatment of animals with intracisternal 6-hydroxydopamine (see Chapters 9 and 10) reduces DOPA-decarboxylase activity but not 5-HTP decarboxylase activity. On the other hand, immune sera raised against the decarboxylase purified from kidney cross-react with both decarboxylases from brain, perhaps indicating that many common antigenic determinants are shared. Since this decarboxylation reaction occurs so rapidly, and since its K_m (5×10^{-6}M) requires less substrate than the preceding steps, tryptophan hydroxylase would appear to be the rate-limiting step in the synthesis of serotonin, providing the added tryptophan is available in the controlled state. Because of this kinetic relationship, drug-induced inhibition of serotonin by interference with the decarboxylation step has never proven particularly effective.

Catabolism

The only effective route of continued metabolism for serotonin is deamination by monoamine oxidase (MAO). The product of this reaction, 5-hydroxyindoleacetaldehyde can be further oxidized to 5-hydroxyindoleacetic acid (5-HIAA) or reduced to 5-hydroxytryptophol depending on the $NAD^+/NADH$ ratio in the tissue. Enzymes have been described in liver and brain by which 5-HT could be catabolized without deamination through formation of a 5-sulfate ester. This could then be transported out of brain, possibly by the acid excretion system handling 5-HIAA.

Brain contains an enzyme that catalyzes the *N*-methylation of 5-HT using *S*-adenosylmethionine as the methyl donor. An *N*-methylating enzyme which uniquely preferred 5-methyl tetrahydrofolate (5MTF) as the methyl donor has been described. Subsequent work by several other groups revealed this reaction to be an artifact of the *in vitro* assay system (see also Chapter 9) in which the 5-MTF is demethylated, releasing formaldehyde that then condenses across the amino-*N* and the *C*-2 of the indole ring to form the cyclic compound tryptoline. Like other derivatives of the 5-HT metabolic product line, the intriguing possibility that one such molecule might be psychotogenic has contributed to as yet unrewarded efforts to find these abnormal metabolites in human psychotics.

Control of Serotonin Synthesis and Catabolism

Although there is a relatively brief sequence of synthetic and degradative steps involved in serotonin turnover, there is still much to be learned regarding the physiological mechanisms for controlling this pathway. At first glance, it seems clear that tryptophan hydroxylase is the rate-limiting enzyme in the synthesis of serotonin, since when this enzyme is inhibited by 80 percent, the serotonin content of the brain rapidly decreases. On the other hand, when the 5-HTP decarboxylase is inhibited by equal or greater amounts, there is no effect upon the level of brain 5-HT. These data could only be explained if the important rate-limiting step were the initial hydroxylation. Since this step also depends upon molecular oxygen, the rate of 5-HT formation could also be influenced by the tissue level of oxygen. In fact, it can be shown that rats permitted to breathe 100 percent oxygen greatly increase their synthesis of 5-HT. It is also of interest that 5-hydroxytryptophan does not inhibit the activity of tryptophan hydroxylase.

If the situation for serotonin were similar to that previously described for the catecholamines, we might also expect that the concentration of 5-HT itself could influence the levels of activity at the hydroxylation step. However, when the catabolism of 5-HT is blocked by monoamine oxidase inhibitors, the brain 5-HT concentration accumulates linearly to levels three times greater than controls, thus sugggesting that end-product inhibition by serotonin is, at best, trivial. Similarly, if the efflux of 5-HIAA from the brain is blocked by the drug probenecid (which appears to block all forms of acidic transport) the 5-HIAA levels also continue to rise linearly for prolonged periods of time, again suggesting that the initial synthesis step is not affected by the levels of any of the subsequent metabolites. Two possibilities therefore remain open: the initial synthesis rate may be limited only by the availability of required cofactors or substrate, such as oxygen, pteridine, and tryptophan from the bloodstream; or the initial synthesis rate may be limited by the other more subtle control features, more closely related to brain activity. In fact, evidence is beginning to accumulate to suggest that impulse flow may, as in the catecholamine systems, initiate changes in the physical properties of the rate-limiting enzyme, tryptophan hydroxylase. Several mechanisms have been postulated for the physiological regulation of tryptophan hydroxylase induced by alterations in neuronal activity within serotonergic neurons. The majority of evidence currently supports the involvement of calcium-dependent phosphorylation in this impulse-coupled regulatory process.

Pineal Body

The pineal organ is a tiny gland (1 mg or less in the rodent) contained within connective tissue extensions of the dorsal surface of the thalamus. While physically connected to the brain, the pineal is cytologically isolated for all intents, since, as one of the circumventricular organs, it is on the "peripheral" side of the blood permeability barriers (see Chapter 2) and its innervation arises from the superior cervical sympathetic ganglion. The pineal is of interest for two reasons. First, it contains all the enzymes required for the synthesis of serotonin plus two enzymes for further processing serotonin, which are not so pronounced in other organs. The pineal contains more than fifty times as much 5-HT (per gram) as the whole brain. Second, the metabolic activity of the pineal 5-HT enzymes can be controlled by numerous external factors, including the neural activity of the sympathetic nervous system operating through release of norepinephrine. As such, the pineal appears to offer us a potential model for the study of brain 5-HT. The pineal also appears to contain, however, many neuropeptides, and thus its secretory role remains as clouded as ever.

Actually, the 5-HT content of the pineal was discovered after the isolation of a pineal factor, melatonin, known to induce pigment lightening effects on skin cells. When melatonin was crystallized and its chemical structure determined as 5-methoxy-*N*-acetyltryptamine, an indolealkylamine, a reasonable extension was to analyze the pineal for 5-HT itself. The production of melatonin from 5-HT requires two additional enzymatic steps. The first is the *N*-acetylation reaction to form *N*-acetylserotonin. This intermediate is the preferred substrate for the final step, the 5-hydroxy-indole-*O*-methyl-transferase reaction, requiring *S*-adenosyl methionine as the methyl donor.

The melatonin content, and its influence in supressing the female gonads, is reduced by environmental light and enhanced by darkness. The established cyclic daily rhythm of both 5-HT and melatonin in the pineal is driven by environmental lighting patterns through sympathetic innervation. In animals made experimentally blind, the pineal enzymes and melatonin content continue to cycle, but with a rhythm uncoupled from lighting cycles. The adrenergic receptors of the pineal are of the β-type, and—as is characteristic of such receptors—their effect on the pineal is mediated by the postjunctional formation of 3′-5′ adenosine monophosphate (cyclic AMP). Elevated levels of cyclic AMP occur within minutes of the dark phase, and lead to an almost immediate activation of the 5-HT-*N*-acetyl transferase. The same receptor action also appears to be responsible on a longer time base for tonic enzyme synthesis. Thus, the proposed use of pineal as a model applicable to brain 5-HT loses its luster since this regulatory step does not seem to be of functional importance in the central nervous system. Furthermore, the "adrenergic" sympathetic nerves also can accumulate 5-HT (which leaks out of pinealocytes) just as they accumulate and bind norepinephrine (NE). Whether this is a functional mistake (i.e., secrete an endogenous false transmitter) or is simply a case of mistaken biologic identity, remains to be shown. It seems likely, however, that it is the NE whose release is required to pass on the intended communications from the sympathetic nervous system, since only NE activates the pinealocyte cyclase to start the enzyme regulation cascade.

Localizing Brain Serotonin to Nerve Cells

While some of these pineal curiosities may prove to be useful for the study of brain 5-HT, they are no substitute for data derived directly from brain. Although most early neurochemical pharmacology assumed that brain 5-HT was a neurotransmitter, more than ten years elapsed before it could be established with certainty that 5-HT in the brain is actually contained within specific nerve circuits. Based on lesion and subcellular fractionation studies, 5-HT content was first related to specific neuronal elements on a very coarse scale. With the introduction of the formaldehyde-induced fluorescence histochemical methodology by Falck and Hillarp, however, it became possible to observe the 5-HT-containing processes directly. Nevertheless, the mapping work has proceeded slowly because the 5-HT fluorophore develops with much less efficiency than that yielded by the condensation with catecholamines and also fades rapidly while viewed in the fluorescence microscope. Although the cell bodies can be seen with relative ease, direct analysis of the nerve terminals of these cell bodies has until recently required extreme pharmacological measures such as combined treatment with MAO inhibitors and large doses of tryptophan; these methods do not lend themselves to discrete mapping of the projections. The situation has been greatly improved for the 5-HT mapping studies, as it has for most other neurochemically defined systems, through the combined application of immunohistochemistry (directed first toward partially purified tryptophan hydroxylase and most recently to the direct localization of 5-HT itself), of orthograde axoplasmic transport of radiolabeled amino acids microinjected into cells identified as containing 5-HT by fluroescence histochemistry, and of retrograde tracing back to known 5-HT cells from suspected terminal fields. (see Chapter 2).

As a result of these extensive efforts by many groups, serotonin-containing neurons are known to be restricted to clusters of cells lying in or near the midline or raphe regions of the pons and upper brain stem (Figure 11-3). In addition to the nine 5-HT nuclei (B_1–B_9) originally described by Dahlström and Fuxe, recent immunocytochemical localization of 5-HT has also detected reactive cells in the area postrema and in the caudal locus ceruleus, as well as in and around the interpeduncular nucleus. The more caudal groups, studied by electrolytic or chemically induced lesions project largely to medulla and spinal cord. The more rostral 5-HT cell groups (raphe dorsalis, raphe medianus, and centralis superior, or B_7–B_9 [Fig. 11-4]) are thought to provide the extensive 5-HT innervation of the telencephalon and diencephalon. The intermediate groups may project into both ascending and descending groups, but since lesions here also interrupt fibers of passage, discrete mapping has re-

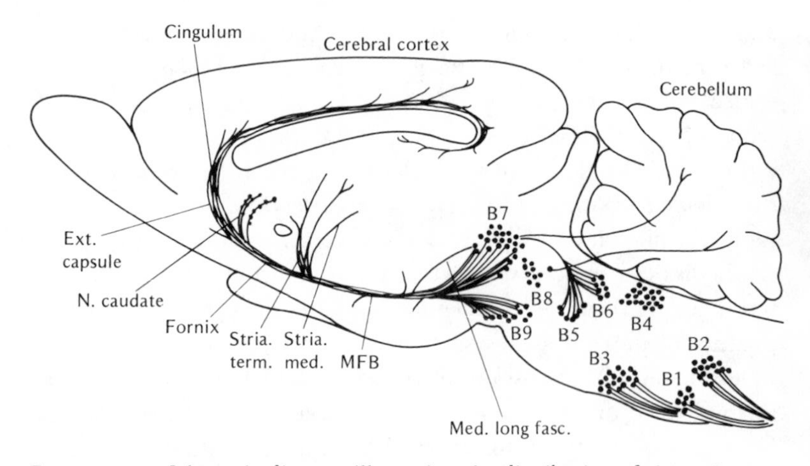

FIGURE 11-3. Schematic diagram illustrating the distribution of the main serotonin-containing pathways in the rat central nervous system. (Modified after G. Breese, *Handbook of Psychopharmacology,* Vol. 1, 1975)

quired the analysis of the orthograde and retrograde methods. The immunocytochemical studies also reveal a far more extensive innervation of cerebral cortex, which unlike the noradrenergic cortical fibers, is quite patternless in general.

In part these studies could be viewed as disappointing in that most raphe neurons appear to innervate overlapping terminal fields, and thus are more NE-like than DA-like in their lack of obvious topography. Exceptions to this generalization are that the B_8 group (raphe medianus) appears to furnish a very large component of the 5-HT innervation of the limbic system, while B_7 (or dorsal raphe) projects with greater density to the neostriatum, cerebral and cerebellar cortices, and thalamus (Fig. 11-5).

Attempts to localize 5-HT-containing terminals in the past relied primarily on the uptake of reactive 5-HT analogues or radiolabeled 5-HT. The use of labeled 5-HT, electron-dense analogues, or 5-HT-selective toxins (like the dihydroxytryptamines) depended for specificity on the selectivity of the uptake process. This situation

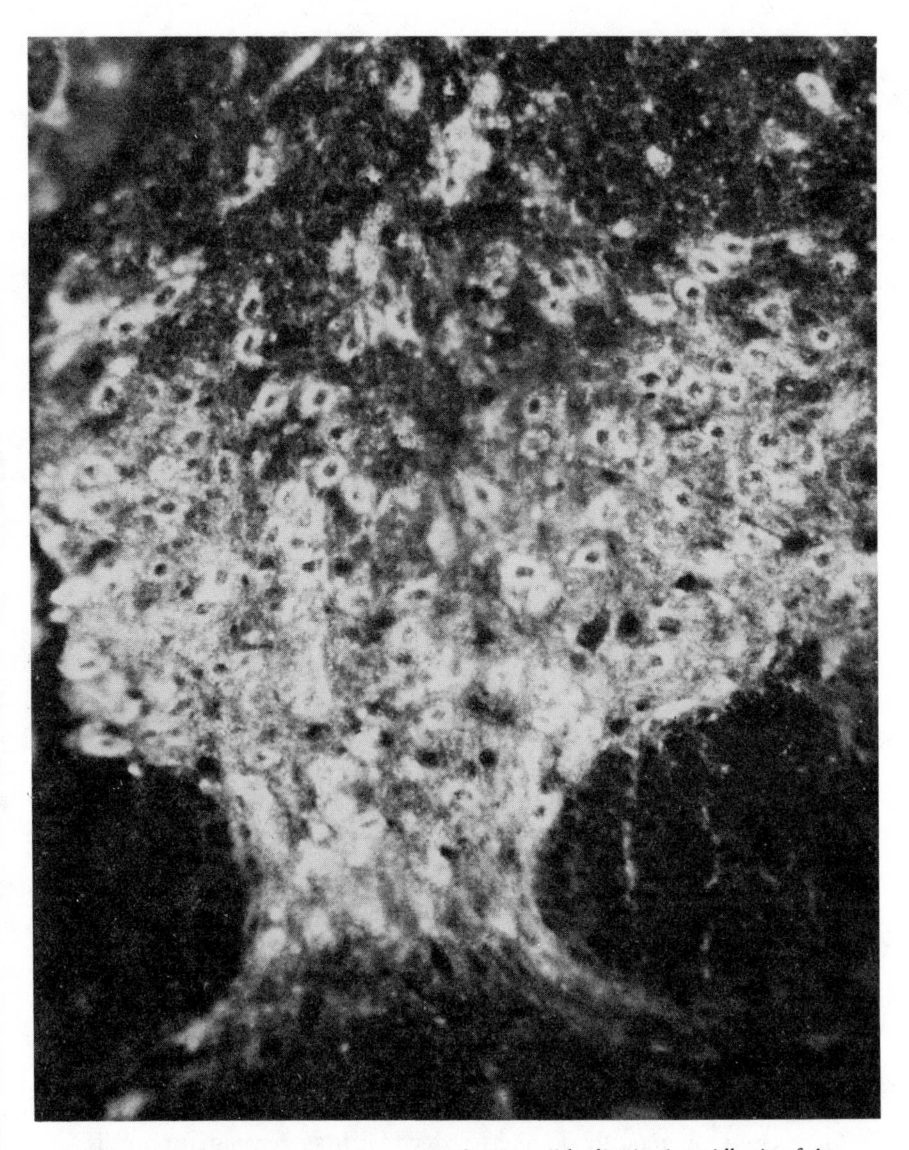

FIGURE 11-4. Fluorescence micrograph of raphe cell bodies in the midbrain of the rat. This rat was pretreated with L-tryptophan (100 mg/kg) one hour prior to sacrifice. (Courtesy of G. K. Aghajanian)

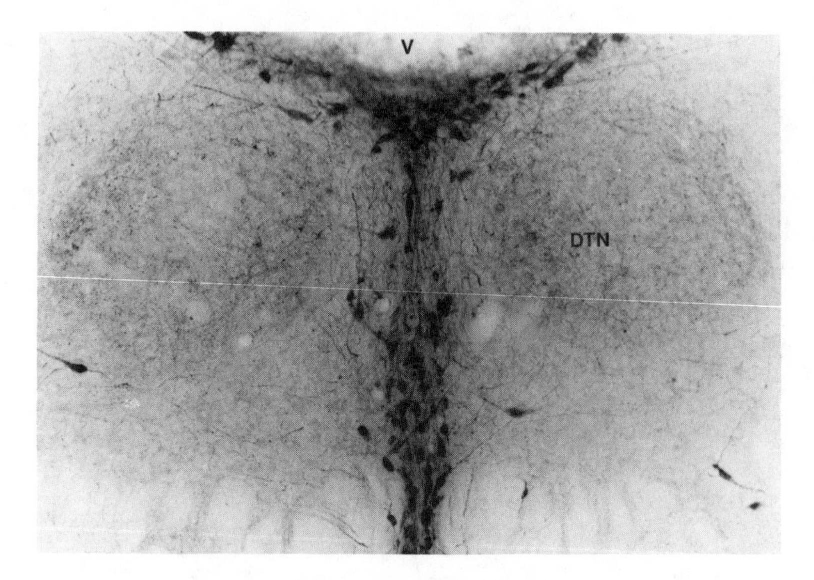

FIGURE 11-5. Photomicrograph of the serotonergic neurons of the caudal portion of the dorsal raphe. Section was stained using an antibody directed against serotonin. The serotonergic innervation of the dorsal tegmental nucleus (DTN) can also be seen. (Courtesy of A. Y. Deutch)

has recently been rectified by immunocytochemistry of endogenous 5-HT (Figs. 11-5, 11-6).

Cellular Effects of 5-HT

From the biochemical and morphological data above we can be relatively certain that the 5-HT of the brain occurs not only within the nerve cells but within specific tracts or projections of nerve cells. We must now inquire into the effects of serotonin when applied at the cellular level. (See Chapter 2.) In those brain areas in which microelectrophoretically administered 5-HT has been tested on cells that exhibit spontaneous electrical activity, the majority of cells decrease their discharge rate. The effects observed typically last much longer than the duration of the microelectrophoretic current. How-

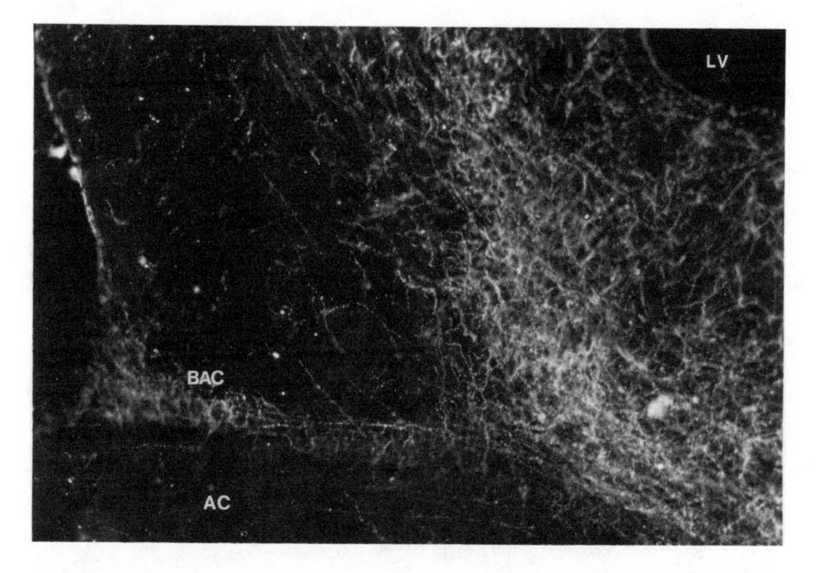

FIGURE 11-6. Darkfield photomicrograph illustrating the serotonergic innervation of the bed nucleus of the stria terminalis, as revealed by immunohistochemical staining with antibody directed against serotonin. The innervation of the bed nucleus of the anterior commissure (BAC) can also be seen above the anterior commissure (AC). LV marks the lateral ventricle. (Courtesy of A. Y. Deutch)

ever, in other regions, 5-HT also causes pronounced activation of discharge rate.

Recent elecrophysiological analyses of 5-HT have focused on neurons of the facial motor nucleus (cranial nerve VII) where the innervation by 5-HT fibers has been well documented. With intracellular recordings from these cells, 5-HT is found to produce a slow, depolarizing action accompanied by a modest increase in membrane resistance. As might then be anticipated on biophysical grounds, the combination of depolarization and increased membrane resistance facilitates the response of these neurons to other excitatory inputs. In a rigorous sense, such effects are not exactly in keeping with the emerging characteristics of modulatory actions (see Chapter 2), since here iontophoretic 5-HT changes both membrane potential and resistance on its own. It will be of interest to determine if activation of a 5-HT pathway to these cells at levels, which do not in themselves directly change membrane properties, will nevertheless modify responses of the target cells to other inputs, analogous to effects described for noradrenergic connections. Unlike the central catecholamines, 5-HT does not activate adenylate cyclase in adult mammalian brain, although the enzyme has been shown to be activated by 5-HT in embryonic or fetal mammalian brain and in invertebrate ganglia. In Aplysia ganglia, where 5-HT does activate adenylate cyclase, known 5-HT-containing neurons produce excitatory actions on certain sensory neurons by activating a cAMP-initiated cascade, and thereby activating a cAMP-dependent protein kinase, that in turn decreases conductance through a specially designated K-channel, called the S-channel. Presumably, the S-channel protein is a substrate for phosphorylation by this cAMP-dependent protein kinase. In the Aplysia, these 5-HT actions occur on the presynaptic terminals of the sensory neurons and play a key role in implementing some simple forms of sensitization, perhaps initiating the sequence of events that lead to longer-term "memory" formation. It is of some phylogenetic interest that many of the functions subserved by 5-HT in invertebrates are subserved by NE systems in vertebrates. The facial nerve and other branchial cleft derived elements of the vertebrate brain appear to be vestiges of this earlier mode of operation, and here a different series of postsynaptic pharmacology seems to hold. For example, many of the drugs developed as peripheral antagonists of 5-HT, and which are without action on 5-HT targets in septum, hippocampus, or amygdala, are nevertheless very good antagonists of 5-HT on facial motoneurons.

Several specific 5-HT-containing tracts, investigated electrophysiologically, indicate that 5-HT produces mainly, if not exclusively, inhibitory effects. While the pathways do conduct slowly, as would be expected for such fine caliber axons, the synaptic mediation process appears to be relatively prompt.

Characterization of 5-HT Receptors

The existence of multiple receptors for serotonin in the central nervous system has been suggested by physiological studies. Radioligand binding studies demonstrating that H^3-5-HT and H^3-spiperone label separate populations of high-affinity binding sites for 5-HT, termed 5-HT_1 and 5-HT_2, respectively, have also provided evidence for multiple receptors in brain tissue. A high correlation between the potencies of 5-HT antagonists in displacing the binding of H^3 spiperone and inhibiting serotonin-induced behavioral hyperactivity, coupled with a lack of such a correlation of H^3-5-HT binding sites, has led to the proposal that inhibition and excitation induced by serotonin are mediated at 5-HT_1 and 5-HT_2 receptors, respectively. Recent binding studies have further subdivided the 5-HT_1 class of recognition sites into 5-HT_{1A} and 5-HT_{1B} subtypes and compounds selective for these "receptor subtypes" have been identified. Autoreceptors for 5-HT appear to fall into the category of 5-HT_{1A} receptor subtypes. Drugs such as 8-OH DPAT and long-chain substituted piperazines appear to have selective 5-HT_{1A} binding activity and these 5-HT autoreceptor agonists are effective in inhibiting midbrain raphe neurons. Ketanserin is a new agent that blocks 5-HT receptors that apparently belong to the 5-HT_2 category without having any significant effect on 5-HT_1 receptors. Because of this discrimination, it has been described as a selective 5-HT_2 blocker. However, it is important to note that ketanserin is nonspecific in the sense that it has appreciable affinity for both α_1-adrenergic receptors and histamine H_1 receptors as well. It does have the advantage that its action on 5-HT receptors is as a pure antagonist.

Historically, the 5-HT responses were originally classified by Gaddum as "D" or "M" receptors, depending upon whether the 5-HT effects (mainly on smooth muscle targets, like stomach and uterus) were blocked by D-LSD or were mediated by release of acetylcholine to produce muscarinic actions. Linking those classical pharmacologic data with modern ligand-binding molecular pharmacology, 5-HT_2 type sites seem most like the D-receptor class, mediating both central and peripheral 5-HT responses that can be blocked by the classical 5-HT antagonists methysergide and cyproheptadine, as well as D-LSD and others.

The reason the old nomenclature has returned from obscurity is the recent description of a new series of 5-HT antagonists that appear best to fit the M class, and are among the most potent drugs ever described. These drugs also represent a success story for those who believe in the powers of molecular drug design; the new antagonists were built upon the demonstration that cocaine can selectively, but weakly, antagonize some 5-HT neuronal responses (in the autonomic system). The drug discovery team first built 5-HT analogues that selectively simulated at higher potency, these peripheral actions, and then set about to use that achievement for still more analogues that were antagonists of the response. One such drug, still highly experimental, acts to antagonize 5-HT actions at 0.01 picomolar concentration. Because the vascular effects of 5-HT may represent one basis for migraine headaches, such drugs may have prompt demonstration of clinical efficacy. The M receptors do not match the 5-HT_1 receptor classification, and may therefore also represent an important understudied central neuronal receptor that could, for example, illuminate the actions of the drugs acting to relieve anxiety or depression, with some hints or hitherto unexplainable 5-HT connection.

Cellular Pharmacology of Brain Serotonin

Let us first consider whether or not a dynamic re-uptake and turnover of serotonin occurs in a fashion similar to that described above for the catecholamines. If we attempt to use the same techniques, namely, to place a small amount of radioactive serotonin, into the ventricle to label the endogeneous stores of serotonin, and to follow the decline in specific radioactivity of serotonin in the brain, we find that the serotonin half-life is very long, on the order of 4 to 5 hrs after the initial acute drop is over. When this ventricular labeling technique is verified by fluorescence histochemistry or light-microscopic or electro-microscopic autoradiography, the majority of the uptake is seen to be into those nerves that normally contain serotonin, but the labeling is mainly over terminals and axons with very little labeling within the cell-body area. Therefore, we know, initially, that the labeling of the serotonin pool by these injection techniques is not complete and that the decline in radioactivity does not reflect total brain 5-HT. Furthermore, if we use any of the other methods for estimating turnover, such as inhibiting monoamine oxidase and following the rate of rise of serotonin, or inhibiting the efflux of 5-HIAA from the brain, we find that the biochemical estimate of this turnover is considerably faster than that determined by the labeling with radioactive 5-HT.

There would seem to be at least two explanations for this kinetic discrepancy. One would be that the uptake for serotonin is, in some respect, less selective or specific than that for the catecholamines. However, the uptake system appears to be similar in that it is energy dependent, that is, it is much more active at 37°C than at 0°C, it requires both glucose and oxygen, and it can be inhibited by ouabain, dinitrophenol, or iodoacetate. Furthermore, the uptake system for 5-HT can be partially inhibited by the same types of drugs that inhibit the uptake of catecholamine into catecholamine-containing nerve endings, such as desmethylimipramine, chlorpromazine, or cocaine, when evaluated in brain slices. However, the tertiary tricyclics appear to be more effective in blocking 5-HT uptake, and one such drug—Cl-imipramine—is very selective for 5-HT uptake.

On the other hand, when the turnover rates of brain serotonin (i.e., complete whole-brain serotonin) are estimated by using precursor labeling (i.e., tryptophan) and determining specific activity changes with time, results are mainly compatible with the fast rather than the slow turnover time of serotonin. The logical tentative conclusion to be reached, then, is that the pool of neuronal serotonin, which the ventricular injection most readily labels, consists mainly of the excess or storage form of serotonin, whereas total endogenous serotonin, in terms of synthesis and utilization, is turned over much more rapidly. These findings would suggest that a great deal of serotonin, which is probably true for other transmitters as well, resides in the storage form, patiently awaiting the call to action. For many years, virtually the only specific pharmacologic tool available for studying the role of 5-HT in the brain was *p*-chlorophenylalanine. In the pursuit of additional specific pharmacological tools, however, it was discovered that 5,6-dihydroxytryptamine can produce a selective long-term destruction of 5-HT axons and terminals in the brain, similar to the effects of 6-OHDA on the catecholamines. 5,6-DHT is more toxic and produces less extensive destruction, and so far it has found its greatest utility after local microinjections and for fluorescence and electron microscopy. Recently, there has been a reawakening of interest in *p*-chloroamphetamine, which for reasons that remain unclear, produces long-lasting 5-HT depletions.

Behavioral Effects of 5-HT Systems

One of the first drugs found empirically to be effective as a central tranquilizing agent, reserpine, was employed mainly for its action in the treatment of hypertension. In the early 1950s when both brain serotonin and the central effects of reserpine were first described, great excitement arose when the dramatic behavioral effects of reserpine were correlated with loss of 5-HT content. However, it was soon found that NE and DA content could also be depleted by reserpine and that all these depletions lasted longer than sedative actions. The brain levels of all three amines remained down for many days while the acute behavioral effects ended in 48 to 72 hr. Hence, it became difficult to determine whether it was the loss of brain catecholamine or serotonin which accounted for the behavioral depression after reserpine.

More extensive experiments into the nature of this problem were possible when the drugs that specifically block the synthesis of catecholamines or serotonin were discovered. After depleting brain serotonin content with parachlorophenylalanine, which effectively removed 90 percent of brain serotonin, investigators observed that no behavioral symptoms reminiscent of the reserpine syndrome appeared. Moreover, when *p*-chlorophenylalanine-treated rats, which were already devoid of measurable serotonin, were treated with reserpine, typical reserpine-induced sedation arose. These results again favor the view that loss of catecholamines could be responsible for the reserpine-induced syndrome. Furthermore, when α-methyl-*p*-tyrosine is given and synthesis of norepinephrine is blocked for prolonged periods of time, the animals are behaviorally sedated and their condition resembles the depression seen after reserpine. Students will find it profitable to review in detail the original papers describing the results just summarized, and the other theories currently in vogue.

Hallucinogenic Drugs

One of the more alluring aspects of the study of brain serotonin is the possibility that it is this system of neurons through which the hallucinogenic drugs cause their effects. In the early 1950s the concept arose that LSD might produce its behavioral effects in the brain by interfering with the action of serotonin there as it did in smooth muscle preparations, such as the rat uterus. However, this theory of LSD action was not supported by the finding that another serotonin blocking agent, 2-bromo LSD, produced minimal behavioral effects in the central nervous system. And, in fact, it was subsequently shown that very low concentrations of LSD itself—rather than blocking the serotonin action—could potentiate it. However, none of these data could be considered particularly pertinent since all the research was done on the peripheral nervous system and all the philosophy was applied to the central nervous system.

Shortly thereafter Freedman and Giarman initiated a profitable series of experiments investigating the basic biochemical changes in the rodent brain following injection of LSD. Although their initial studies required them to use bioassay for changes in serotonin, they were able to detect a small (on the order of 100 ng/gm, or less) increase in the serotonin concentration of the rat brain shortly after the injection of very small doses of LSD. Subsequent studies have shown that a decrease in 5-HIAA accompanies the small rise in 5-HT. Although the biochemical effects are similar to those that would be seen from small doses of monoamine oxidase inhibitors, no direct monoamine oxidase inhibitory effect of LSD has been described. This effect was generally interpreted as indicating a temporary decrease in the rate at which serotonin was being broken down, and it could also be seen with higher doses of less effective psychoactive drugs. In related studies by Costa and his co-workers, who estimated the biochemical turnover of brain serotonin, prolonged infusion of somewhat larger doses of LSD clearly promoted a decrease in the turnover rate of brain serotonin.

The next advance in the explanation of the LSD response was made when Aghajanian and Sheard reported that electrical stimulation of the raphe nuclei would selectively increase the metabolism of 5-HT to 5-HIAA. This finding suggested that the electrical activity of the 5-HT cells could be directly reflected in the metabolic turnover of the amine. Subsequently, the same authors recorded single raphe neurons during parenteral administration of LSD and observed that they slowed down with a time course similar to the effect of LSD. Thus, following LSD administration both decreased electrical activity of these cells and decreased transmitter turnover occur.

Of the several explanations originally proposed for this effect, the data now support the view that LSD is able to depress directly 5-HT-containing neurons at receptors that may be the sites where raphe neurons feed back 5-HT messages to each other through recurrent axon collaterals or dendrodendritic interactions; at these receptive sites on raphe neurons, LSD is a 5-HT agonist. In other tests on raphe neurons, iontophoretic LSD will antagonize activity of raphe neurons that show excitant responses to either NE or 5-HT. However, at sites in 5-HT terminal fields, when LSD is evaluated as an agonist or antagonist of 5-HT, its effects are considerably weaker than on the raphe neurons. These observations have led Aghajanian to suggest that LSD begins its sequence of physiological and neurochemical changes by acting on 5-HT neurons. In this view, the psychedelic actions of LSD must entail a primary or perhaps total reliance upon decreased efferent activity of the raphe neurons.

As Freedman has indicated, however, several points need further clarification before the description of LSD-induced changes in cell firing and cell chemistry can be incorporated in an explanation of the pharmacological and behavioral effects of LSD. If the effects of LSD could be "simply" equated with silencing of the raphe and subsequent disinhibition of raphe synaptic target cells, then many aspects of LSD-induced behavioral changes should be detectable in raphe-lesioned animals, but they are not. Typical LSD effects are also seen when raphe-lesioned animals are administered LSD. When normal animals are pretreated with *p*-Cl-phenylalanine (in dose schedules that antagonize raphe-induced synaptic inhibition) and given LSD, the effect is accentuated. In further contrast to the predictions of the 5-HT silencing effect, the LSD response is further accentuated with concomitant treatment with 5-HTP. The physiologic manipulation that most closely simulates behavioral actions of LSD is stimulation of the raphe, leading to decreased habituation to repetitive sensory stimuli. These data are difficult to reconcile with the view that LSD-induced behavior results from inactivation of tonic 5-HT-mediated postsynaptic actions. An important aspect to be evaluated critically in the continued search for the cellular changes that produce the behavioral effects is the issue of tolerance: LSD and other indole psychotomimetic drugs show tolerance and cross-tolerance in humans, but these properties have not been seen in animal studies at the cellular level.

The student must realize that to track down all of the individual cellular actions of an extremely potent drug like LSD and fit these effects together in a jigsaw puzzlelike effort to solve the question of how LSD produces hallucinations is extremely difficult. Similar jigsaw puzzles lie just below the surface of every simple attempt to attribute the effects of a drug or the execution of a complex behavioral task like eating, sleeping, mating, and learning (no rank ordering of author's priorities intended) to a single family of neurotransmitters like 5-HT. While it is clearly possible to formulate hypothetical schemes by which divergent inhibitory systems like the 5-HT raphe cells can become an integral part of such diverse behavioral operations as pain suppression, sleeping, thermal regulation, and corticosteroid receptivity, a very wide chasm of unacquired data separates the concept from the documentary evidence needed to support it. The gap is even wider than it seems since we do not at present have the slightest idea of the kinds of methods that can be used to convert correlational data (raphe firing associated with sleeping-stage onset or offset) into proof of cause and effect. While previous editions of this guide to self-instruction in cellular neuropharmacology have attempted superficial overviews of the behavioral implication of 5-HT neuronal actions, we now relinquish that effort until the cellular bases become more readily perceptible.

Actions of Other Psychotropic Drugs

Subsequent studies by Aghajanian and his colleagues have indicated that other drugs which can alter 5-HT metabolism can also disturb the discharge pattern of the raphe neurons. Thus, monoamine oxidase inhibitors and tryptophan, which would be expected to elevate brain 5-HT levels, also slow raphe discharge; while clinically effective tricyclic drugs (antidepressants such as imipramine, chlorimipramine, and amitriptyline) also slow the raphe neurons and could elevate brain 5-HT levels locally by inhibition of 5-HT reuptake. At the present time it remains unclear whether these metabolic changes in 5-HT after psychoactive drugs are administered reflect changes in the cells that receive 5-HT synapses or in the raphe neurons themselves. A large question is whether these circuits mediate the therapeutic responses of the drugs or are the primary pathologic site of the diseases the drugs treat. One etiologic view of schizophrenia, for example, might be based on the production of an abnormal indole, such as the hallucinogenic dimethyltryptamine, a compound that can be formed in human plasma from tryptamine. Some evidence indicates that schizophrenic patients have abnormally low monoamine oxidase activity in their platelets, which could permit the formation of abnormal amounts of plasma tryptamine and consequently abnormal amounts of dimethyltryptamine as well. Figure 11-7 summarizes possible sites of drug interaction in a hypothetical serotonin synapse in the CNS.

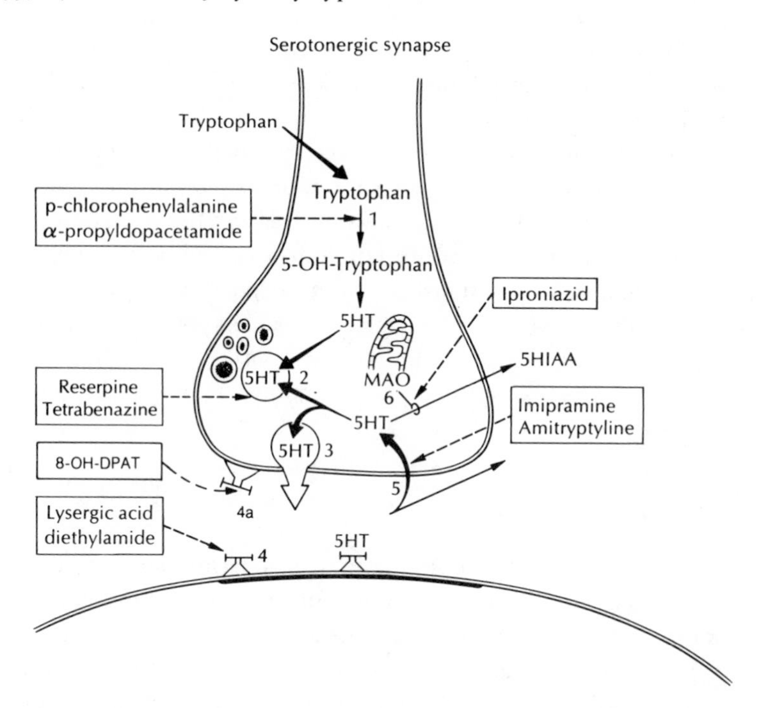

FIGURE 11-7. Schematic model of a central serotonergic neuron indicating possible sites of drug action.

Site 1: *Enzymatic synthesis:* Tryptophan is taken up into the serotonin-containing neuron and converted to 5-OH-tryptophan by the enzyme tryptophan hydroxylase. This enzyme can be effectively inhibited by *p*-chlorophenylalanine and α-propyldopacetamide. The next synthetic step involves the decarboxylation of 5-OH-tryptophan to form serotonin (5-HT).

Site 2: *Storage:* Reserpine and tetrabenazine interfere with the uptake–storage mechanism of the amine granules, causing a marked depletion of serotonin.

Site 3: *Release:* At present there is no drug available which selectively blocks the release of serotonin. However, lysergic acid diethylamide, because of its ability to block or inhibit the firing of serotonin neurons, causes a reduction in the release of serotonin from the nerve terminals.

Site 4: *Receptor interaction:* Lysergic acid diethylamide acts as a partial agonist at serotonergic synapses in the CNS. A number of compounds have also been suggested to act as receptor blocking agents at serotonergic synapses, but direct proof of these claims at the present time is lacking.

Site 4a: *Autoreceptor interaction:* Autoreceptors on the nerve terminal appear to play a role in the modulation of serotonin release. 8-Hydroxy-diproplaminotetra-

In this chapter we have encountered one of the more striking examples of an intensively studied brain biogenic amine for which there is every reason to believe that it is an important synaptic transmitter. Still to be determined are the precise synaptic connections at which this substance accomplishes the transmission of information and the functional role these connections play in the overall operation of the brain with respect to both effective and other multicellular interneuronal operations. The central pharmacology of serotonin, while intensively investigated, remains poorly resolved. Specific inhibition of uptake and synthesis are possible, but truly effective and selective postsynaptic antagonists remain to be developed.

Histamine

The challenge posed by histamine to neuropharmacologists has led to a vigorous chase across meadows of enticing hypotheses surrounded by bogs of confusion and dubious methodology. At last, more than sixty years after its isolation from pituitary by J. J. Abel, the role of histamine in the brain seems to be approaching a resolution that this amine occurs in two types of cells: mast cells and as yet unidentified hypothalamic neurons.

Most of what has been known about the synthesis and degradation of histamine in brain was based upon attempts to simulate in brain tissue, data obtained from more or less homogeneous samples of peritoneal mast cells (Fig. 11-8). But since mast cells were not

---

line (8-OH-DPAT) acts as an autoreceptor agonist at serotonergic synapses in the CNS.

Site 5: *Re-uptake:* Considerable evidence now exists to suggest that serotonin may have its action terminated by being taken up into the presynaptic terminal. The tricyclic drugs with a tertiary nitrogen such as imipramine and amitryptyline appear to be potent inhibitors of this uptake mechanism.

Site 6: *Monoamine oxidase. (MAO):* Serotonin present in a free state within the presynaptic terminal can be degraded by the enzyme MAO, which appears to be located in the outer membrane of mitochondria. Iproniazid and clorgyline are effective inhibitors of MAO.

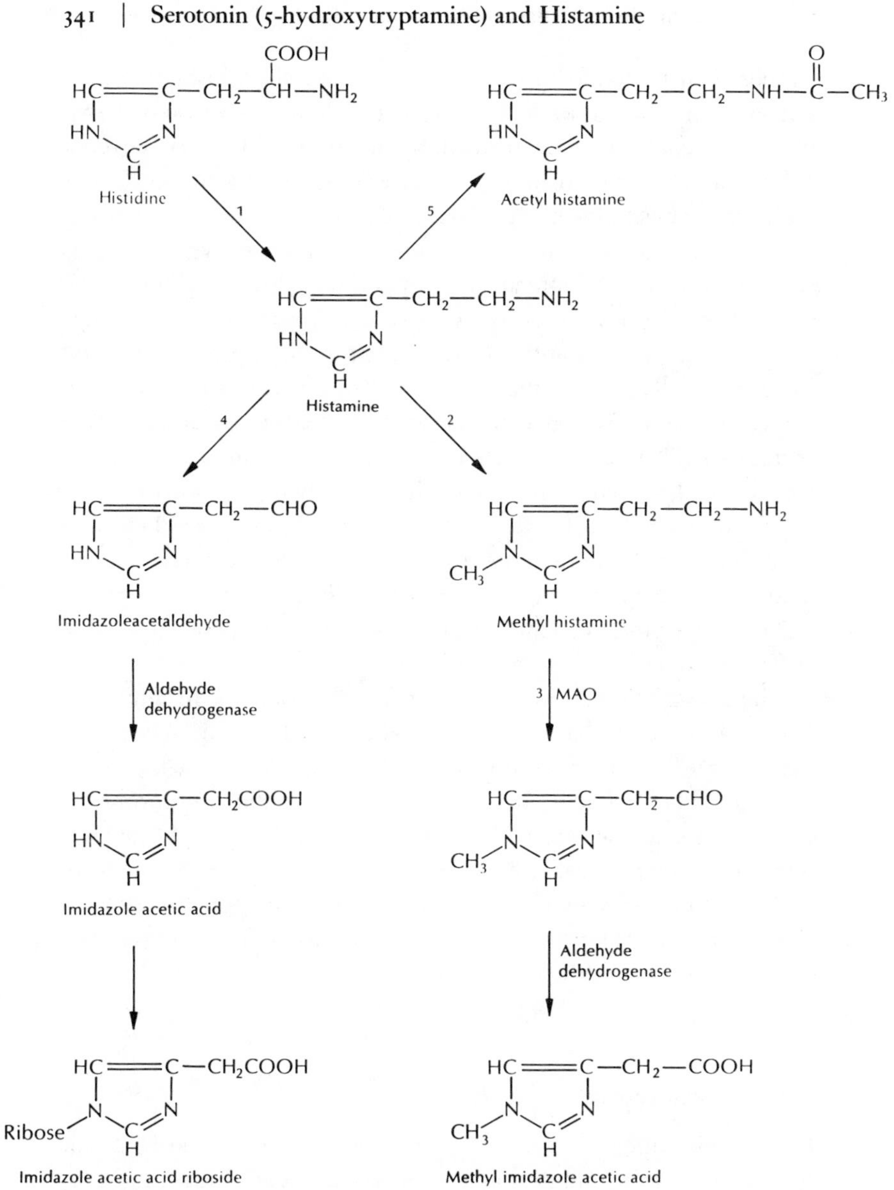

FIGURE 11-8. Metabolism of histamine. (1) Histidine decarboxylase. (2) Histamine methyl transferase. This is the major pathway for inactivation in most mammalian species. (3) Monoamine oxidase. (4) Diamine oxidase (Histaminase). (5) Minor pathway of histamine catabolism.

supposed to be found in healthy brains, and since histamine does not cross the blood–brain barrier, it had long been assumed that histamine could also be formed by neurons and hence be considered as a neurotransmitter. In fact, attempts to develop drugs which were histamine antagonists, in the mistaken belief that battle field shock was compounded by histamine release, were a key to the subsequent development of antipsychotic drugs. Furthermore, as every hayfever sufferer knows, present-day antihistamine drugs clearly produce substantial CNS actions such as drowsiness and hunger. Finally, we should realize that in retrospect the increased content of histamine found in transected degenerating sensory nerve trunks was likely to be an artifact of mast cell accumulation, rather than a peculiar form of neurochemistry. Viewing the other data on CNS histamine through that same retrospectroscope, as they have been illuminated by the innovative experiments of Schwartz and his associates, we now have a rather compelling case that histamine qualifies as a putative central neurotransmitter in addition to the role the same amine plays in mast cells.

The major obstacle in elucidating the functions of histamine in brain has been the absence of a sensitive and specific method to demonstrate putative histaminergic fibers *in situ*, the lack of suitable methods to measure histamine and its catabolites and problems in the characterization of histamine receptors in the nervous system. Some progress has been made in recent years to overcome these impediments and evidence has accumulated to support the hypothesis made more than two decades ago that histamine functions as a neurotransmitter in the brain.

Histamine Synthesis and Catabolism

Because histamine penetrates the brain from blood so poorly, brain histamine arises from histamine synthesis *in situ* from histidine. Active transport of histidine by brain slice has been demonstrated and because histidine loading has been shown to elevate brain histamine, histidine transport could be a controlling factor in brain histamine synthesis.

Two enzymes are capable of decarboxylating histidine *in vitro:* L-aromatic amino acid decarboxylase (i.e., DOPA decarboxylase) and the specific histidine decarboxylase. The pH optimum, affinity for histidine, effects of selective inhibitors, and the regional distribution of histamine synthesizing activity indicate that the specific histidine decarboxylase is responsible for histamine biosynthesis in brain (Fig. 11-8). The instability of histidine decarboxylase and its low activity in adult brain have precluded purification of the enzyme from this source, but fetal liver histidine decarboxylase has been purified to near homogeneity. Studies of the pH optimum, cofactor (pyridoxal phosphate), requirements, inhibitor sensitivity, and antigenic properties have demonstrated that the brain and fetal enzyme have similar properties. Antibodies to histidine decarboxylase have been used to map for the distribution of this enzyme in brain by employing immunohistochemical techniques.

The estimates of the K_m of histidine for brain histidine decarboxylase are much higher than the concentration of histidine in plasma or brain, suggesting that this enzyme is not saturated with substrate *in vivo* and consistent with the observations that administration of L-histidine increases brain histamine levels.

There are suprisingly few selective inhibitors of histidine decarboxylase. Most of the effective inhibitors act at the cofactor site and, although they produced a reduction in histamine formation, they also inhibit other pyridoxal requiring enzymes.

Histamine is metabolized by two distinct enzymatic systems in mammals. It is oxidized by diamine oxidase to imidazoleacetaldehyde and then to imidazoleacetic acid as well as methylated by histamine methyl transferase to produce methyl histamine. Mammalian brain lacks the ability to oxidize histamine and nearly quantitatively methylates it. Methyl histamine undergoes oxidative deamination by either diamine oxidase or monoamine oxidase. The affinity is higher for diamine oxidase, but in brains which lack diamine oxidase, methyl histamine is oxidized primarily by monoamine oxidase type B.

The major route of catabolism of orally injested histamine is via *N*-acetylation by bacteria in the gastrointestinal tract to form *N*-acetylhistamine.

Histamine-Containing Cells

Although there are fluorogenic condensation reactions that can detect histamine in mast cells by cytochemistry, the method has never been able to demonstrate histamine-containing nerve fibers or cell bodies. Because the histamine content of mast cells is quite substantial, however, it has been possible to use the cytochemical method to detect mast cells in brain and peripheral nerve. From such studies, Schwartz has estimated that the histamine content of brain regions, and nerve trunks that show about 50 ng histamine/gm (i.e., every place except the diencephalon) could all be explained on the basis of mast cells. Moreover, the once inexplicable rapid decline of brain histamine in early postnatal development can also be attributed to the relative decline in the number of mast cells in the brain during development.

Mast cell histamine shows some interesting differences from what we shall presume is neuronal histamine: in mast cells, the histamine levels are high, the turnover relatively slow, and the activity of histidine decarboxylase relatively low; moreover mast cell histamine can be depleted by mast cell-degranulating drugs (48/80 and Polymixin B). In brain, only about 50 percent of the histamine content can be released with these depletors, and for that which remains, the turnover is quite rapid. The activity of histidine decarboxylase is also much greater. Even better separation of the two types of cellular histamine dynamics comes from studies of brain and mast cell homogenates. In adult brain, a significant portion of the histamine is found in the crude mitochondrial fraction that is enriched in nerve endings and is released from these particles on hypo-osmotic shock. In cortical brain regions and in mast cells, most histamine is found in large granules that sediment with the crude nuclear fraction, and this histamine, unlike that in the hypothalamic nerve endings, has the slow half-life characteristic of mast cells.

Despite the encouraging result that histamine may be present in fractions of brain homogenates containing—among other broken cellular elements—nerve endings, it is difficult to parlay that information into a direct documentation of intraneuronal storage. More promising steps in that direction were taken in studies where specific brain lesions were made. The student will recall that lesions of the medial forebrain bundle region produced a loss of forebrain norepinephrine and 5-HT along a time course parallel to nerve fiber degeneration, in experiments that were done at a time when our understanding of CNS monoaminergic systems was not so far along as it is for histamine today. When such lesions were placed unilaterally, and specific histidine decarboxylase activity followed, a 70 percent decline in forebrain histidine decarboxylase activity was found at the end of one week. The decline was not due to the concomitant loss of monoaminergic fibers since lesions made by hypothalamic injections of 6-OHDA or 5,7-dihydroxytryptamine did not result in histidine decarboxylase loss. The most reasonable explanation would be that the lesion interrupted a histamine-containing diencephalic tract; the histamine and decarboxylating activity on the ipsilateral cortex was reduced only about 40 percent, suggesting that the pathway may make diffuse contributions to nondiencephalic regions. However, subsequent studies of completely isolated cerebral cortical "islands" show a complete loss of histamine content. As mentioned earlier, antibodies to histidine decarboxylase have also been utilized to map the distribution of this enzyme in rodent brain by immunohistochemical techniques. Recently Steinbusch and co-workers have developed an immunohistochemical method for the visualization of histamine in the central nervous system using an antibody raised against histamine coupled to a carrier protein. Results obtained with this technique are in general agreement with studies on the immunohistochemical localization of histidine decarboxylase. The highest density of histamine positive fibers are found in brain regions known to contain high histamine levels such as the median eminence and the premammillary regions of the hypothalamus.

Histamine Receptors

Direct physiologic evidence indicating the existence of specific histamine receptors in brain remains rather meager. Intraventricular

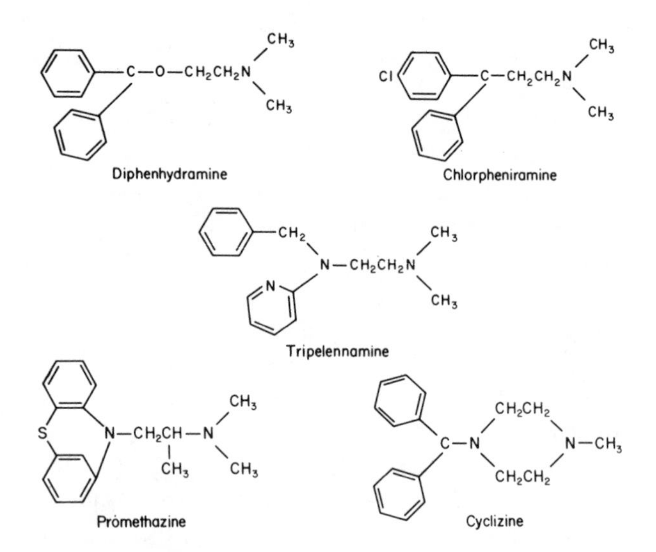

FIGURE 11-9. Representative H_1-receptor blocking drugs.

histamine will alter the cortical EEG and produce a sedated state; in iontophoretic tests of unidentified hypothalamic spinal and brain stem neurons, histamine shows depressant activity but also excites certain hypothalamic and cortical cells. The latter effects show some features of specificity, as the effects could be blocked only with so-called "classical" antihistamine drugs, such as mepyramine and promethazine (Fig. 11-9), although the local anesthetic activity of these agents casts some doubts on their specificity. More detailed studies of histamine receptors on smooth-muscle preparations have revealed that they exist in two forms: an H_1 receptor, with which the well-known histamine blockers interact, and an H_2 receptor, where these classical antagonists do not work but other antagonists do (such as metiamide and burimamide) (Fig. 11-10). These two classes of histamine receptors also reveal themselves by their differential responses to a number of histamine-like agonists. For example, 2-methylhistamine preferentially elicits responses mediated by

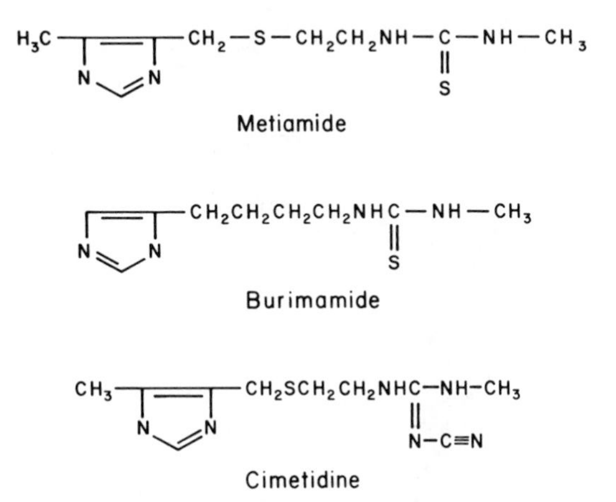

FIGURE 11-10. Representative H_2-receptor blocking drugs.

H_1 receptors whereas 4-methyl histamine and dimaprit have preferential effects mediated through H_2 receptors. The structures of some of these histamine agonists are illustrated in Figure 11-11.

Attempts to characterize the electrophysiological effects elicited by histamine on neurons in the central nervous system by employing selective histamine antagonists have been hampered by the local anesthetic properties of histamine–H_1 antagonists and the ability of H_2 antagonists themselves to depress the spontaneous firing rate of neurons. However, electrophysiological studies reveal that stimulation of histamine receptors can cause stimulation or inhibition of subpopulations of neurons in the CNS. In general, the excitatory actions of histamine are most often related to an interaction with H_1 receptors, whereas the inhibitory actions appear to be related to H_2 receptors. Most recently, a histamine-sensitive adenylate cyclase has been described in both rat and guinea pig brain. This receptor is by pharmacological criteria an H_2 receptor; the response to adenylate cyclase is most pronounced in neocortex and hippocampus, but is not seen in hypothalamus. The ability of histamine to activate metabolic processes in neocortex by a presumed cyclic

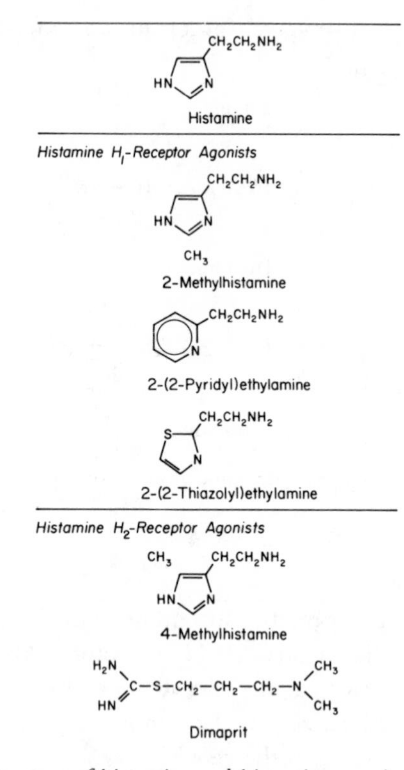

FIGURE 11-11. Structure of histamine and histamine agonists that act at H_1 and H_2 receptors.

AMP-mediated activation of glycogenolysis further supports a histaminergic physiology. Correlative physiological studies separating histamine receptors pharmacologically and attempting to test their possible cyclic AMP mediation could offer much support to the now reasonable hypothesis that brain histamine occurs in neurons as well as mast cells.

In summary, compelling evidence has been generated to indicate that histamine, in addition to its presence in mast cells, occurs in neurons where it probably functions as a neurotransmitter in the CNS. Brain has a similar non-uniform distribution of histamine, a specific histidine decarboxylase, and methyl histamine (the major metabolite of brain histamine). The cerebellum has the lowest levels, whereas the hypothalamus is most enriched. Hypothalamic synaptosomes contain histamine and histidine decarboxylase suggesting a localization in nerve endings. Depolarization of hypothalamic slices causes a calcium-dependent release of histamine, although histamine release from neurons has not been convincingly demonstrated in *in vivo*. The turnover of brain histamine is quite rapid similar to other biogenic amine transmitters. Electrophysiological and biochemical experiments have demonstrated the presence of H_1 and H_2-histamine receptors in mammalian brain and indicated that these receptors can stimulate the formation of cAMP. Finally, lesions of the midbrain or caudal hypothalamus cause a progressive reduction in forebrain histamine and histidine decarboxylase consistent with the existence of an ascending histamine-containing system, which has now been directly visualized by employing immunohistochemical techniques. Continued application of these immunohistochemical techniques for visualization of histidine decarboxylase-like and histamine-like activity in brain should provide detailed maps of histamine-containing neuron systems in the brain in the near future. However, we are still far from being able to speak with conviction of specific histaminergic circuits and their relation to behavior, and thus the functions of histamine in brain are highly speculative.

The high levels of histamine and the presence of histamine receptors in the hypothalamus taken with the known ability of histamine to alter food and water intake, thermoregulation, autonomic activity, and hormone release implicate a possible role for histamine in these vegetative functions. Also, because of the presence of ascending histamine-containing fibers originating from the reticular formation coupled with the sedative nature of H_1 antagonists and the activating effects of histamine observed on some brain cells. A possible role for histamine in arousal has been suggested.

Selected References

Aghajanian, G. K. (1981). The modulatory role of serotonin at multiple receptors in brain. In *Serotonin Neutotransmission and Behavior* (B. L. Jacobs and A. Gelperin, eds.), 156. MIT Press, Cambridge, MA.

Fuller, R. W. (1980). Pharmacology of central serotonin neurons. *Ann. Rev. Pharmacol. Toxicol. 20,* 111.

Hamon, M., S. Bourgoin, S. El Mestikaway, and C. Goetz (1984). Central serotonin receptors. In *Handbook of Neurochemistry*, Receptors in the Nervous System, Vol. 6 (A. Lajtha, ed.), p. 107. Plenum, New York.

Hough, L. B., and J. P. Green (1984). Histamine and its receptors in the nervous system. In *Handbook of Neurochemistry*, Receptors in the Nervous System, Vol. 6 (A. Lajtha, ed.), p. 145. Plenum, New York.

Parent, A., L. Descarries, and A. Beaudet (1981). Organization of ascending serotonin systems in the adult rat brain. A radioautographic study after intraventricular administration of ^{3}H 5-hydroxytryptamine. *Neuroscience 6,* 115.

Richardson, B. P., G. Engel, P. Donatsch, and P. A. Stadler, (1985). Identification of serotonin M-receptor subtypes and their specific blockade by a new class of drugs. *Nature 316,* 126.

Schwartz, J. C., G. Barbin, M. Baudry, M. Gargarg, M. P. Martres, H. Pollard, and M. Verdiere (1979). Metabolism and functions of histamine in the brain. In *Current Developments in Pharmacology* (W. B. Essman, ed.), p. 173. Spectrum Publications, New York.

Sills, M. A., B. B. Wolfe, and A. Frazer (1984). Determination of selective and nonselective compounds for the 5-HT_{1a} and 5-HT_{1b} receptor subtypes in rat frontal cortex. *J. Pharmacol. and Exp. Therapy, 231,* 480.

Steinbusch, H. W. M., and R. Nieuwenhuys (1983). The raphe nuclei of the rat brainstem: A cytoarchitectonic and immunohistochemical study. In *Chemical Neuroanatomy* (P. C. Emson, ed.), p. 131. Raven Press, New York.

Steinbusch, H. W. M. (1984). Serotonin-immunoreactive neurons and their projections in the CNS. In *Handbook of Chemical Neuroanatomy*, Classical Transmitters and Transmitter Receptors in the CNS, Vol. 3, Part II (A. Björklund, T. Hökfelt, and M. Kuhar, eds.). Elsevier, New York.

Steinbusch, H. W. M., and A. H. Mulder (1984). Immunohistochemical localization of histamine in neurons and mast cells in the rat brain. In *Handbook of Chemical Neuroanatomy*, Classical Transmitters and Transmitter Receptors in the CNS, Vol. 3, Part II (A. Björklund, T. Hökfelt, and M. Kuhar, eds.), p. 126. Elsevier, New York.

Wamsley, J. K., and J. M. Palacios (1984) Histaminergic receptors. In *Handbook of Chemical Neuroanatomy*, Classical Transmitters and Transmitter Receptors in the CNS, Vol. 3, Part II (A. Björklund, T. Hökfelt, and M. Kuhar, eds.), p. 386. Elsevier, New York.

12 | Neuroactive Peptides

A very large number of peptides made by neurons have now been shown to affect specific target cells of the central and peripheral nervous systems: neuronal, glial, smooth muscle, glandular, and vascular. Although the number of known endogenous peptides already exceeds several dozen, with no obvious end in sight, the actual molar concentrations of any given peptide in the brain is maximally two to three orders of magnitude lower than that of the monoamines, acetylcholine, and amino acids and with potencies at correspondingly lower concentrations. The specific signaling advantages that peptides may hold for the operation of neuronal circuits is not yet obvious. Furthermore, no specific peptides have yet been linked selectively to a given functional system in the brain, nor correlated in any exclusive manner with any pathological states. Thus, the challenge of devising new drugs for neurological or psychiatric disorders based on these peptides is considerably more difficult than *just* designing potent congeners that can act as their agonists or antagonists. Nevertheless, there are strong implications that the biologic properties evoked by peptide-mediated signals will influence future pharmacological development.

Some Basic Questions

What's Different about Peptides?

Peptide-secreting neurons differ in their biology from neurons using amino acids and monoamines in ways other than the molecular structure of their transmitter. The amino-acid or monoamine transmitters are formed from dietary sources by one or two intracellular enzymatic steps; the end product of these enzymatic actions is the active transmitter molecule which is then stored in the nerve terminal until release. After release, the transmitter (or choline in the case of ACh) can be reaccumulated back into the nerve terminal by the energy-dependent active re-uptake property, thus conserving the requirement for *de novo* synthesis. Peptide-secreting cells employ a somewhat more formidable approach: Synthesis is directed by messenger RNA on ribosomes, and thus this synthesis can only take place in the perikaryon or dendrites of a neuron. Furthermore, all peptides studied so far are synthesized as part of a much larger precursor molecule (or prohormone) which has no biological activity and from which the active peptide is cleaved by special processing peptidases. The process for peptides thus starts with ribosomal synthesis of the prohormone, which is then packaged into vesicles in the smooth endoplasmic reticulum, and transported from the perikaryon to the nerve terminals for eventual release. These steps in the molecular processing of a neuropeptide are described in greater detail in Chapter 3.

Insofar as present data reveal, peptide-releasing neurons share certain basic properties with all other chemically characterized neurons: transmitter release is Ca-dependent, and some of the postsynaptic effects may be mediated by directly altering ion channel conductances or indirectly regulating ion channels through second messengers such as Ca^{++}, cyclic nucleotides or inositol triphosphates (see Chapter 6). Furthermore, like the monoamines, peptides identical to those extracted from nervous system are often made by nonneural secretory cells in endocrine glands and the mucosa of the hollow viscera. Some observers of the peptide scene would like to emphasize characteristics of neuropeptide action which are extrapolations from the actions of endocrine peptide hormones: targets that are distant from release sites, release into portal or other circulatory systems, higher potency and longer time course of effects. Although it is clear that endocrine hormones, peptides, or amines can often act in concentrations in the nanomolar range or below, it remains to be determined how time course and thresholds vary when the same neuropeptides act on their target cells in brain or ganglia. In amphibian autonomic ganglia, at least one peptide does act in such a local endocrine, or *paracrine*, fashion, producing synaptic potentials that last for hundreds of milliseconds. These effects of an endogenous peptide very similar to hypothalamic luteinizing hormone releasing hormone (LHRH) are produced on a nonsynaptic neuronal target by diffusing at least 10 to 100 μ from preganglionic synaptic release sites, without apparently acting at all on the immediate postsynaptic neuron. Whether these particular paracrine properties are ways around the simple amphibian autonomic anatomy or a prototype for all peptide actions peripherally and centrally remains debatable.

How Are New Peptides Found?

Before considering the interesting relationships among the many known peptides, we can anticipate future events by examining the recurrent pattern of events that seems to underlie every wave of peptide discovery and characterization. From this historical overview, the student can see that new peptides do not just happen into existence. The primary necessity for the discovery of a new peptide has classically been the recognition of a biological function controlled by an unknown factor; this biological action is then employed as the bioassay for the subsequent steps of the discovery process. Although this approach was used for the characterization of gastrointestinal peptide hormones in the 1940s and 1950s, the process was perfected and greatly increased in power by Guillemin and Schally and their teams during the 1960s and 1970s as they pursued the hypophysiotrophic hormones by which the hypothalamus regulates the secretion of adenohypophysial (anterior pituitary) hormones for control of the major peripheral endocrine tissues. The starting bioassay is used as a screen for chemical manipulations of tissue extracts, and through chemical manipulations of the extract (i.e., differential extractions, molecular sieves, etc.) the active factor is enriched and ultimately purified. For some factors, the purification of nanograms of active peptide has required hundreds of thousands of brains. The peptidic nature of the purified material is inferred by the loss of biological activity after treatment with peptidases (carboxypeptidases or endopeptidases). The cleaved peptide fragments are then sequenced and the entire factor is eventually reconstructed.

When a sequence has been achieved and matched against the amino-acid composition of the pure factor, it is then necessary to synthesize the peptide and determine if the synthetic replicate matches actions and potencies with the purified natural factor. Only when the replicate's properties match those of the factor can the natural peptide be considered to have been "identified." At this point the factor is given a name which reflects its actions in the original bioassay.

However, this is far from the end of the discovery process. When the chemical structure of a peptide has been identified and synthetically replicated, two new opportunities arise: large amounts of the synthetic material can be prepared for examination of physiological and pharmacological properties; at the same time, antibodies can be prepared against the synthetic material. The antibodies are used for radioimmunoassay and for immunocytochemistry. The quantitative measurements made possible by the radioimmunoassay, and the qualitative distribution of the peptide assessed by microscopic cytochemical localizations then frequently reveals that the peptide has a much broader distribution, and, therefore, presumably has a much more general series of actions than was originally presumed. Autoradiographic mapping of ligand binding sites for peptides, generally the last step in the molecular and cellular characterization are then defined. Curiously, these ligand binding maps show a flagrant disregard for the cellular distribution of the recognized forms of the endogenous peptides; this mismatch strengthens the concept of paracrine diffusion for peptide actions in some minds. Thereafter, the endogenous peptide or an analog of it is ordinarily proposed to be the cause or cure of a major mental illness. In many of the more recent reports, it is frequently observed that the peptide isolated and found to be active is in fact not the only active form contained in the tissue: In many cases better extraction methods confirmed by radioimmunoassay reveal larger molecules with full or even greater potency. Although the general rule is that active peptides are synthesized from larger precursor hormones, processing can sometimes lead to active larger peptides that (then) cannot really be viewed as "precursors" since they have potent activity in their own right.

As more and more peptides have been identified within the nervous system and from other sources, their internal structural similarities allow us to consider them as members of one or another family grouping of peptides. The family groupings are useful because they indicate that certain sequences of amino acids have had considerable evolutionary conservation, presumably because they provide unambiguous signals between secreting and responding cells. Two aspects of peptide family groupings deserve recognition: families in which separate propeptide genes or mRNAs sharing one or more short peptide sequences are expressed by different cell groups in the same species and families in which the peptides expressed by homologous cell groups share generally identical peptide sequences with minor to moderate variation. The first category, exemplified by the opioid peptide and the tachykinin peptide families, implies that the natural ligand's receptors may diverge into subtypes that can discriminate the fine differences in peptide sequences. Although some individual peptides have already earned their own receptor subtypes (vasopressin, somatostatin), in other cases, such as the tachykinins and the opioid peptides, the multiple receptor subtypes predicted the later definition of multiple natural ligands. In those cases in which the propeptide form may give rise to multiple different agonists, it is possible—but by no means yet broadly supported—that more than one peptide messenger is released from the possible array. Some dreamy-eyed pundits have theorized that different branches of a neuron might have the option of changing the mix of cleavage products available for release.

The latest additions to the list of neuroactive peptides have been discovered by strategies that depart from the classical bioassay strategy purification schemes and instead rely either on specific common general features, such as a C-terminally amidated sequence or on the detection of potential cleavage fragments when the propeptide sequence has been deduced by molecular cloning of the mRNA or gene. Although these historical discovery cycles are individually entertaining stories, the punch lines—determining what the peptides do and how to make drugs based upon these actions—still remain elusive.

Transmitters or Not?

In Chapter 3 we considered the kinds of evidence needed to demonstrate that a substance found in nerve cells is the transmitter whose cells secrete at their synapses. To recapitulate, it must be shown that the substance is present within the nerve cell and specifically in its presynaptic terminals, that the nerve cell can make or accumulate the substance, and that it can release that substance when activated; when the substance is released, it must be shown to mimic in every aspect the functional activity following stimulation of the nerves that released it, including the magnitude and quality of changes in postsynaptic membrane conductances. A criterion especially pertinent to the objectives of this book is that drugs which can simulate or antagonize the effects of nerve fiber activation must have an identical influence on the effects of the substance applied exogenously to the target cell.

In the case of peptides, the opportunities for supplying these data would at first glance seem better than they are with the many categories of small molecules we have discussed in previous chapters. Once the peptide has been isolated and its structure determined, immunocytochemical localization of the peptide usually follows promptly, as does the demonstration that with the right sensitive assay, the peptide can be shown to be released from brain slices by a Ca^{++}-dependent, depolarization process. The development of a suitable synthetic ligand helps define receptor distribution, possible actions on presumptive synaptic targets, and on potential mechanisms of cellular regulation, and in a few cases, the development of receptor subtype specific antagonist analogs. In several peptide families, synthetic agonist and antagonist peptide analogs have been employed to document behavior-altering actions of the exogenous forms, and by inference a role of the endogenous form in behavior, and possibly in disease states. Although these molecular tools have accelerated the early partial fulfillment of transmitter criteria for peptides generally, there are as yet no cases within the CNS in which the complete array of neuronal actions transmitted by a peptide-containing circuit have been shown to be completely attributable to the peptide. There are important reasons for this unsatisfying state of peptide-transmitter documentation, and lack of effort is not one of them.

They Are Not Alone

Earlier editions of this book were organized according to a then-prevalent concept in neuropharmacology: a given neuron operated by one and only one transmitter. According to this concept, one aspect of a neuron's phenotype, as classic as its size, shape, location, circuitry, and synaptic function was its neurochemical designation as "GABA-ergic" or "cholinergic" or "adrenergic" or whatever. All card-carrying pharmacologists knew that autonomic neurons came in only two color-coded categories: adrenergic and cholinergic. And then a curious finding began to repeat itself: Neuroactive peptides started showing up in autonomic neurons where there was no need for additional transmitters. Initially viewed by many as an oddity, the idea of coexisting peptides in central as well as peripheral neurons already occupied by an amino acid or monoamine or even another neuropeptide, has now gained considerable acceptance. (see Table 12-1).

The notion of coexistence of other transmitters with peptides leads one to ask whether there is any such thing as a truly "peptidergic" neuron, or rather only neurons in which the peptide there simply expands the armamentarium of messenger molecules available for transmitting *an enriched* message that requires more than one type of transmitter for its fully refined signal content. Although the latter concept is far from a neuroscientific consensus, the data in hand are certainly compatible, recognizing the enormous void in our knowledge in which the majority of neurons have not yet had either their peptidic or nonpeptidic transmitters identified.

Such a revolutionary concept then forces us to ask: If the peptide

TABLE 12-1. Neuroactive peptides coexist with other transmitters

Transmitter	*Peptide*	*Location*
GABA	Somatostatin	Cortical and hippocampal neurons
	Cholecystokinin	Cortical neurons
Acetylcholine	VIP	Parasympathetic and cortical neurons
	Substance P	Pontine neurons
Norepinephrine	Somatostatin	Sympathetic neurons
	Enkephalin	Sympathetic neurons
	NPY	Medullary and pontine neurons
	Neurotensin	Locus ceruleus
Dopamine	CCK	Ventrotegmental neurons
	neurotensin	Ventrotegmental neurons
Epinephrine	NPY	Reticular neurons
	Neurotensin	Reticular neurons
Serotonin	Substance P	Medullary raphe
	TRH	Medullary raphe
	Enkephalin	Medullary raphe
Vasopressin	CCK, dynorphin	Magnocellular hypothalamic neurons
Oxytocin	Enkephalin	Magnocellular hypothalamic neurons

is not the sole signaling molecule, but rather a minor, fractional percent of messenger signaling capacity of the nerve terminal, how might peptides affect signals transmitted by the coexisting amino acid or monoamine? The honest answer now is "We don't know," at least for the CNS. However, in the autonomic nervous system, where Tomas Hökfelt and his peptide liberation team have managed to find at least one peptide in every sympathetic, parasympathetic, or enteric neuron they have studied, some solid peptide–monoamine interactions are emerging. Vasoactive intestinal peptide (VIP), released when cholinergic nerves discharge at high frequency, augments parasympathetic control of salivation by increasing glandular blood flow. Neuropeptide Y (NPY), found in many sympathetic neurons, sensitizes smooth muscle target cells to adrenergic signals. Thus, peptide agonist or antagonist actions might well alter neuronal information flow by disturbing the normal symbiotic relationships between the signals.

Other peptide actions, although not necessarily on coexisting amino-acid or monoamine transmitters, have been reported with irregularity: increasing or decreasing the amount of transmitter released by a given stimulus, or enhancing or diminishing the response of a common target cell (in either amplitude or duration or both) to combined transmitter signals. For example, VIP greatly accentuates the response of cortical neurons to subthreshold doses of norepinephrine as indicated by the amount of cyclic AMP generation and the potency on altering neuronal activity. VIP and some other peptides have been reported to alter the affinity of postsynaptic receptors for their nonpeptidic ligands.

One way to recast these players in the proper perspective might be to consider that a peptide usually embellishes what the primary transmitter for a neuronal connection seeks to accomplish. Such an effect may be to strengthen or prolong the primary transmitter's actions, especially when the nerve is called upon to fire at higher than normal frequencies. The peptide may provide a part of the intraneuronal signal to alter the rate of production of the primary transmitter, and there may still be places in which the peptide itself can fulfill all of the effects of a primary transmitter.

Accepting this concept for the moment, several other important concepts for pharmacology emerge: the anticipated effects of peptide-directed drugs will not only depend on the location of the receptors and their actions, but on the context of the signals with which the peptides normally cooperate. Thus, the pharmacology of NPY or neurotensin may be most appropriately understood as a part of the total picture of central noradrenergic pharmacology, given the degree to which these two peptides may participate in that monoaminergic neuron's transmission. Furthermore, one may imagine a relatively modest number of response mechanisms on which peptide messengers may operate (e.g., direct regulation of Na, K, or Ca channels, or second messenger-mediated regulation of those or other channels), but with a great enrichment of signaling possibilities attainable through the interplay between frequency-dependent and diffusion-dependent release and response sites. However, before we all grow obese on the meringue of what may only be a lemon pie-in-the-sky, there are some facts worth having. The rest of this chapter reviews pharmacologically relevant data for the grand peptide families (oxytocin/vasopressin, the tachykinins, glucagon-related peptides, pancreatic polypeptide-related peptides, and opioid peptides,) and for a few individual peptides (CCK, somatostatin, angiotensin, calcitonin-gene related peptide, and CRF). Future editions will certainly require an expansion of this roster.

In terms of the development of drugs active on the nervous system, research on the peptides has given us a humbling view of the riches that may await mining in the inner depths of the brain.

The Grand Peptide Families

Vasopressin and Oxytocin

These two highly similar nonapeptides (see Fig. 12-1) with internal 1,6 disulfide bridges are the original mammalian peptide "neurohormones": They are synthesized in the neuronal perikarya found in the large neurons of the supraoptic and paraventricular nuclei, and stored in the axons of these neurons in the neurohypophysis from which they are released into the bloodstream. Each peptide is synthesized as part of a larger propeptide (see Chap. 3) with which it is stored and released and from which it is cleaved as part of the release process. In the periphery, vasopressin (also known as antidiuretic hormone) facilitates distal tubular water reabsorption, while

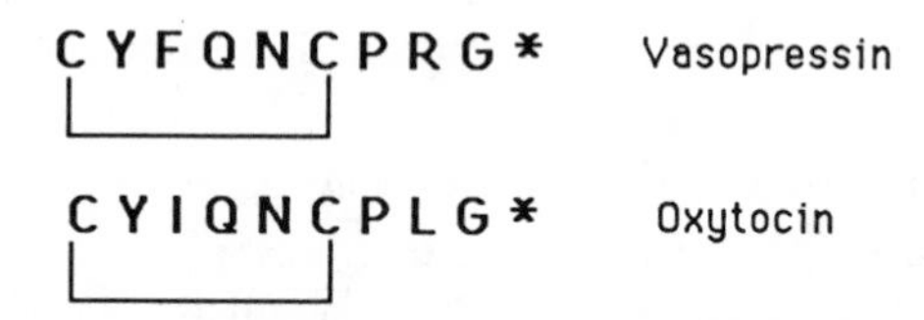

FIGURE 12-1. Molecular sequence of oxytocin and vasopressin, in which amino acid names are symbolized by the standard single letter symbols, and * indicates an amidated C-terminus. (Key to amino acid single letter symbols: A = Ala; R = Arg; N = Asn; D = Asp; B = Asparagine; C = Cys; Q = glutamine; E = Glu; G = Gly; H = His; I = Ile; L = Leu; K = Lys· M = Met; F = Phe; P = Pro; S = Ser; T = Thr; W = Trp; Y = Tyr; V = Val.).

oxytocin stimulates epididymal and uterine muscle contraction. In the hypothalamus, the neurons of the supraoptic and paraventricular nuclei have been found to give off axon collaterals that project within the nuclei, between the nuclei and to the median eminence. When these so-called "magnocellular" neurons are recorded from and identified on the basis of antidromic stimulation of their axons in the neurohypophysis, a recurrent inhibition is seen. Tests with antagonists of all other presumed neurotransmitters fail to explain the chemical mediator of this inhibition, and since both oxytocin and vasopressin were found to have potent inhibitory actions on neurons projecting to the neurohypophysis, Nicoll and Barker proposed that the same peptide released from the pituitary into the blood might also function as the transmitter for the recurrent pathway. If this could be demonstrated convincingly, that would considerably defuse the sham conflict between neurohormones and neurotransmitters. Some parvocellular neurons of the supraoptic and paraventricular nuclei also contain oxytocin or vasopressin. Their circuits project either to the median eminence where at least vasopressin acts as a coreleasing hormone of ACTH, or to other distant targets in the diencephalon, pons, and spinal cord, including the locus ceruleus, and the nucleus of the solitary tract. Additional vasopressin-containing neurons have been identified outside of the hypothalamus in the anterior pro-optic area, and these may be the source of the minute quantities of vasopressin and oxytocin reported within the limbic system.

The behavioral actions of vasopressin are quite impressive, although the mechanisms accounting for these effects remain unclear. In microgram quantities, subcutaneously administered vasopressin or synthetic analogs with minimal renal actions alter the behavior of rats by generally delaying the extinction of learned aversive or appetitive tasks. Humans given analogs of vasopressin by nasal spray also show enhanced performance of attention-related memory tasks. Even though they are minute, the doses of vasopressin required for behavioral effects after subcutaneous injection are 1000 times higher than the doses that are effective when given into the cerebral ventricles of a rat. Because of this greatly lowered threshold, some presume that the peripherally injected peptides somehow penetrate into and act upon the CNS. An alternative view is that peripheral target systems send to the brain neuronal information that induces a state of arousal sufficient to alter behavior. The circumventricular organs located within the ventricles, but on the blood side of the blood–brain barrier, may also detect the circulating hormones. Present data indicate that the effects of the peptides on behavior persist far longer than any detectable change in circulating peptide, and even after direct peripheral effects such as elevation of blood pressure have been normalized. Analogs of vasopressin and oxytocin have now been developed with selective agonist or antagonist properties; these molecules should make useful pharmacological probes. Based on the selective actions of vasopressin analogs that act as agonists or antagonists, two subtypes of peripheral receptors have been characterized: the V_1 receptors mediate responses of some analogs on arteriolar walls, and the V_2 receptors mediate the effects of other analogs on the renal tubules. The effects of vasopressin analogs on either memory or learning behaviors or on secretory responses mediated through the median eminence do not adhere to either receptor class and may indicate that more subtypes will be characterized.

The Tachykinin Peptides

In 1931, von Euler and Gaddum discovered an unexpected pharmacologically active substance in extracts of brain and intestine,

The TACHYKININ Peptides

R P K P E E F F G L M *	Substance P
D V **P K** S D **E F** V **G L M** *	Kassinin
H **K** T D S **F** V **G L M** *	Substance K
pE **P** S K D A **F** I **G L M** *	Eledoisin
D M H D F **F** V **G L M** *	Neuromedin K
pE L Y E N K P R R P Y I L	Neurotensin

FIGURE 12-2. Structural homologies between the peptides of the tachykinin family, presented according to the schema of Figure 12-1. The sequence of neurotensin, unrelated except by the discoverer Susan Leeman, is also shown.

which they later named Substance P because it was present in the dried acetone powder of the extract. Although studied intermittently through bioassays, Substance P remained somewhat shaded in obscurity until almost forty years later, when Leeman and her coworkers purified a sialogogic peptide from hypothalamic extracts, while looking for the still-elusive corticotropin-releasing factor. This sialogogic peptide turned out to have an amino-acid content and pharmacological profile identical to Substance P, and when finally purified, sequenced, and synthesized was identified as Substance P, a undecapeptide. (Fig. 12-2) The knowledge of the structure and the availability of large amounts of the synthetic material permitted the development of radioimmunoassays and immunocytochemical tests that were then used to map the brain and assay its Substance P content. Substance P is present in small neuronal systems in many parts of the CNS, and on subcellular fractionation it is found in the vesicular layer. It is especially rich in neurons projecting into the substantia gelatinosa of the spinal cord from the dorsal root ganglia and has even been proposed as the transmitter for primary afferent sensory fibers. While it is very potent as a depolarizing substance in direct tests of spinal cord excitability (about 200 times stronger on a molar basis than GLU), its long duration of action prompted caution in accepting Substance P as the primary sensory transmitter. Other peptides have also been identified in dorsal root ganglion cells, and GLU released from cultures of these cells can also produces a prompt powerful excitatory action on spinal dorsal horn neurons. If GLU and Substance P were cotransmitters for some sensory fibers, Substance P could be viewed as prolonging and intensifying the transmission into the cord.

Radioimmunoassays and immunocytochemistry also show brain regions other than spinal cord to be rich in Substance P, especially the substantia nigra, caudate-putamen, amygdala, hypothalamus, and cerebral cortex. Tests with iontophoretic application in these regions generally indicate excitatory actions, again with a long duration. The presence of high levels of Substance P in the striatonigral fiber system and its disappearance there after interruption of the system have led to the proposal that Substance P is the transmitter for this pathway. This nearly accepted view came into question when maps of autoradiographically localized Substance P binding sites showed no presumptive receptors in the substantia nigra, but many binding sites in parts of brain with much lower content of Substance P. This was taken as another example of a mismatch between endogenous ligand maps and ligand binding site maps, and a log on the fire of those favoring peptide actions through diffusion. In the human neurologic disease, Huntington's chorea, characterized by profound movement disorders and psychological changes, Substance P levels in the substantia nigra are considerably reduced. Substance P coexists with 5-HT in some raphe neurons projecting to spinal cord, but there is no 5-HT in the dorsal root ganglia.

Based on smooth muscle bioassays of N-terminally shortened Substance P analogs, and early antagonist analogs, an initial subtyping of receptors suggested a P-type, where Substance P was as potent or more potent than other nonmammalian analogs (see Fig. 12-2), and an E-type receptor where one such amphibian peptide, *eledoisin*, was potent, and *kassinin*, an even more potent agonist. Search for kassinin-like peptides using such smooth muscle receptor subtypes as a bioassay revealed two forms, that were clearly not Substance P, and have been termed Substance K, and neuromedin K. Molecular cloning of the pro-Substance P mRNA from bovine striatum has partly clarified some of the mysteries here: Two mRNA forms have been sequenced (see Fig. 12-2), one of which contains only Substance P, while the other contains both Substance P and Substance K. Ligand binding and radioimmunoassays for Substance K have helped square the accounts between the apparent ligand versus binding site maps, and substantia nigra seems now to be rich in both the peptide and suitable receptive sites, although direct functional tests are still needed. No precursor for neuromedin K has yet been identified, nor is the genomic relationship between the three yet clear.

Neurotensin

Strictly speaking, neurotensin is not part of the tachykinin family of peptides and probably deserves independent billing. However, there is a tenuous historical relationship in that Leeman and coworkers isolated and sequenced the tridecapeptide immediately after their success with Substance P while still looking for the corticotropin-releasing factor. When injected intracisternally, the synthetic neurotensin lowers body temperature, accentuates barbiturate sleeping time, and stimulates the release of growth hormone and prolactin. Again, the availability of large amounts of the synthetic material has provided radioimmunoassays and immunocytochemistry which indicate that the largest amounts of this peptide are to be found in the anterior and basal hypothalamus, in the nucleus accumbens and septum, and in the spinal cord and brainstem within small interneurons of the substantia gelatinosa and motor trigeminal nucleus. Neurotensin colocalizes to some dopamine and some adrenergic neurons. Cellular tests of neurotensin action suggest it may produce excitation, but the mechanisms are not yet defined. In the various peripheral, whole-animal bioassays, the blood pressure receptors and the hyperglycemia receptors appear to read the

Arg-Arg dipeptide near the middle of neurotensin. Neurotensin produces diminished responsiveness to painful stimuli and hypothermia after central injection. Studies in humans and dogs show marked increases in neurotensin-like material circulating in plasma after fatty meals.

Glucagon-related Peptides

Vasoactive intestinal polypeptide (VIP), a peptide composed of twenty-nine amino acids, was originally isolated from porcine intestine and named for its ability to alter enteric blood flow. Establishment of its sequence revealed marked structural similarities to glucagon, secretin, and another peptide of gastric origin which inhibits gastric muscular contraction (GIP). These structural similarities constitute the basis of what is presumed to be a peptide family. (See Fig. 12-3.) The development of synthetic VIP for preparation of immune assays and cellular localization revealed that VIP exists independently of its cousins in gut and pancreas and showed that VIP was prominent in many regions of the autonomic and central nervous system. In parasympathetic nerves to the cat salivary gland, VIP coexists with acetylcholine and is apparently released with ACh as part of an integrated command to activate secretion and to increase blood flow through the gland. In the CNS,

FIGURE 12-3. The glucagon-related peptide family represented by their single-letter amino-acid symbols, and in which the sequences of PHI-27, PHM-27, growth hormone releasing hormone 1–24, glucagon and secretin, that match those of VIP are indicated in bold letters. For an interesting exercise, the reader may wish to construct a complementary table in which matches to glucagon are highlighted.

GLUCAGON RELATED PEPTIDES

HSDAVFTDNYTRLRKOMAVKKYLNSILN*	VIP
HADGVFTSDFSRLLGQLSAKKYLESLI*	PHI 27
HADGVFTSDFSKLLGQLSAKKYLESLM*	PHM 27
YADIFTNSYRKVLGQLSARKLLQD ---	$GHRH_{1-24}$
HSQGTFTSDYSKYLDSRRAQDFVQWLMNT	Glucagon
HSDGTFYSELSRLRDSARLQRLLOGLV*	Secretin

FIGURE 12-4. Neurons in rat cerebral cortex which contain vasoactive intestinal polypeptide, as visualized by immunoperoxidase staining. Note the precise vertical orientation of the smooth thick-branching dendrites, and their rather precise lateral separation by almost identical distances. In this field, only one of the processes can be seen connecting with its perikaryon, which is located in lamina IV. (Unpublished micrograph provided by Dr. John Morrison) × 650

VIP-reactive neurons are among the most numerous of the chemically defined cells of the neocortex, and there exhibit a very narrow radial orientation, suggesting they innervate cellular targets located wholly within a single cortical column assembly. (See Fig. 12-4.) In peripheral structures, VIP-reactive nerves innervate gut, especially sphincter regions, as well as lung and possibly even thyroid gland. Preliminary tests on neuronal activity suggest VIP excites; biochemical tests show it is able to activate synthesis of cyclic AMP, an action that in the gut is blocked by somatostatin. A very pernicious gut cancer that actively secretes VIP has been described, but no other known disease states involving this peptide have yet been documented.

Additional members were added to this important family of peptides when Guillemin and colleagues isolated the long-sought growth hormone releasing hormone (GHRH) peptide and recognized the homologous shared amino-acid sequences. The latest members of the family are termed PHI-27 and PHM-27, names that reflect their size, twenty-seven amino acids long, and the one letter initials of their N-terminal (H = Histidine) and C-terminal (isoleucine or methionine) peptides (where P stands for "peptide" and not proline). PHI-27 was initially isolated by Mutt and associates from gut extracts searching for other C-terminally amidated peptides, a property all active members of this family share. PHM-27 was identified from the deduced sequence of the pro-VIP mRNA in cloning experiments. GHRH has a relatively limited neuronal distribution concentrated in the hypothalamus and median eminence. No cellular maps for PHI or PHM in brain have yet been reported.

Pancreatic Polypeptide-related Peptides

The pancreatic polypeptides were recognized in extracts of pancreatic islets in the mid-1970s, and those from pigeon, pig, and human pancreas were all found to be highly homologous thirty-six residue peptides with amidated C-termini. (see Fig. 12-5.) Antisera against these peptides recognized cells and fibers in the autonomic and central nervous system, but these immunocytochemical observations were rightly qualified as "pancreatic polypeptide-like im-

FIGURE 12-5. The pancreatic polypeptide family represented by their single-letter amino-acid symbols, and in which the sequences of peptide YY (PYY), avian pancreatic polypeptide (APP), and human pancreatic polypeptide (HPP) that match those of NPY are indicated in bold letters.

PANCREATIC POLYPEPTIDES

YPSKPDNPGEDAPAEDLARYYSALRHYINL ITRQRY*	NPY
YPAKPEAPGQNASPQQLSRYYASLRHYLNLVTRQRY*	PYY
GPSQPTYPGDDAPVEDLI RFYDNLQQYLNVVTRHRY*	APP
APLEPVYPGDNATPEQMAQYAADLRRYINMLTRPRY*	HPP

munoreactivity" because when the same sera were used in their highly dilute form for radioimmunoassay, only scant extractable material was detected. When Mutt and Tatemoto implemented their chemical isolation strategy centered on the detection of novel C-terminally amidated gut and brain peptides, they soon were able to announce the amino-acid sequences of two more members of this family. The first one found in gut was given the name PYY (a neuropeptide with tyrosine (single-letter amino-acid symbol "Y") residues at both the N- and C-termini). However, the brain one also had the same length, and the same N- and C-terminal tyrosines, with slight internal amino-acid substitutions, and has been termed NPY. In rapid order, this peptide's distribution was described in rodent and human brains, and its regional content quantified, showing NPY to form one of the most extensive of the peptide systems, with very high amounts in the hypothalamus, limbic system, and neocortex, not to mention some loosely correlated losses in some schizophrenics. In many places, including the autonomic nervous system, NPY coexists with either norepinephrine or epinephrine,. NPY increases the sensitivity of sympathetically innervated smooth muscle to norepinephrine and is one of the most potent natural vasoconstrictors known.

Opioid Peptides

An "endorphin" is any "endogenous substance" (i.e., one naturally formed in the living animal) which exhibits pharmacological properties of morphine. When this term was first coined in mid-1975, it was a useful abstraction to cover "morphine-like factors," from brain extracts and spinal fluids that were active in opiate ligand displacement assays or opiate-sensitive smooth muscle assays. Within a year, a highly competitive effort resulted in the purification, isolation, sequencing, and synthetic replication not only of one but of nearly a half-dozen peptides that deserved the term endorphin. Subsequent research has greatly clarified the molecular and genomic relationships between the three major branches of the opioid peptide greater family (see Fig. 12-6): the proopiomelanocortin

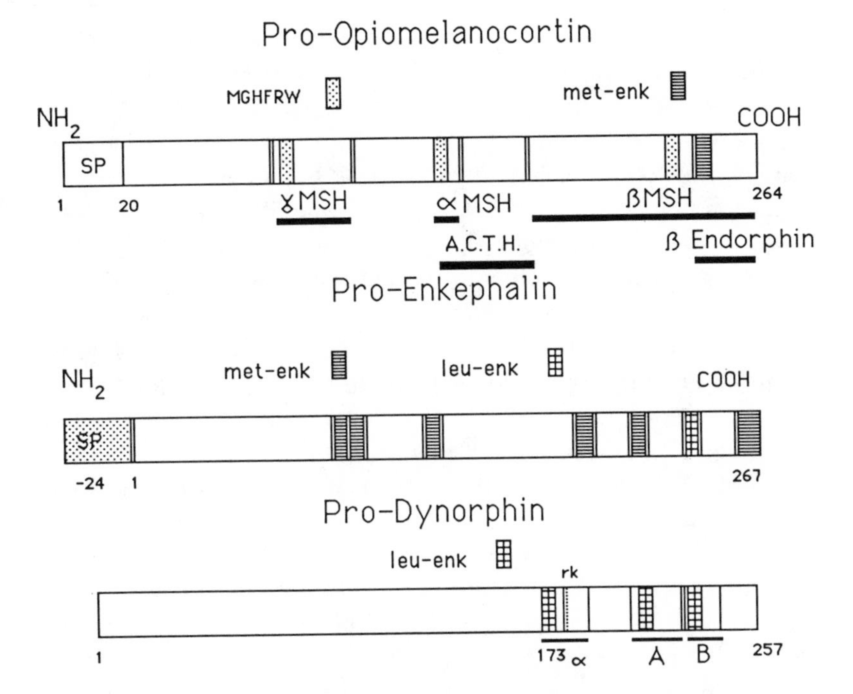

FIGURE 12-6. Structural relationships among the pro-hormone, precursor forms of the three major branches of the opioid peptides, depicted as a bar diagram, whose length in amino-acid residues is indicated by the small number at the corresponding C-termini. The location of the repeating peptide sequences are indicated. Basic amino-acid sequences that constitute consensus cleavage sites for processing are indicated as single or double vertical lines within the bar. Proopiomelanocortin and proenkephalin have consensus signal peptides indicated at their N-termini.

(POMC)-derived peptides, the proenkephalin-derived peptides, and the prodynorphin-derived peptides. Other structurally related natural peptides lack opioid receptor activity (such as the invertebrate cardioacceleratory peptide FMRF-amide); furthermore, some opioid-acting peptides have been found "exogenously" in milk and in plant proteins and have been called "exorphins." These developments have lead activists in the field to give the entire group of structures a more general name, such as endogenous opioid peptides. This also spares those who prefer to study enkephalin and dynorphin peptides from the burden of seeming to carry the endorphin banner.

The incredible explosion of work on this class of peptides began with attempts to isolate and characterize the receptor to opiates as part of a molecular biological approach to the question of narcotic addiction. When specific binding assays were developed, substantial evidence was accumulated to indicate that a high-affinity binding site in synaptic membranes showed stereo-selective opioid recognition properties. The receptor showed Na-sensitive allosteric-like affinity shifts (i.e., in the presence of Na, agonists show reduced affinity) but was not altered in number or affinity during narcotic dependence or withdrawal.

A whole new approach to neurotransmitter identification took shape when Hughes, Kosterlitz, and their colleagues in Aberdeen demonstrated that extracts of brain contain a substance that can compete in the opiate receptor assays and show opioid activity in *in vitro* smooth-muscle bioassays. In late 1975, this endogenous opioid activity was attributed to two pentapeptides, named enkephalins, which shared a common tetrapeptide sequence, YGGF, varying only in the C-terminal position: hence called Met^5-enkephalin and Leu^5-enkephalin. Perhaps even more dramatic than the announcement that brain contained not one but two opioid peptides was the realization that the entire structure of Met^5-enkephalin is contained within a 91 amino acid pituitary hormone, β-lipotropin (β-LPH), whose isolation and sequence was reported several years earlier by C. H. Li but whose function was never really very clear.

Several groups then reported isolation, purification, chemical structures, and synthetic confirmation of three additional endorphin peptides: α-endorphin (β-LPH $_{61-76}$), γ-endorphin (β-LPH $_{61-77}$) and β-endorphin (β-LPH $_{61-91}$; also called "C-fragment"). Numerous claims and counterclaims were parried across the symposium stages for several months as to one person's peptide being another person's artifact or precursor, eventually clarifying.

CHEMICAL AND CELLULAR RELATIONSHIPS AMONG THE OPIOID PEPTIDES

1. The proopiomelanocortin (POMC) peptides are expressed independently in the anterior pituitary, intermediate lobe of the pituitary, and one main cluster of neurons in the area of the arcuate nucleus of the hypothalamus. The major endorphin agonist produced from POMC is the thirty-one amino-acid C-terminal fragment, β-endorphin, the most potent of the natural opioids. N-terminal fragments of β-endorphin are much less potent, and analogs with no N-terminal tyrosine (so-called des-Tyr versions) lose all opioid activity, although in some hands, such des-Tyr peptides are reputed to be active behaviorally. In the corticotropin-secreting cells of the anterior pituitary, POMC is processed largely to corticotropin, and to an inactive form of β-endorphin, whereas in intermediate lobe cells and arcuate neurons, the same precursor is processed to α-MSH and active β-endorphin. A third MSH-containing heptapeptide component, discovered during the cloning and sequencing of the POMC mRNA, suggests that yet another end product may be possible.

2. *The enkephalin pentapeptides*, Met^5-enkephalin and Leu^5-enkephalin are expressed in wholly separate neuronal systems from the POMC neurons and are much more pervasively distributed throughout the central and peripheral nervous systems, including the adrenal medulla and enteric nervous system. The cloning and sequencing of the mRNA for the proenkephalin, starting with mRNA from adrenal medullary tissue, produced an unexpected dividend in that the precursor exhibits multiple copies of the two peptides in almost exactly the 6:1 ratio of Met^5- to Leu^5-enkephalins that had been described in regional brain and gut assays, and thus solving the mystery of the two similar peptides.

3. *The prodynorphin peptides* consist of C-terminally extended forms of Leu^5-enkephalin arising from a different gene and from a different mRNA that encodes for production of four major peptides, termed Dynorphin A, Dynorphin B, and two Neoendorphins, α and β. These C-terminally extended peptides all act as potent opioid agonists without cleavage down to the enkephalin pentapepide form, and on mapping they were found to represent a third separated series of rather generally distributed central and peripheral neurons.

Each of the separate classes of neurons containing β-endorphin, enkephalin or dynorphin peptides have distinct morphological features (see Fig. 12-7). The β-endorphin containing neurons are long projection systems that fall within the general endocrine-oriented

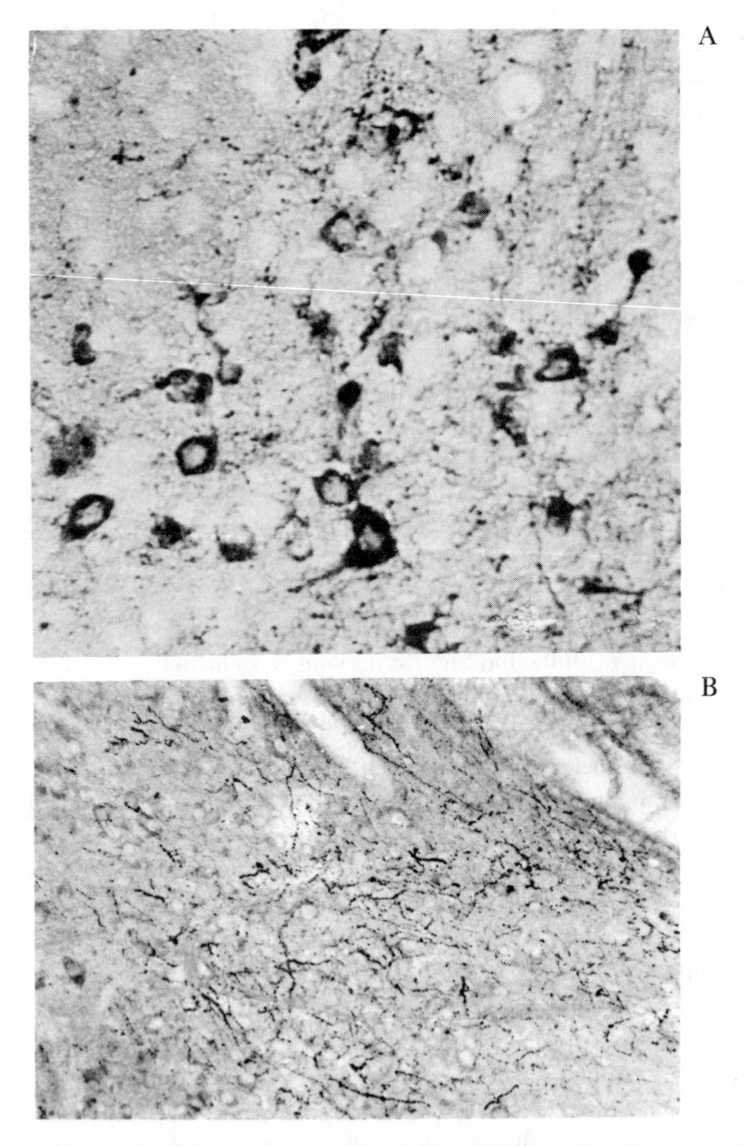

FIGURE 12-7. Brain immunohistochemistry. (A) Cell bodies and varicose processes of immunoperoxidase-stained rat hypothalamus using antiserum to β-endorphin. Note the granular deposits within the neuronal cell bodies and the long, thick varicose fibers (Bloom, unpublished). (B) Nerve fibers stained by immunoperoxidase with antiserum to β-endorphin; these fibers are in the nucleus locus ceruleus. (Bloom, unpublished) ×300

systems of the medial hypothalamus, diencephalon, and pons. The proenkephalin-derived peptides and the prodynorphin-derived peptides are generally found in neurons with modest to short projections, groups of which are widespread. In some regions the enkephalin-derived and dynorphin-derived peptides show intriguing relationships. For example, enkephalin—containing neurons project from the entorhinal cortex to the molecular layer of the dentate gyrus of the hippocampus, while dynorphin-containing-neurons project from the dentate gyrus to the CA3 pyramidal cells. In the spinal cord, intrinsic interneurons contain dynorphin peptides, while descending long axons from the pons and medulla contain enkephalin peptides. In addition, all these peptide systems presumably contain other nonpeptidic transmitters as well.

Cellular Effects

Given these twists and turns at the molecular level, analysis of the effects of synthetic endorphins at the cellular level may well be expected to be preliminary since it remains unclear just which peptides should be tested on which target cells. As this chapter is being assembled, there are at least four schemes of endorphin–receptor classifications based on the comparative actions of morphine or of morphine-like drugs with mixed agonist–antagonist actions on the dog spinal cord (the μ, κ, and σ scheme of Martin and Gilbert), or on the relative effects of endogenous and synthetic opioids on various smooth muscle *in vitro* assays (the μ and δ scheme of Kosterlitz, or the μ, δ; and ϵ scheme of Herz), or on, alternatively, GTP-modifiable ligand binding. One of the most remarkable of these features of the partially characterized endorphin "receptor" schemes is how poorly they equate with the endorphin or enkephalin-like peptide circuits, and how little any of them are affected by chronic exposure to morphine.

The best concordance between receptor subtypes and opioid peptide systems stems from the studies of dynorphin-derived peptides by Goldstein and Chavkin, which emphasized the ability of these peptides to act on the κ type of opioid receptor site. This is most apparent in studies of the gut receptors, but is supported by studies on central neurons as well. Naloxone, the workhorse antagonist for all opioid receptors, has far more potent ability to reverse μ and δ opioid actions, but at 10-fold higher doses also will antagonize κ sites. When the prodynorphin-derived peptides found within the hippocampus are evaluated for their relative content and potency, the most potent are the least prevalent and vice versa. Although none of the receptors has yet been purified completely, a 500-fold enrichment of the μ-receptor from rat brain membranes discriminates its macromolecular properties from the δ-receptor. Chronic exposure of nonaddicted rats to naltrexone, the long acting form of naloxone, increases the number of both μ- and δ receptors.

The membrane mechanisms underlying the activation of the various classes of opioid receptors are not well understood. Some studies suggest they may block the activation of adenylate cyclase, and that such effects may underlie the chronic tolerance that develops with repeated dosing. Intracellular recordings of myenteric gangionic neurons and of locus ceruleus neurons *in vitro* indicate that these cells respond by hyperpolarization to acute administration of opioids mediated by an increased K^+ conductance, that secondarily depresses Ca-dependent spike activity. In dorsal horn, neither membrane potential nor conductance are impressively altered, but responsiveness to depolarizing synaptic potentials is reduced.

Iontophoretic tests with enkephalins and β-endorphin suggest that neurons throughout the CNS can be influenced by these peptides at naloxone-sensitive receptors; in general, responses tend to be depressant except in the hippocampus where excitatory actions are so profound that hippocampal seizures can be induced. However, the mechanism of this excitation is the exception which proves the rule, as it is based on inhibition on *inhibitory* interneurons and produces excitation by disinhibition. In such tests, all endorphins show qualitatively equivalent effects and similar onset of actions; effects of β-endorphin appear to be more potent and longer lasting, which may be due to its slower hydrolysis. Analogs in which cleavage of the N-terminal dipeptide bond is delayed by substitution of a D-Ala are more long lasting, while peptides with no N-terminal Tyr are much weaker.

Behavioral Effects

It is in terms of behavior that the effects of these peptides have attracted growing attention. Early efforts were made to answer the question of morphine substitution and were somewhat disappointing in showing that the enkephalins produced only transient analgesia after direct intracerebroventricular injection; in such tests β-endorphin is 50 to 100 times more potent than morphine on a molar basis. In mice, β-endorphin produces some significant analgesia after intravenous injection, but the dose required is about 1000 times that needed on central injection, and the duration of effect after peripheral injection is much shorter.

Morphine-dependent rats show cross-tolerance to all opioid peptides; similarly, animals given continuous injections of endorphin or enkephalin show tolerance and also show withdrawal signs when given naloxone. The fact that blood-borne peptides do not penetrate into brain well or survive the gauntlet of peptidases to which they are exposed in the process has spurred the search for more effective analogs. However, the real power of such analogs may not be realized until the cellular mechanisms of the several opioid peptide-containing systems are better evaluated.

The behavioral effects of the endorphins which first drew attention required the central injection of 1 to 10 μg of synthetic peptide, and far more for those presumed breakdown products of β-endorphin (i.e., α-endorphin and γ-endorphin), which have almost no opioid activity on central injection. However, a second wave of behavioral interest in endorphins has arisen from studies using peripherally injected peptides in active or passive avoidance tests. As with the behavioral actions of vasopressin, submicrogram amounts of the endorphins or of corticotrophin, or of any of their less well known fragments such as α-MSH, $ACTH_{4-10}$, or the endorphins α-, γ-, or des-tyrosine gamma,—none of which have any known targets—all produce similar effects in delaying extinction of active or passive avoidance for hours after subcutaneous injection. Since all these peptides produce similar effects, it will come as no shock to realize that they do not act on "typical" opiate receptors: the effects are not antagonized by naloxone, nor are they replicated by morphine. The explanation of these low-dose behavioral effects of peripheral peptides, and their even lower dose effectiveness, which give extremely flat dose-response curves on intracerebral injection, probably will reside in mechanisms similar to those potential explanations previously discussed for vasopressin.

The sheer number of whole animal effects that have been interpreted as endorphinergic physiology on the basis of naloxone effects has grown considerably but remains open to more comprehensive analysis. Among the proposed physiological properties that may be regulated by one or another of the endorphin substances are blood pressure, temperature, feeding, sexual activity, and lymphocyte mitosis, along with pain perception and memory. All this seems like a lot to ask of one peptide. But then, who are we to ask what a peptide may—can—do for us? As we said before, don't blink or you'll miss the next blazing development.

Individual Peptides Worth Tracking

We conclude this chapter with brief comments on a few of the other neuroactive peptides that are under active investigation. Around those we have selected, bodies of research are consolidating and the neuropharmacological implications seem strong.

Somatostatin (Somatotropin Release-Inhibiting Factor)

In 1971, other workers began to look for other goodies in their hard-won hypothalamic extracts; Vale, Brazeau, and Guillemin tried their extracts for potency in releasing growth hormone from long-term cultured anterior pituitary cells and found to their amazement a factor that inhibited even basal growth hormone release in minute amounts. They named this factor somatostatin. Isolation and purification studies eventually culminated in the characterization of this molecule as a tetradecapeptide with disulfide bridge between Cys^3 and Cys^{14} (Fig. 12-8). Radioimmunoassays, immunocytochemistry, and whole-animal tests of the synthetic peptide made it clear that so-

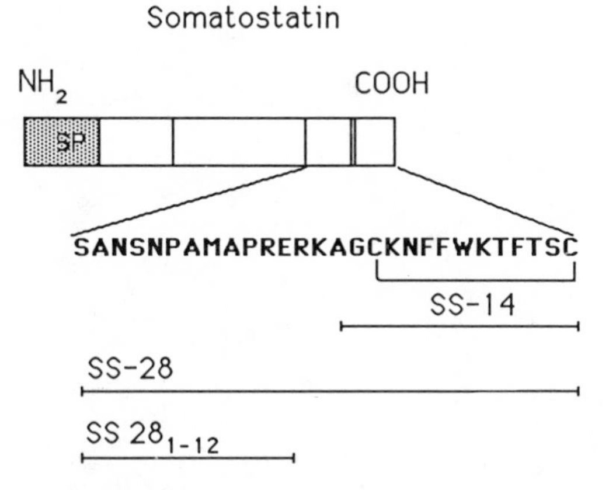

FIGURE 12-8. Structure of the prohormone form of somatostatin, with the location and sequence of the twenty-eight amino acid residue form at the C-terminus, and the relative sequences and locations of the different forms of somatostatin indicated.

matostatin was doing more than "just" inhibiting growth hormone release. Somatostatin was found to be widely distributed in the gastrointestinal tract and pancreatic islets, localized in the latter tissue to the delta cells by immunocytochemistry. Somatostatin in islets can apparently suppress the release of both glucagon and insulin, and in diabetic humans, who have no insulin, suppression of glucagon can be an important element in determining insulin requirement. The mechanisms of this suppression are not known but may be similar to the effect of somatostatin on growth hormone release from pituitary, an action accompanied by suppression of TSH release.

When somatostatin is injected intracerebroventricularly, animals show decreased spontaneous motor activity, reduced sensitivity to barbiturates, loss of slow wave and REM sleep, and increased appetite. Radioimmunoassays and immunocytochemistry of rat brain regions show that somatostatin is largely concentrated in the mediobasal hypothalamus, with much smaller amounts present in a few other brain regions (see Fig. 12-9). Immunocytochemical studies led

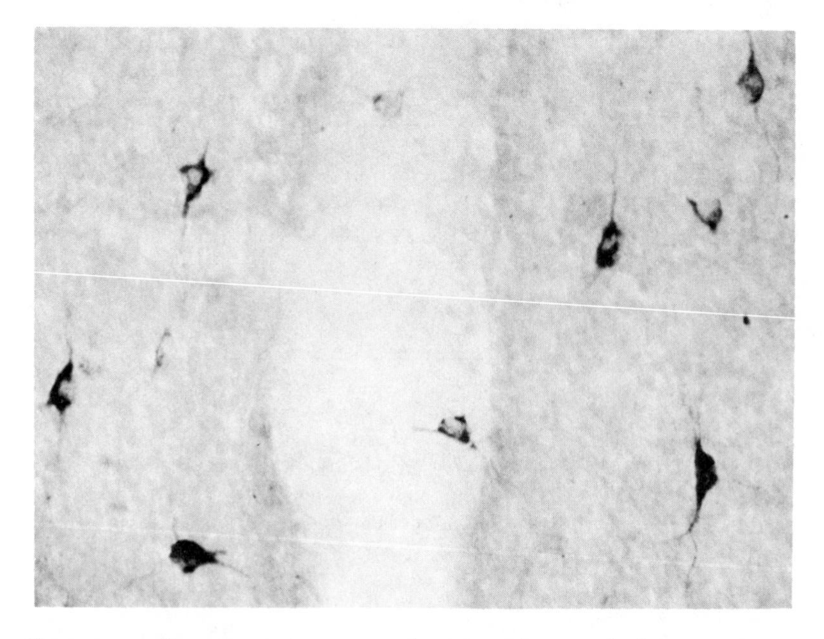

FIGURE 12-9. Somatostatin immunoreactive neurons in rat cerebral cortex contrast with the VIP cells by having no precise laminar orientation and having multiple shorter processes. Nerve cell in center appears to overlay a small blood vessel that runs vertically through the field. (Unpublished micrograph contributed by Dr. John Morrison) ×410

to the surprising finding that somatostatin reactive cells and fibers are also present in dorsal root ganglia, in the autonomic plexi of the intestine, and in the amygdala and neocortex. Avian and amphibian brains contain considerably more somatostatin than mammalian brains: physiological tests on the isolated frog spinal cord suggest that it may facilitate the transmission of dorsal root reflexes after a long latent period. Iontophoretic tests on rat neurons indicate a relatively common depressant action that is brisk in onset and termination. It has been suggested that somatostatin shares a pharmacological profile with the anticonvulsant drug, phenytoin. An endogenous large somatostatin, extended at the N-terminus by an additional 14 amino acids also exists in both brain and gut, and shares equipotency for most somatostatin actions. In rodent and primate neocortex, the N-terminally extended forms reveal a far more extensive neuronal density, which at least overlaps with the intrinsic GABA-ergic cortical neurons. In early onset Alzheimer's disease, cortical somatostatin content is depleted, and somatostatin 28 immunoreactive neuritic processes are involved in the formation of the hallmark Alzheimers lesion, the plaque. Cysteamine, a drug developed to treat the metabolic disorder known as cystinosis, will selectively deplete Somatostatin-14 without altering brain levels of Somatostatin 28 or Somatostatin 28 [1–12], suggesting that these peptides may be stored in separable subcellular compartments.

Cholecystokinin (CCK)

Among the gut hormones with the longest history are gastrin and cholecystokinin (CCK). Subsequently, when their molecular structures were determined, a striking homology was revealed between the C-termini (Fig. 12-10). After some confusing interludes, it became recognized that the gastrinlike material extractable from nervous tissue was due to antisera cross reacting with the C-terminal octapeptide of CCK whose tyrosine version is sulfated. Although

FIGURE 12-10. The homologies between ovine and rodent corticotropin releasing factor and sauvagine are indicated by bold single letter amino acid symbols. Below are indicated the amino acid sequences of calcitonin gene-related peptide (CGRP) and of the C-terminal octapeptide of cholecystokinin, with a sulfated tyrosine at position 2.

SQEPPISLDLTFHLLREVLEMTKADQLAQQAHSNRKLDIA CRF (OVINE)
SEEPPISLDLTFHLLREVLEMARA**EQLAQQAHSNRK**MEII CRF (RAT)
pEG**PPISI DL**SL**E**LL**R**KMIEIEKQEKEK**QQA**AN**NR**LLDTI SAUVAGINE

SCNTATCVTHRLAGLLSRSGGVVKDNFVPTNVGSEAF* "CGRP"

DYMGWMDF* CCK_8

the exact species of the peptide present in brain is still in some chaos, much evidence favors the view that the C-terminal octapeptide exists as a separate entity and is released from nerves. CCK-like material coexists with dopamine in nerve cells of the ventrotegmental area and in their terminals in nucleus accumbens. A minimum amount of data exist as to cellular action which generally causes responsive cells to fire more rapidly. Immune staining patterns also reveal this peptide to exist within numerous small GABA-containing intracortical neurons. Because CCK is released from distended intestines, a minor controversy has developed around its ability to reduce appetite in experimental animals and its possible role in gut and in cortex to regulate eating. Although the picture is yet far from complete, it is of interest that the "satiety" effects of peripherally injected CCK are blocked when the vagus nerve is cut below the diaphragm. While an alternative drug approach to dietary control is clearly a desirable goal, it has not yet become clear how these data fit that need. Two interesting, but as yet unexplained, pharmacological interactions deserve continued monitoring. In iontophoretic tests on spontaneously active hippocampal pyramidal cells, CCK-8 shows potent ability to activate firing, an action that is compatible with the CCK contained within the dentate granule cell mossy fiber synapses to these neurons. Low doses of systemic or iontophoretically administered benzodiazepines selectively block CCK induced activation without altering the excitatory effects of enkephalin or acetylcholine. In separate lines of investigation, largely at the behavioral level, systemic CCK proved to have opiate antagonistic actions depressing opiate analgesia, but without any direct effects on opioid receptors. When rats were actively immunized against CCK, exogenous opiates had enhanced initial potency, but tolerance still developed with repeated doses of opiates.

Angiotensin II

Present evidence suggests that we should keep an open mind as to just which Roman numerals are used to refer to the "naturally stored" and "naturally released" forms of this peptide, whose metabolism

Angiotensinogen (liver)

D R V Y I H P F H L L V Y S (Protein)
Renin ↑ (kidney)

Angiotensin I

D R V Y I H P F H L
Angiotensin-Converting ↑ (plasma, liver,
Enzyme nerve)

Angiotensin II

D R V Y I H P F
↑ Aspartate Amino Peptidase

Angiotensin III

R V Y I H P F

FIGURE 12-11. The angiotensin family and the relationships between specific cleavage enzyme actions and derived peptide products.

comes across as a shaggy dog story yet vividly illustrates the mode in which at least one group of active peptides is synthesized and released. Until a few years ago, the role of angiotensin was thought to be mainly the peripheral modulation of blood pressure. Angiotensin (Fig. 12-11) was the name given to a vasoconstricting decapeptide formed in blood from an α-2 hepatic globulin by the proteolytic enzyme renin; as its name suggests, renin is made and secreted by specialized cells of the kidney. However, the decapeptide, now termed Angiotensin I, is converted to an octapeptide Angiotensin II with a similar pharmacological profile of higher intrinsic activity through a "converting enzyme" that selectively cleaves off the C-terminal dipeptide. Subsequently Angiotensin III (a natural N-des-Asp Angiotensin II) was identified, is even more potent, and may more readily traverse biological membranes. All of the substrates and enzymes needed for these conversions of angiotensin have been described as occurring in brain. Radioimmunoassays, immunocytochemistry, and receptor displacement studies all indicate the presence of neural systems containing or responding to one or another of these peptides. The angiotensin system appears to be relatively concentrated in the periventricular zone, and the specialized innervated organs that characterize this zone—the area postrema, the pineal, the subfornical, and subcommisural organs as well as the intermediate lobe of the pituitary and others. These organs lie within the recesses of the brain but are on the blood side of the blood–brain barrier.

Circulating Angiotensin II can indeed modify blood pressure and induce secretion of aldosterone. The angiotensin-converting enzyme inhibitors have been able to realize one of the major ambitions of future drug development research by computer modeling drugs to fit the active catalytic pockets of purified peptide processing enzymes. This class of drugs has been shown to be very effective, alone or with low doses of diuretics, in regulating hypertension through direct vascular actions. The central effect that has generally caught the attention of pharmacologists is its ability to produce profound drinking behavior when injected in physiological (i.e., picomol) amounts into the third ventricle. Although angiotensin can elicit drinking, it is probably not the "drinking" hormone since circulating levels do not correlate with drinking activity. Iontophoretic studies have shown that angiotensin excites the neuronal-like cells of the subfornical organ. Neurons in the supraoptic nucleus are also excited by Angiotensin II, which may in part explain how central or peripheral administration of angiotensin can result in vasopressin release from the neurohypophysis. In the long-term nephrectomized rat and dog, immunocytochemically detectable Angiotensin II is present in small amounts throughout the diencephalon, pons, and brainstem, but the possible role these presumed neurons may have in central cardiovascular regulation or other functions has only recently been suggested on the basis of long-acting inhibitors of angiotensin-converting enzyme.

Calcitonin Gene-related Peptide

Calcitonin is a thirty-two amino-acid peptide hormone secreted by special cells in the thyroid gland whose function is to oppose parathormone and increase the urinary excretion of Ca^{++} and PO_4. The cells that make this hormone, the C-cells, are embryologically regarded as arising from neural crest. However, this ancestral lineage would not have merited inclusion of the hormone in this edition were it not for the results of studies originally carried out on a very undifferentiated thyroid carcinoma as part of a plan to characterize the genomic relationships between the segments of DNA encoding the calcitonin prohormone. Under certain conditions of growth in culture, the thyroid carcinoma cells were observed to produce additional mRNA forms that contained part of a series of amino acids encoded by the procalcitonin mRNA, but that differed at their carboxy terminus where the calcitonin occurs. Analysis of the genomic relationship between the exons encoding the segments of the mRNA revealed that the new form of the calcitonin-gene related prohormone contained a potentially novel peptide, termed calcitonin gene-related peptide (CGRP) which bore no significant structural similarity to calcitonin. Antisera were then raised to a synthetic version of the predicted thirty-seven amino-acid long peptide (see Fig. 12-10). The antisera as well as the cDNA forms of the CGRP mRNA were then used to determine whether this previously unexpected peptide was actually expressed in normal tissues. Tests on hypothalamus and on the trigeminal sensory ganglion were positive, as they were in other neuronal systems when the rat nervous system was mapped more completely. The exact mechanisms of mRNA processing by which the same primary RNA transcript of the calcitonin-CGRP gene can give rise to calcitonin in thyroid C cells and to CGRP in neurons remain unclear. However, the concept of "alternative" RNA processing may underlie one part of the peptide coexistence puzzle and clearly represents a mechanism by which a limited number of peptide-encoding gene sequences can be variably arranged into different end products in different cell types. Early physiological tests of CGRP indicate that it is an extremely potent direct vasodilator (at femtomolar doses) but inactive on calcitonin-responsive peripheral cells. No actions on central neurons have yet been reported.

Corticotropin Releasing Factor

The search for the corticotropin releasing factor was completed in 1981 when Wylie Vale and colleagues completed the purification and sequence analysis of the peptide taken from extracts of sheep hypothalamus. Thus ended a three-decade-long search for a mystery factor that put the secretion of corticotropin and other anterior pituitary hormones under neuronal control and allowed brain to replace pituitary as the master gland of the body. The remaining member of the original four horsemen of the hypophysiotrophic hypothalamic control system was lassoed a year later when Vale and Roger Guillemin independently reported the isolation, sequencing, and synthetic replication of growth hormone releasing hormone. Nonparticipants in this highly competitive struggle probably had mixed emotions when these quests ended, since along the way the world of neuropeptide discovery was at least triply blessed by first Substance P, then somatostatin, and eventually neurotensin, all of which were discovered by teams intent on CRF.

When CRF was in hand, it turned out to be an interesting forty-one amino-acid peptide (see Fig. 12-10) structurally related to a rather obscure frog skin peptide, sauvagine, and to another obtained from an obscure peripheral organ, the urohypophysis of certain fish, termed *urotensin*. The cellular origin of the mammalian CRF was traced to that portion of the paraventricular nucleus that projects to the median eminence, but not into the posterior pituitary. Later, these and other CRF-containing circuits were identified as innervating a very extensive group of neurons in the pons and medulla, perhaps representing a stress-related circuitry. CRF is an extremely potent ACTH secretogogue, but its effects are significantly augmented by vasopressin (produced by the adjacent magnocellular paraventricular neurons) and by norepinephrine and angiotensin.

In addition to its premier action in regulation of corticotropin secretion, synthetic CRF has equally potent effects on neurons *in vitro* and *in vivo*, increasing the frequency of action potentials in cells that fire in bursts, like the hippocampal pyramidal cells. Even modest doses of the peptide can induce seizure like activity within the limbic system. At still smaller doses given intracerebroventricularly, CRF is a potent activator of spontaneous locomotion, and can also produce an "anxiogenic" response (opposite to the effects of antianxiety drugs like benzodiazepines and ethanol) in different behavior tests. Clearly, the complete physiological effects of CRF at the pituitary and at other neuronal sites represent a much more comprehensive basis for the mobilization of bodily systems during stress. Early antagonists of CRF may provide some means of reducing the sequence of secretory events that until now has characterized the "stress" reaction.

A Reader's Guide to Peptide Poaching

Until more is learned about the functions, sites, and mechanisms of action of any one peptide, we will have trouble formulating even tentative roles for these substances in neurotransmitter-neurohormonal regulation of central drug actions.

As the story of neuroactive peptides unfolds, (the sheer number of peptides available for pursuit makes the going slower owing to the division of the available work force), interested students should be alert for answers to the following questions: Are there general patterns of circuitry? Are specific peptidases amenable to selective pharmacological intervention, or do neuronal and glial peptidases read only the dipeptides, whose bonds they are about to cleave? Do the peptides act presynaptically, postsynaptically, or at both sites? How are peptidases specifically activated either to release neuroactive peptides from precursors or to terminate the activity of the peptide? Can the peptides modulate the release or response to transmitters of other neurons or those with which they coexist in a given neuron. Can the promise of peptide chemistry deliver useful antagonists to prove identity of receptors with nerve pathway stimulation? Does the presence of receptors to other centrally active drugs indicate that still more endogenous peptides should be sought? (A novel eighteen amino-acid neuropeptide identified as the natural ligand for at least one of the benzodiazepine ligand-binding sites in brain was described just as these revisions were being prepared; unexpectedly, the natural ligand acts in the inverse fashion of the benzodiazepines—causing, rather than relieving, anxiety and reducing, rather than lengthening, GABA effects.) Can theorists of brain function find a useful purpose for so many excitatory and inhibitory substances? Stay tuned, the data flow fast.

Selected References

General Sources

Hökfelt, T., O. Johansson, A. Ljüngdahl, J. M. Lundberg, and M. Schültzberg (1980). Peptidergic neurones. *Nature 284*, 515–521.

Krieger, D. T., (1983). Brain peptides: What, where, and why? *Science 222*, 975–985.

Krieger, D. T., M. J. Brownstein, and J. B. Martin, eds. (1983). *Brain Peptides*. Wiley, New York.

Palkovits, M. (1984). Distribution of neuropeptides in the central nervous system: A review of biochemical mapping studies. *Prog. Neurobiol. 23*, 151–189.

Snyder, S. H. (1980). Brain peptides as neurotransmitters. *Science 290*, 976–983.

Angiotensin II

Bennett, J. P., Jr., and S. H. Snyder (1976). Angiotensin II: Binding to mammalian brain membranes. *J. Biol. Chem. 251*, 7423.

Fitzsimons, J. T. (1975). The renin-angiotensin system and drinking behavior. *Prog. Brain Res. 42*, 215.

Phillips, M. I., J. Weyhenmeyer, D. Felix, and D. Ganten (1979). Evidence for an endogenous brain renin angiotensin system. *Fed. Proc. 38*, 2260–2266.

Oxytocin and Vasopressin

de Wied, D. (1983). The importance of vasopressin memory. Trends in *Neuroscience* 7, 62–64.

Doris, P. A. (1984). Vasopressin and central integrative processes. *Neuroendocrinology 38*, 75–85.

Gash, D. M., and G. J. Thomas (1983). What is the importance of vasopressin in memory processes? *Trends in Neuroscience, 6* 197–198.

Hagler, A. T., D. J. Ogusthorpe, Dauber-Osguthorpe, and J. C. Hempel (1985). Dynamics and conformational energetics of a peptide hormone: Vasopressin. *Science 227*, 1309–1315.

Hayward, J. N. (1975). Neural control of the posterior pituitary. *Ann. Rev. Physiol. 37*, 191.

Koob, G. F., R. Dantzer, F. Rodriguez, F. E. Bloom, and M. Le Moal (1985). Osmotic stress mimics the effects of vasopressin on learned behaviour. *Nature 315*, 750–752.

Nicoll, R. A., and J. L. Barker (1971). The pharmacology of recurrent in-

Nicoll, R. A., and J. L. Barker (1971). The pharmacology of recurrent inhibition in the supraoptic neurosecretory system. *Brain Res. 35*, 501–518.

Swanson, L. W., and P. E. Sawchenko (1983). Hypothalamic integration: Organization of the paraventricular and supraoptic nuclei. *Ann. Rev. Neurosci. 6*, 269–324.

Neurotensin

Carraway, R., and S. E. Leeman (1975). The amino acid sequence of a hypothalamic peptide, neurotensin. *J. Biol. Chem. 250*, 1907–1912.

Kobayashi, R. M., M. Brown, and W. Vale (1977). Regional distribution of neurotensin and somatostatin in rat brain. *Brain Res. 126*, 584–590.

Rosell, S. (1980). Experimental evidence for Neurotensin or a metabolite being a hormone. *Acta Physiol. Scand. 110*, 325–326.

Tachykinin Peptides

Jessel, T. M., and M. D. Womack (1985). Substance P and the novel mammalian tachykinins: A diversity of receptors and cellular actions. *Trends in Neuroscience 8*, 43–45.

Lundberg, J. M., A. Saria, E. Brodin, S. Rosell, and K. Folkers (1983). A substance P antagonist inhibits vagally induced increase in vascular permeability and bronchial smooth muscle contraction in the guinea pig. *Proc. Nat. Acad. Sci. (USA) 80*, 1120–1124.

Nawa, H., H. Kotani, and S. Nakanishi (1984). Tissue specific generation of two preprotachykinin mRNAs from one gene by alternative RNA splicing. *Nature. 312*, 729–734.

Nicoll, R. A., C. Schenker, and S. E. Leeman (1980). Substance P as a transmitter candidate. *Ann. Rev. Neurosci. 3*, 227–268.

Nilsson, G., T. Hökfelt, and B. Pernow (1974). Distribuiton of Substance P-like immunoreactivity in the rat central nervous system as revealed by immunohistochemistry. *Med. Biol. 52*, 424–448.

Substance P in the Nervous System (1982). *Ciba Foundation Symposium 91*. Pitman Publishers, London, 349 pp.

Glucagon-related Peptides

Guillemin, R., P. Brazeau, P. Bohlen, F. Esch, N. Ling, and W. B. Wehrenberg (1982). Growth hormone releasing factor from a human pancreatic tumor that caused acromegaly. *Science 218*, 585–587.

Lundberg, J. M., B. Hedlund, and T. Bartfai (1982). Vasoactive intestinal polypeptide enhances muscarinic ligand binding in cat submandibular gland. *Nature 295*, 147–149.

Magistretti P. J., and M. Schorderet (1985). Norepinephrine and histamine potentiate the increases in cyclic adenosine 3′:5′-monophosphate elicited by vasoactive intestinal polypeptide in mouse cerebral cortical slices: mediation by $\alpha 1$-adrenergic and H_1-histaminergic receptors. *J. Neurosci. 5*, 362–368.

Morrison, J. H., P. J. Magistretti, R. Benoit, and F. E. Bloom (1984). The distribution and morphological characteristics of the intracortical VIP-positive cell: An immunohistochemical analysis. *Brain Res. 292*, 269–282.

Rivier, J., J. Spiess, M. Thorner, and W. Vale (1982). Characterization of a growth hormone releasing factor from a human pancreatic islet tumor. *Nature 300*, 276–278.

Said, S., and V. Mutt (1972). Isolation from porcine intestinal wall of a vasoactive octacosapeptide related to secretin and glucagon. *Eur. J. Biochem. 28*, 199–206.

Tapia-Arancibia, L., and S. Reichlin (1985). Vasoactive intestinal polypeptide and PHI stimulate somatostatin release from cat cerebral cortical and diencephalic cells in dispersed cell culture. *Brain Res. 336*, 67–72.

Tatemoto K., and V. Mutt (1981). Isolation and characterization of the interstitual peptide porcine PHI (PHI-27), a new member of the glucagon-secretin family. *Proc. Nat. Acad. Sci. (USA) 78*, 6603–6607.

Pancreatic Polypeptide-related Peptides

Allen, Y. S., T. E. Adrian, J. M. Allen, K. Tatemoto, T. J. Crow, S. R. Bloom, and J. M. Polak (1983). Neuropeptide Y distribution in the rat brain. *Science 221*, 877–879.

Emson, P. C., and M. E. de Quidt (1984). NPY—a new member of the pancreatic polypeptide family. *Trends in Neuroscience 7*, 31–33.

Mutt, V. (1983). New approaches to the identification and isolation of hormonal polypeptides. *Trends in Neuroscience 6*, 357–360.

Tatemoto, K. (1982). Neurpeptide Y: Complete amino acid sequence of the brain peptide. *Proc. Nat. Acad. Sci. (USA) 79*, 5485–5480.

Opioid Peptides

Bloom, F. E. (1983). The endorphins: A growing family of pharmacologically pertinent peptides. *Ann. Rev. Pharmacol. Toxicol. 23*, 151–170.

Bradbury, A. F., D. G. Smyth, C. R. Snell, N. J. M. Birdsall, and E. C. Hulme (1976). C-fragment of lipotropin has a high affinity for brain opiate receptors. *Nature 260*, 793.

Civelli, O., J. Douglass, A. Goldstein, and E. Herbert (1985). Sequence and expression of the rat prodynorphin gene. *Proc. Nat. Acad. Sci. (USA) 82*, 4291–4295.

Goldstein, A., S. Tachibana, L. I. Lowney, and L. Hold (1979). Dynorphin—(1–13), an extraordinarily potent opioid peptide. *Proc. Nat. Acad. Sci. (USA) 76*, 6666–6670.

Hughes, J., T. W. Smith, H. W. Kosterlitz, L. A. Fothergill, G. A. Morgan, and H. R. Morris (1975). Identification of two related pentapeptides from the brain with potent opiate agonist activity. *Nature 258*, 577.

Maneckjee, R., S. R. Zukin, S. Archer, and J. Michael (1985). Purification and characterization of the μ opiate receptor from rat brain using affinity chromatography. *Proc. Nat. Acad. Sci. (USA) 82*, 594–598.

Nakanishi, S., A. Inoue, T. Kita, A. C. Y. Chang, S. Cohen, and S. Numa (1979). Nucleotide sequence of cloned cDNA for bovine corticotropin-β-lipotropin precursor. *Nature 257*, 238–240.

Nicoll, R. A., B. E. Alger, and C. E. Jahr (1980). Enkephalin blocks inhibitory pathways in vertebrate CNS. *Nature 287*, 22–25.

Nicoll, R. A., G. R. Siggins, N. Ling, F. E. Bloom, and R. Guillemin (1977). Neuronal actions of endorphins and enkephalins among brain regions: A comparative microiontophoretic study. *Proc. Nat. Acad. Sci. (USA) 74*, 2584.

Schaefer, M., M. R. Picciotto, T. Kreiner, R.-R. Kaldany, R. Taussig, and R. H. Scheller (1985). Aplysia neurons express a gene encoding multiple FMRFamide neuropeptides. *Cell 41*, 457–467.

Zukin, S. R., and S. R. Zukin (1984). The case for multiple opiate receptors. *Trends in Neuroscience 7*, 160–164.

Somatostatin

Benoit, R., N. Ling, B. Alford, and R. Guillemin (1982). Seven peptides derived from prosomatostatin in rat brain. *Biochem. Biophys. Res. Commun. 107*, 944–950.

Effendic, S., and R. Luft (1980). Somatostatin: A classical hormone, a locally active polypeptide, and a neurotransmitter. *Ann. Clin. Med. 12*, 87–94.

Johansson, O., T. Hökfelt, and R. P. Elde (1984). Immunohistochemical distribution of somatostatin-like immunoreactivity in the central nervous system of the adult rat. *Neuroscience 13*, 265–339.

Morrison, J. H., R. Benoit, P. M. Magistretti, and F. E. Bloom (1983). Immunohistochemical distribution of pro-somatostatin related peptides in cerebral cortex. *Brain Res. 262*, 344–351.

Reichlin, S. (1983). Somatostatin. *N. Engl. J. Med. 309*, 1495–1501, 1556–1563.

Tran, V. T., M. F. Beal, and J. B. Martin (1985). Two types of somatostatin receptors differentiated by cyclic somatostatin analogs. *Science 228*, 492–494.

Vale, W., P. Brazeau, C. Rivier, M. Brown, B. Boss, J. Rivier, R. Burgus, N. Ling, and R. Guillemin (1975). Somatostatin. *Recent Prog. Horm. Res. 31*, 365.

Cholecystokinin

Beinfeld, M. C. (1985). Cholecystokinin (CCK) gene-related peptides: distribution, and characterization of immunoreactive pro-CCK, and an amino terminal pro-CCK fragment in rat brain. *Brain Res. 344:* 351–355.

Dodd, J., and J. S. Kelly (1981). The actions of cholecystokinin and related peptides on pyramidal neurones of the mammalian hippocampus. *Brain Res. 205*, 337–350.

Hendry, S. H. C., E. G. Jones, and M. C. Beinfeld (1983). Cholecystokinin-immunoreactive neurons in rat and monkey cerebral cortex make symmetric synapses and have intimate associations with blood vessels. *Proc. Nat. Acad. Sci. (USA) 80*, 2400–2404.

Innis, R. B., R. M. Correa, G. R. Uhl, B. Schneider, and S. H. Snyder (1979). Cholecystokinin octapeptide-like immunoreactivity: Histochemical localization in rat brain. *Proc. Nat. Acad. Sci. (USA) 76*, 521:525.

Rehfeld, J. F., N. R. Goltermann, L.-I. Larsson, P. M. Emson, and C. M. Lee (1979). Gastrin and cholecystokinin in central and peripheral neurones. *Fed. Proc. 38*, 2325–2329.

White, F. J., and R. Wang (1984). Interactions of cholecystokinin octapeptide and dopamine on nucleus accumbens neurons. *Brain Res. 300*, 161–166.

Calcitonin Gene-related Peptide

Brain, S. D., T. J. Williams, J. R. Tippins, H. R. Morris, and I. MacIntyre (1985). Calcitonin-gene related peptide is a potent vasodilator. *Nature 313*, 54–56.

Rosenfeld, M. G., J.-J. Mermod, S. G. Amara, L. W. Swanson, P. E.

Sawchenko, J. Rivier, W. W. Vale, and R. M. Evans (1983). Production of a novel neuropeptide encoded by the calcitonin gene via tissue-specific RNA processing. *Nature 304*, 129–135.

Rosenfeld, M. G., S. G. Amara, and R. M. Evans (1984). Alternative RNA processing: determining neuronal phenotype. *Science 225*, 1315–1320.

Vale, W., and M. Greer (1985). *Corticotropin Releasing Factor*—A conference report. *Fed. Proc. 44*, 145–263.

Vale, W., J. Spiess, J. Rivier, and C. Rivier (1981). Characterization of a 41-residue ovine hypothalamic peptide that stimulates secretion of corticotropin and beta-endorphin. *Science 213*, 1394–1397.

Vale, W., C. Rivier, M. R. Brown, J. Spiess, G. Koob, L. Swanson, L. Bilezikjian, F. Bloom, and J. Rivier (1983). Chemical and biological characterization of corticotropin releasing factor. *Rec. Progr. Horm. Res. 39*, 245–270.

Index

Acetylcholine, 173–201
  acetylcholinesterase, 180–85
  assays, 173
  choline acetyltransferase, 178
  choline transport, 176
  cholinergic drugs, structure of, 198
  cholinergic pathways, 190–92
  cholinergic receptor, 193
  synthesis, 175
  uptake, synthesis and release, 186–88
Action potential, 21–22
  analysis of, 21
  thresholds of, 22
Adenosine, 121
α-Adrenergic receptors, 80, 255–56
β-Adrenergic receptors, 80, 256–58
Adrenergic systems, 263
  pharmacology, 303–4
Alzheimer's disease, 196
γ-Aminobutyrric acid (GABA), 124–55
  distribution, 125–55
  metabolism, 127–35
  as neurotransmitter, 145–50
  transaminase, 131–32
γ-Aminobutyrobetaine, 137
γ-Aminobutyrylcholine, 137
γ-Aminobutyrylhistidine (homocarnosine), 137
γ-Aminobutyrllysine, 137
γ-Amino-β-hydroxybutyric acid (GABOB), 137
Aminoxyacetic acid, 134, 150
Amphetamine, 293–95, 305
Angiotensin, 382
Antianxiety drugs, 289–91
Antidepressants, 291–92
Antipsychotic drugs, 152, 283–89
Apomorphine, 294
Aromatic amino acids,
Astrocyte, 12
Autoreceptors, 33, 87
Autoreceptors and transmitter release, 241
Axon, 9
Axoplasmic transport, 247–49

Benzodiazepines, 137–39
Bicuculline, 137, 139–42
Brain permeability barriers, 13
Brocresine (NSD 1055), and histidine decarboxylase,
Bufotenine, 316

Calcitonin gene-related peptide, 384
Carboxypeptidases, 354
CCK, and GABA, 382
Catecholamines, 203–311
  and affective disorders, 304–9
  assay of
    electronmicroscopy, 209–10
    gas chromatography, 203–4
    GC-mass fragmentography, 203–4
    high performance liquid chromatography, 205–6
histochemical fluorescence microscopy and

microspectrofluoremetry, 207–9
radioisotopes, 205
voltammetry, 207
axonal transport, 247–49
biosynthesis, 215–31
coexistence, 269–70
distribution, 212–15
electron microscopy of,
metabolism, 242–45, 271–74
as neurotransmitters, 249–52
pharmacology, 252–56
antianxiety drugs, 289–91
antidepressant drugs, 291–92
antipsychotic drugs, 288–89
effects on DA neurons, 296–98
effects on NE neurons, 299–302
stimulants, 293
physiological function, 264, 267
prostaglandins and, 241
release, 239–41
storage, 235–38
turnover, 232–38
uptake, 245–47
Catechol-O-methyltransferase (COMT), 242–45
Chlorpromazine, 297
Cholecystokinin, 381
Chromaffin granules, 235–38
Chromogranin, 237
Clonidine, 290, 300
Clorgyline, 242
Cocaine, 256, 307
Colony hybridization, 61
Corticotropin releasing factor, 381, 385

Deprenyl, 242
Desipramine, 256
DFP, 182, 184, 185, 191, 197
Diethyldithiocarbamate, 223
Dihydropteridine reductase, 220, 221
Dihydroxyphenylacetic acid, 242–45, 272, 273
Dimaprit, 346
DNA sense strand, 54
DOPA decarboxylase, 221–22
Dopamine, 265–68
autoreceptors, 277–81
and cyclic AMP,
distribution, 282–83
function, 278, 285
pharmacology, 279, 287
Dopamine hypothesis of schizophrenia, 309–11
Dopaminergic neuron systems, 265–67
incerto-hypothalamic neurons, 265
interplexiform neurons, 265
medullary periventricular neurons,
mesocortical neurons, 267, 284–87
mesolimbic neurons, 267, 281–84
nigrostriatal neurons, 266, 267, 281–84
periglomerular neurons, 265
pharmacology, 281–83
tuberohypophysial neurons, 265–66
DSP 4, 211

Eicosanoids, 119
Eledoisin, 364, 365
EMD-23-448, 280–87
En passant terminals, 11
Endocytosis, 32
Endopeptidases, 354
Epinephrine, 268–69
Excitatory amino acids, 161–70
agonists, 163, 164, 168, 169
antagonists, 163, 164, 168, 169
receptors, 167–69

Exocytotic release, 32, 189
Exons, 56

False neurotransmitters, 254
Familial dysautonomia, 196
Femoxetine, 292
FLA-63, 223
Fluorescence microscopy, of catecholamines, 207–9

GABA, 124–55
agonists, 141–45
alternative metabolic pathways, 135
antagonists, 139–41
distribution, 125–27
metabolism, 127–35
pharmacology, 139–45
receptors, 135–39
shunt, 128–29
turnover, 151–53
Gabaculine, 132, 144
GABARINS, 145
GABA-Transaminase (GABA-T), 131–32
Ganglia, 9
Genetic code, 55
Genotype, 50
GHRH, 369
GIP, 367
Glia, 12
Glucagon-related peptides, 367
Glutamate, 161–70
Glutamic acid decarboxylase (GAD), 129–31
Glutamine, 128
Glutathione, 79
Glycine, 155–61
distribution, 158
metabolism, 156, 157
neurotransmitter role, 157–60
Golgi apparatus, 52

Hallucinogenic drugs, 335–38
Haloperidol, 297
Hemicholinium, 180
Histamine, 340–49
agonists, 346–47
antagonists, 346–48
catabolism, 341–43
and cyclic AMP, 347
receptors, 345–49
synthesis of, 342–43
Histamine-containing cells, 344–45
Histidine decarboxylase, 343
inhibitors of, 343
Homovanillic acid, 272–74
5 HT-receptors, 331–32
Huntington's chorea, 365
Hydrazinopropionic acid, 135
γ-Hydroxybutyric acid, 275, 297
Hydroxylamine, 134
6-Hydroxydopamine, 210
5-Hydroxytryptamine (Serotonin), 315–40
biosynthesis, 316–21
catabolism of, 318, 321
and cyclic AMP, 330
decarboxylation of, 320
5-Hydroxytryptophan decarboxylase, 320

Ibotinic acid, 163–65
Idazoxane, 300
Immunocytochemistry, 355
Introns, 56
Iodotyrosines, 219, 220
Ion pump inhibitors, 20
Iprindole, 292, 307
Iproniazid, 304

Junctional transmission, 24

Kainic acid, 163–64
Kassinin, 364, 365

Leakage channels, 23
Leu5-enkephalin, 372
Lipids, 76–78
β-Lipotropin, 372
Lithium, 114, 306, 319
Local anesthetics, 22
Locus coeruleus, 262–64
  functional hypothesis, 264, 302–3
LSD, explanations of action, 335–38

Mast cells and 48/80, 344
Mast cells and polymixin B, 344
Membrane ion pumps, 19
Membrane potentials, 16
Met5-enkephalin, 372
3-Methoxy-4-hydroxyphenethy-
    leneglycol (MHPG), 271–72
Methyl-histamine, 341, 343
α-Methyl-5-hydroxytryptophan, 220
α-Methyl-3-iodotyrosine, 219
α-Methyl-norepinephrine, 222
α-Methyl-ρ-tyrosine (AMPT), 220
Methyl-tetrahydrofolate, 224
Microelectrophoresis, 38
Modulation, 106–23
  definitions, 106
  second messngers, 109
Molecular biological methods,
    power of, 49
Monoamine oxidase, 242–44
Monoamine oxidase inhibitors,
Morphine, 153
MPTP, 211–12
Muscimol, 142–43
Mutagenesis, 69
Myasthenia Gravis, 194
Myelin, 13

N-acetylaspartate, 80
Nerve cell, parts of, 9, 10
Neuroactive agents, definition of, 4–6
Neuroglia, types of, 12
Neuromedin K, 364
Neurons, cellular variation of, 49
Neuropeptide Y (NPY), 360
Neurotensin, 366
Neurotoxins, 210–12
N-methyl-D-aspartate, 164, 167–70
Node of Ranvier, 13
Noradrenergic neurons, the lateral
    tegmental system, 264
Noradrenergic systems, 261, 263
  correlation between activity and
    MHPG levels,
  locus coeruleus, 262–64
Norepinephrine
  catabolism, 242–45, 271–72
  pharmacology, 252–56, 290–91
  release, 239–40
  storage, 235–38
  synthesis, 216–27
  turnover, 232–35
  uptake, 245–47
Northern blot, 65
Nucleic acid, base pairing
    complementarity of, 53
Nucleic acid, sequence determination
    of, 57

Ohm's law, 16
Oligodendrocyte, 12
Oligodeoxynucleotide probe, 63
Opioid peptides, 370
  behavioral effects of, 377

and CCK, 382
cellular effects of, 375
receptor classes of, 375

Pancreatic polypeptide-related peptides, 369
Paracrine, actions of peptides, 353
Pargyline, 290, 294
Parkinson's Disease, 212
Peptides
actions of, 357
and coexisting transmitters, 269, 358
discovery methods of, 354
special properties of, 352
synthetic ligands of, 357
Perikaryon, 9
Permeability barriers, 14
Perphenazine, 294
Phenotype, 50
Phenylethanolamine-N-methyl-transferase, 224
Phosphoinositide hydrolysis, 113–16
lithium, 114
Phospholipid methylation, 118
Picrotoxin, 139–40
Pineal body, 323–24
Piperoxane, 300
Postsynaptic potentials, 24
Post-translational processing, 52
3-PPP, 280
Presynaptic dense projections, 12
Prodynorphin peptides, 373
Progabide, 127
Prolactin, 288
Proopiomelanocortin, 370
Prostaglandins and norepinephrine release, 241
Protein carboxylmethylation, 117–18
Protein phosphorylation, 109–13
Pyrogallol, 256
2-Pyrrolidonone, 127

Quisqualic acid, 163–64

Radioimmunoassay, 355
Receptors, 86–105
assays, 91
classification, 88–91
definition, 87
identification, 93
kinetics, 95
Reserpine, 290
Restriction endonucleases, 58
Restriction fragment length polymorphisms, analysis of, 67
Reuptake of transmitter, 256
Reverse transcriptase, 58
Ribosomes, 52
RNA, 52

S-adenosylmethionine and serotonin, 321
Schwann cell, 12
Serotonin (5-hydroxytryptamine)
biosynthesis of, 316–21
catabolism, 318, 321
cellular effects of,
cellular pharmacology, 332–34
cytochemistry of, 324, 328
hallucinogenic drugs, 335–38
indolealkylamines, structure, 316
psychotropic drugs and, 338–39
receptor pharmacology, 331–32, 338
regulation of synthesis, 321–23
SIF cells, 214
Signal peptide, 55
Slot blots, 65

Slow postsynaptic potentials, 28
Somatostatin, 378
Southern blot, 65
Specialized contact zone of synapses, 11
Spikeless neurons, 27
Strychnine, 34
Substance K, 364
Substance P, and 5-HT, 364, 365
Succinic semialdehyde dehydrogenase, 132–33
Sulpiride, 287
Synapse, 10
Synaptic effects, pharmacology of, 39
Synaptic transmitters, identification of, 35
Synaptic vesicles, 10

Tachykinin peptides, 363
Tetraethylammonium, 23
Tetrahydro-β-carbolines, 224–25
Tetrahydroisoquinolines, 225
Tetrodotoxin, 23
Thiosemicarbazide, 134
THIP, 143
Transcription, 51, 54
Transgenic expression, 69
Translation, 52, 54
Transmitters
  conditional actions of, 29
  identification criteria, 36
  in vitro analysis of, 41
  localization of, 36
  modulators and neurohormones, 4, 45
  patch clamp analysis of, 42
  secretion, 31
  synaptic mimicry of, 37
Tryptophan, uptake of, 317
Tryptophan hydroxylase, 317–19
Tyrosine, 218, 230, 231
Tyrosine hydroxylase, 218–19
  inhibitors, 219–20
  regulation, 224–29

Varicosities, 10
Vasoactive intestinal peptide, 360
Vasopressin and Oxytocin, 361
VIP, 367
  and acetylcholine, 367
  and cyclic AMP, 360
Voltage clamp, 41

Watson-Crick double helix, 53
Western blot, 65

Yohimbine, 231, 300